D1201894

The Finite Element Method
Fifth edition
Volume 3: Fluid Dynamics

Professor O.C. Zienkiewicz, CBE, FRS, FREng is Professor Emeritus and Director of the Institute for Numerical Methods in Engineering at the University of Wales, Swansea, UK. He holds the UNESCO Chair of Numerical Methods in Engineering at the Technical University of Catalunya, Barcelona, Spain. He was the head of the Civil Engineering Department at the University of Wales Swansea between 1961 and 1989. He established that department as one of the primary centres of finite element research. In 1968 he became the Founder Editor of the *International Journal for Numerical Methods in Engineering* which still remains today the major journal in this field. The recipient of 24 honorary degrees and many medals, Professor Zienkiewicz is also a member of five academies – an honour he has received for his many contributions to the fundamental developments of the finite element method. In 1978, he became a Fellow of the Royal Society and the Royal Academy of Engineering. This was followed by his election as a foreign member to the U.S. Academy of Engineering (1981), the Polish Academy of Science (1985), the Chinese Academy of Sciences (1998), and the National Academy of Science, Italy (Academia dei Lincei) (1999). He published the first edition of this book in 1967 and it remained the only book on the subject until 1971.

Professor R.L. Taylor has more than 35 years' experience in the modelling and simulation of structures and solid continua including two years in industry. In 1991 he was elected to membership in the U.S. National Academy of Engineering in recognition of his educational and research contributions to the field of computational mechanics. He was appointed as the T.Y. and Margaret Lin Professor of Engineering in 1992 and, in 1994, received the Berkeley Citation, the highest honour awarded by the University of California, Berkeley. In 1997, Professor Taylor was made a Fellow in the U.S. Association for Computational Mechanics and recently he was elected Fellow in the International Association of Computational Mechanics, and was awarded the USACM John von Neumann Medal. Professor Taylor has written several computer programs for finite element analysis of structural and non-structural systems, one of which, FEAP, is used world-wide in education and research environments. FEAP is now incorporated more fully into the book to address non-linear and finite deformation problems.

Front cover image: A Finite Element Model of the world land speed record (765.035 mph) car THRUST SSC. The analysis was done using the finite element method by K. Morgan, O. Hassan and N.P. Weatherill at the Institute for Numerical Methods in Engineering, University of Wales Swansea, UK. (see K. Morgan, O. Hassan and N.P. Weatherill, 'Why didn't the supersonic car fly?', *Mathematics Today, Bulletin of the Institute of Mathematics and Its Applications,* Vol. 35, No. 4, 110–114, Aug. 1999).

Plate 1 Analysis of subsonic flow around an aircraft (Dassault Falkon)

Courtesy of Prof. Ken Morgan, Department of Civil Engineering, University of Wales Swansea. Source: J. Peraire, J. Peiro and K. Morgan, 'Multigrid solution of the 3-D compressible Euler equations on unstructured tetrahedral grids', *Int. J. Num. Meth. Eng.* 36, 1029–1044, 1993.

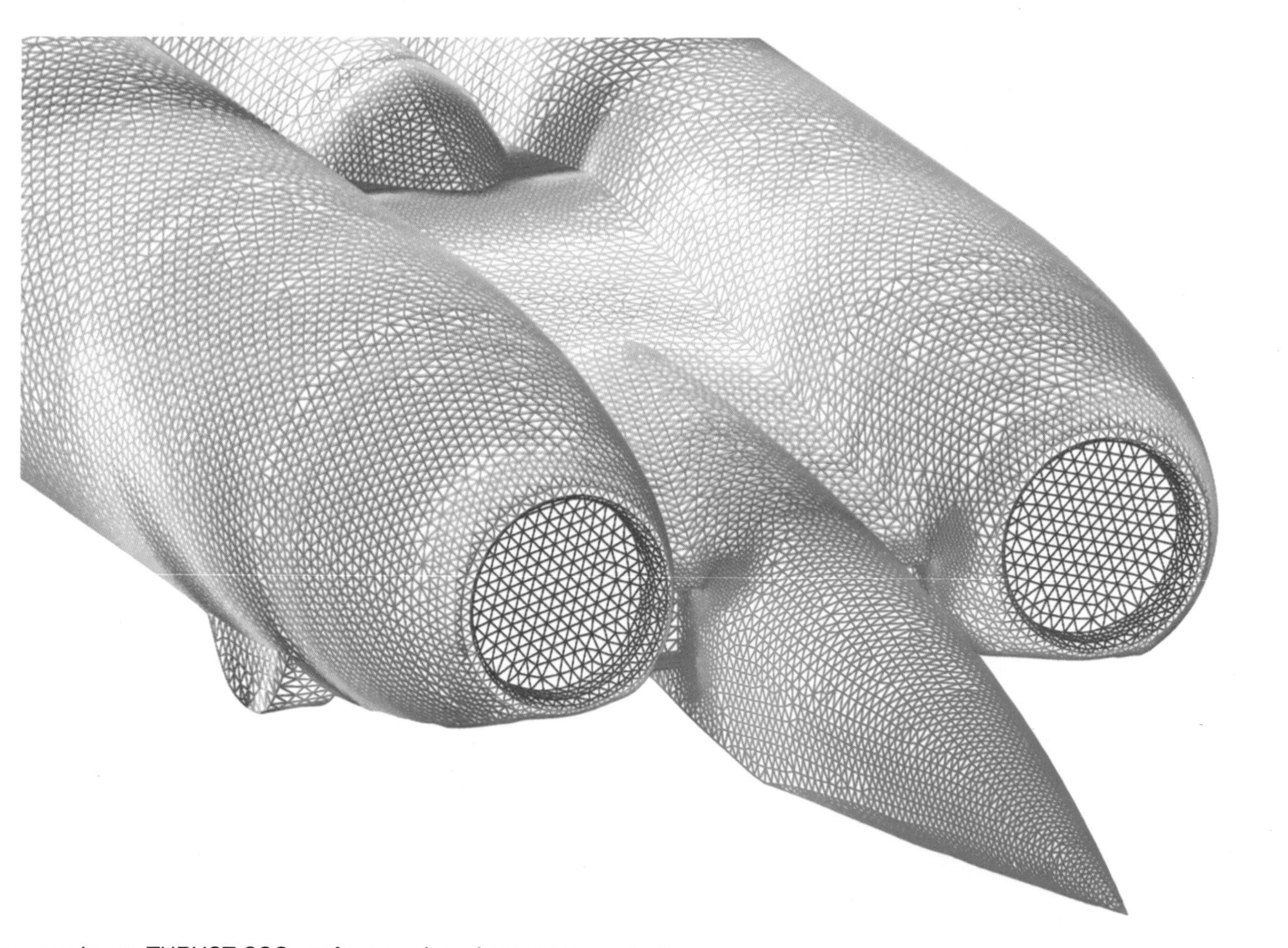

Plate 2 Supersonic car, THRUST SSC, surface mesh and pressure contours

Details near the nose. Complete car is shown on the book cover. Courtesy of Prof. Ken Morgan, Department of Civil Engineering, University of Wales Swansea (Nodes: 39, 528 and Elements: 79060). Source: K. Morgan, O. Hassan and N. P. Wetherill, 'Why didn't the supersonic car fly?', *Mathematics Today, Bulletin of the Institute of Mathematics and its Applications*, 35, 110–114, August 1999.

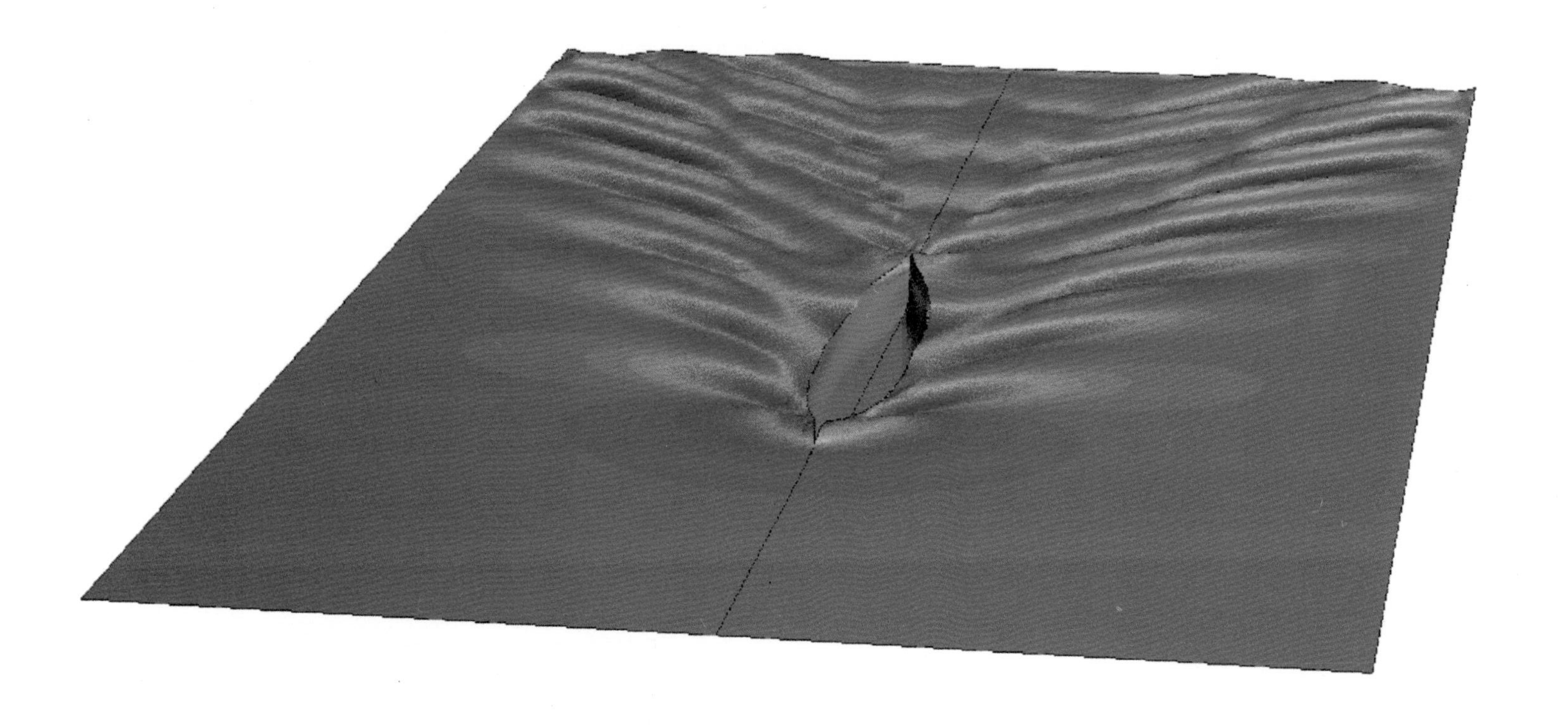

Plate 3 Wave elevation pattern behind a ship
C60 hull for Froude number 0.238. Courtesy of Prof. E. Oñate, CIMNE, Barcelona. Source: E. Oñate and J. Garcia, 'A stabilized finite element formulation for fluid-structure interactions problems', *Comp. Meth. Appl. Mech. Eng.* (to be published)

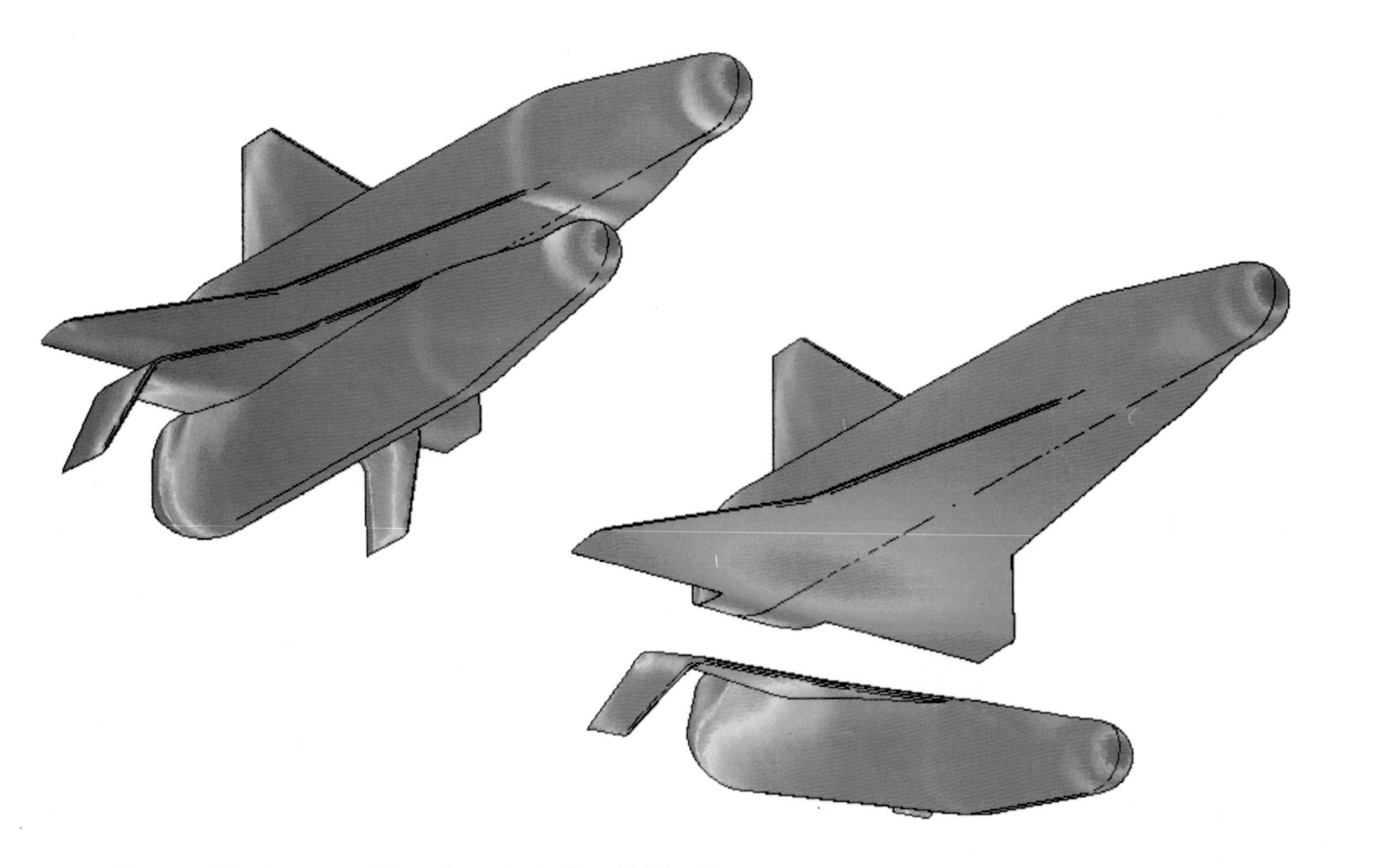

Plate 4 Transient finite element modelling of generic shuttle vehicle booster

Pressure contours. Courtesy of Dr. O. Hassan, Department of Civil Engineering, University of Wales Swansea. Source: O. Hassan, L.B. Bayne, K. Morgan and N. P. Wetherill, 'An adaptive unstructured mesh method for transient flows involving moving boundaries', *ECCOMAS' 98*. John Wiley.

The Finite Element Method

Fifth edition

Volume 3: Fluid Dynamics

O.C. Zienkiewicz, CBE, FRS, FREng
UNESCO Professor of Numerical Methods in Engineering
International Centre for Numerical Methods in Engineering, Barcelona
Emeritus Professor of Civil Engineering and Director of the Institute for Numerical Methods in Engineering, University of Wales, Swansea

R.L. Taylor
Professor in the Graduate School
Department of Civil and Environmental Engineering
University of California at Berkeley
Berkeley, California

OXFORD AUCKLAND BOSTON JOHANNESBURG MELBOURNE NEW DELHI

Butterworth-Heinemann
Linacre House, Jordan Hill, Oxford OX2 8DP
225 Wildwood Avenue, Woburn, MA 01801-2041
A division of Reed Educational and Professional Publishing Ltd

A member of the Reed Elsevier plc group

First published in 1967 by McGraw-Hill
Fifth edition published by Butterworth-Heinemann 2000

British Library Cataloguing in Publication Data
A catalogue record for this book is available from the British Library

Library of Congress Cataloguing in Publication Data
A catalogue record for this book is available from the Library of Congress

ISBN 0 7506 5050 8

Published with the cooperation of CIMNE,
the International Centre for Numerical Methods in Engineering,
Barcelona, Spain (www.cimne.upc.es)

Typeset by Academic & Technical Typesetting, Bristol
Printed and bound by MPG Books Ltd

Dedication

This book is dedicated to our wives Helen and Mary Lou and our families for their support and patience during the preparation of this book, and also to all of our students and colleagues who over the years have contributed to our knowledge of the finite element method. In particular we would like to mention Professor Eugenio Oñate and his group at CIMNE for their help, encouragement and support during the preparation process.

Contents

Volume 1: The basis

1. Some preliminaries: the standard discrete system
2. A direct approach to problems in elasticity
3. Generalization of the finite element concepts. Galerkin-weighted residual and variational approaches
4. Plane stress and plane strain
5. Axisymmetric stress analysis
6. Three-dimensional stress analysis
7. Steady-state field problems – heat conduction, electric and magnetic potential, fluid flow, etc
8. 'Standard' and 'hierarchical' element shape functions: some general families of C_0 continuity
9. Mapped elements and numerical integration – 'infinite' and 'singularity' elements
10. The patch test, reduced integration, and non-conforming elements
11. Mixed formulation and constraints – complete field methods
12. Incompressible problems, mixed methods and other procedures of solution
13. Mixed formulation and constraints – incomplete (hybrid) field methods, boundary/Trefftz methods
14. Errors, recovery processes and error estimates
15. Adaptive finite element refinement
16. Point-based approximations; element-free Galerkin – and other meshless methods
17. The time dimension – semi-discretization of field and dynamic problems and analytical solution procedures
18. The time dimension – discrete approximation in time
19. Coupled systems
20. Computer procedures for finite element analysis

Appendix A. Matrix algebra
Appendix B. Tensor-indicial notation in the approximation of elasticity problems
Appendix C. Basic equations of displacement analysis
Appendix D. Some integration formulae for a triangle
Appendix E. Some integration formulae for a tetrahedron
Appendix F. Some vector algebra
Appendix G. Integration by parts
Appendix H. Solutions exact at nodes
Appendix I. Matrix diagonalization or lumping

Volume 2: Solid and structural mechanics

Preface to Volume 3

This volume appears for the first time in a separate form. Though part of it has been updated from the second volume of the fourth edition, in the main it is an entirely new work. Its objective is to separate the fluid mechanics formulations and applications from those of solid mechanics and thus perhaps to reach a different interest group.

Though the introduction to the finite element method contained in the first volume (the basis) is general, in it we have used, in the main, examples of elastic solids. Only a few applications to areas such as heat conduction, porous media flow and potential field problems have been presented. The reason for this is that all such problems are self-adjoint and that for such self-adjoint problems Galerkin procedures are optimal. For *convection dominated problems* the Galerkin process is no longer optimal and it is here that most of the fluid mechanics problems lie.

The present volume is devoted entirely to fluid mechanics and uses in the main the methods introduced in Volume 1. However, it then enlarges these to deal with the non-self-adjoint problems of convection which are essential to fluid mechanics problems.

It is our intention that the present volume could be used by investigators familiar with the finite element method in general terms and introduce them to the subject of fluid mechanics. It can thus in many ways stand alone. However, many of the general finite element procedures available in Volume 1 may not be familiar to a reader introduced to the finite element method through different texts and therefore we recommend that this volume be used in conjunction with Volume 1 to which we make frequent reference.

In fluid mechanics several difficulties arise. (1) The first is that of dealing with *incompressible* or *almost incompressible* situations. These, as we already know, present special difficulties in formulation even in solids. (2) Second and even more important is the difficulty introduced by the convection which requires rather specialized treatment and stabilization. Here particularly in the field of compressible high-speed gas flow many alternative finite element approaches are possible and often different algorithms for different ranges of flow have been suggested. Although slow creeping flows may well be dealt with by procedures almost identical to those of solid mechanics, the high-speed range of supersonic and hypersonic flow may require a very particular treatment. In this text we shall generally use only one algorithm the so-called *characteristic based split* (*CBS*), introduced a few years ago by the authors. It turns out that this algorithm is applicable to all ranges of flow and indeed gives results which are at least equal to those of specialized methods. We shall therefore stress its development and give details of its use in the third chapter dealing with discretization.

We hope that the book will be useful in introducing the reader to the complex subject of fluid mechanics and its many facets. Further we hope it will also be of use to the experienced practitioner of computational fluid dynamics (CFD) who may find the new presentation of interest and practical application.

Acknowledgements

The authors would like to thank Professor Peter Bettess for largely contributing the chapter on waves (Chapter 8) in which he has made so many achievements† and to Dr. Pablo Ortiz who, with the first author, was the first to apply the CBS algorithm to shallow-water equations. Our gratitude also goes to Professor Eugenio Oñate for adding the section on free surface flows in the incompressible flow chapter (Chapter 5) documenting the success and usefulness of the procedure in ship hydrodynamics. Thanks are also due to Professor J. Tinsley Oden for the short note describing the discontinuous Galerkin method and to Professor Ramon Codina whose participation in recent research work has been extensive. Thanks are also due to Drs Joanna Szmelter and Jie Wu who both contributed in the early developments leading to the final form of the CBS algorithm.

The establishment of finite elements in CFD applications to high-speed convection-dominated flows was first accomplished at Swansea by the research team working closely with Professor Ken Morgan. His former students include Professor Rainald Löhner and Professor Jaime Peraire as well as many others to whom frequent reference is made. We are very grateful to Professor Nigel Weatherill and Dr. Oubay Hassan who have contributed several of the diagrams and colour plates and, in particular, the cover of the book. The recent work on the CBS algorithm has been accomplished by the first author with substantial support from NASA (Grant NAGW/2127, Ames Control Number 90-144). Here the support, encouragement and help given by Dr. Kajal K. Gupta is most gratefully acknowledged.

Finally the first author (O.C. Zienkiewicz) is extremely grateful to Dr. Perumal Nithiarasu who worked with him for several years developing the CBS algorithm and who has given to him very much help in achieving the present volume.

OCZ and RLT

† As already mentioned in the acknowledgement of Volume 1, both Peter and Jackie Bettess have helped us by writing a general subject index for Volumes 1 and 3.

‡ Complete source code for all programs in the three volumes may be obtained at no cost from the publisher's web page: http://www.bh.com/companions/fem

1

Introduction and the equations of fluid dynamics

1.1 General remarks and classification of fluid mechanics problems discussed in this book

The problems of solid and fluid behaviour are in many respects similar. In both media stresses occur and in both the material is displaced. There is however one major difference. The fluids cannot support any deviatoric stresses when the fluid is at rest. Then only a pressure or a mean compressive stress can be carried. As we know, in solids, other stresses can exist and the solid material can generally support structural forces.

In addition to pressure, deviatoric stresses can however develop when the fluid is in motion and such motion of the fluid will always be of primary interest in fluid dynamics. We shall therefore concentrate on problems in which displacement is continuously changing and in which velocity is the main characteristic of the flow. The deviatoric stresses which can now occur will be characterized by a quantity which has great resemblance to shear modulus and which is known as dynamic viscosity.

Up to this point the equations governing fluid flow and solid mechanics appear to be similar with the velocity vector $\mathbf{u}$ replacing the displacement for which previously we have used the same symbol. However, there is one further difference, i.e. that even when the flow has a constant velocity (steady state), *convective acceleration* occurs. This convective acceleration provides terms which make the fluid mechanics equations non-self-adjoint. Now therefore in most cases unless the velocities are very small, so that the convective acceleration is negligible, the treatment has to be somewhat different from that of solid mechanics. The reader will remember that for self-adjoint forms, the approximating equations derived by the Galerkin process give the minimum error in the energy norm and thus are in a sense optimal. This is no longer true in general in fluid mechanics, though for slow flows (creeping flows) the situation is somewhat similar.

With a fluid which is in motion continual preservation of mass is always necessary and unless the fluid is highly compressible we require that the divergence of the velocity vector be zero. We have dealt with similar problems in the context of elasticity in Volume 1 and have shown that such an incompressibility constraint introduces very serious difficulties in the formulation (Chapter 12, Volume 1). In fluid mechanics the same difficulty again arises and all fluid mechanics approximations have to be such that even if compressibility occurs the limit of incompressibility can be modelled. This precludes the use of many elements which are otherwise acceptable.

In this book we shall introduce the reader to a finite element treatment of the equations of motion for various problems of fluid mechanics. Much of the activity in fluid mechanics has however pursued a *finite difference* formulation and more recently a derivative of this known as the *finite volume* technique. Competition between the newcomer of finite elements and established techniques of finite differences have appeared on the surface and led to a much slower adoption of the finite element process in fluid mechanics than in structures. The reasons for this are perhaps simple. In solid mechanics or structural problems, the treatment of continua arises only on special occasions. The engineer often dealing with structures composed of bar-like elements does not need to solve continuum problems. Thus his interest has focused on such continua only in more recent times. In fluid mechanics, practically all situations of flow require a two or three dimensional treatment and here approximation was frequently required. This accounts for the early use of finite differences in the 1950s before the finite element process was made available. However, as we have pointed out in Volume 1, there are many advantages of using the finite element process. This not only allows a fully unstructured and arbitrary domain subdivision to be used but also provides an approximation which in self-adjoint problems is always superior to or at least equal to that provided by finite differences.

A methodology which appears to have gained an intermediate position is that of finite volumes, which were initially derived as a subclass of finite difference methods. We have shown in Volume 1 that these are simply another kind of finite element form in which subdomain collocation is used. We do not see much advantage in using that form of approximation. However, there is one point which seems to appeal to many investigators. That is the fact that with the finite volume approximation the local conservation conditions are satisfied within one element. This does not carry over to the full finite element analysis where generally satisfaction of all conservation conditions is achieved only in an assembly region of a few elements. This is no disadvantage if the general approximation is superior.

In the reminder of this book we shall be discussing various classes of problems, each of which has a certain behaviour in the numerical solution. Here we start with incompressible flows or flows where the only change of volume is elastic and associated with transient changes of pressure (Chapter 4). For such flows full incompressible constraints have to be applied.

Further, with very slow speeds, convective acceleration effects are often negligible and the solution can be reached using identical programs to those derived for elasticity. This indeed was the first venture of finite element developers into the field of fluid mechanics thus transferring the direct knowledge from structures to fluids. In particular the so-called linear Stokes flow is the case where fully incompressible but elastic behaviour occurs and a particular variant of Stokes flow is that used in metal forming where the material can no longer be described by a constant viscosity but possesses a viscosity which is non-newtonian and depends on the strain rates.

Here the fluid (flow formulation) can be applied directly to problems such as the forming of metals or plastics and we shall discuss that extreme of the situation at the end of Chapter 4. However, even in incompressible flows when the speed increases convective terms become important. Here often steady-state solutions do not exist or at least are extremely unstable. This leads us to such problems as eddy shedding which is also discussed in this chapter.

The subject of turbulence itself is enormous, and much research is devoted to it. We shall touch on it very superficially in Chapter 5: suffice to say that in problems where turbulence occurs, it is possible to use various models which result in a flow-dependent viscosity. The same chapter also deals with incompressible flow in which free-surface and other gravity controlled effects occur. In particular we show the modifications necessary to the general formulation to achieve the solution of problems such as the surface perturbation occurring near ships, submarines, etc.

The next area of fluid mechanics to which much practical interest is devoted is of course that of flow of gases for which the compressibility effects are much larger. Here compressibility is problem-dependent and obeys the gas laws which relate the pressure to temperature and density. It is now necessary to add the energy conservation equation to the system governing the motion so that the temperature can be evaluated. Such an energy equation can of course be written for incompressible flows but this shows only a weak or no coupling with the dynamics of the flow.

This is not the case in compressible flows where coupling between all equations is very strong. In compressible flows the flow speed may exceed the speed of sound and this may lead to shock development. This subject is of major importance in the field of aerodynamics and we shall devote a substantial part of Chapter 6 just to this particular problem.

In a real fluid, viscosity is always present but at high speeds such viscous effects are confined to a narrow zone in the vicinity of solid boundaries (*boundary layer*). In such cases, the remainder of the fluid can be considered to be inviscid. There we can return to the fiction of so-called ideal flow in which viscosity is not present and here various simplifications are again possible.

One such simplification is the introduction of potential flow and we shall mention this in Chapter 4. In Volume 1 we have already dealt with such potential flows under some circumstances and showed that they present very little difficulty. But unfortunately such solutions are not easily extendable to realistic problems.

A third major field of fluid mechanics of interest to us is that of shallow water flows which occur in coastal waters or elsewhere in which the depth dimension of flow is very much less than the horizontal ones. Chapter 7 will deal with such problems in which essentially the distribution of pressure in the vertical direction is almost hydrostatic.

In shallow-water problems a free surface also occurs and this dominates the flow characteristics.

Whenever a free surface occurs it is possible for transient phenomena to happen, generating waves such as for instance those that occur in oceans and other bodies of water. We have introduced in this book a chapter (Chapter 8) dealing with this particular aspect of fluid mechanics. Such wave phenomena are also typical of some other physical problems. We have already referred to the problem of acoustic waves in the context of the first volume of this book and here we show that the treatment is extremely similar to that of surface water waves. Other waves such as electromagnetic waves again come into this category and perhaps the treatment suggested in Chapter 8 of this volume will be effective in helping those areas in turn.

In what remains of this chapter we shall introduce the general equations of fluid dynamics valid for most compressible or incompressible flows showing how the particular simplification occurs in each category of problem mentioned above. However, before proceeding with the recommended discretization procedures, which we present in Chapter 3, we must introduce the treatment of problems in which convection and diffusion occur simultaneously. This we shall do in Chapter 2 with the typical convection–diffusion equation. Chapter 3 will introduce a general algorithm capable of solving most of the fluid mechanics problems encountered in this book. As we have already mentioned, there are many possible algorithms; very often specialized ones are used in different areas of applications. However the general algorithm of Chapter 3 produces results which are at least as good as others achieved by more specialized means. We feel that this will give a certain unification to the whole text and thus without apology we shall omit reference to many other methods or discuss them only in passing.

1.2 The governing equations of fluid dynamics[1–8]

1.2.1 Stresses in fluids

The essential characteristic of a fluid is its inability to sustain shear stresses when at rest. Here only hydrostatic 'stress' or pressure is possible. Any analysis must therefore concentrate on the motion, and the essential independent variable is thus the velocity $\mathbf{u}$ or, if we adopt the indicial notation (with the x, y, z axes referred to as x_i, $i = 1, 2, 3$),

$$u_i, \qquad i = 1, 2, 3 \tag{1.1}$$

This replaces the displacement variable which was of primary importance in solid mechanics.

The rates of strain are thus the primary cause of the general stresses, σ_{ij}, and these are defined in a manner analogous to that of infinitesimal strain as

$$\dot{\varepsilon}_{ij} = \frac{\partial u_i/\partial x_j + \partial u_j/\partial x_i}{2} \tag{1.2}$$

This is a well-known tensorial definition of strain rates but for use later in variational forms is written as a vector which is more convenient in finite element analysis. Details of such matrix forms are given fully in Volume 1 but for completeness we mention them here. Thus, this strain rate is written as a vector ($\dot{\boldsymbol{\varepsilon}}$). This vector is given by the following form

$$\dot{\boldsymbol{\varepsilon}}^{\mathrm{T}} = [\dot{\varepsilon}_{11}, \dot{\varepsilon}_{22}, 2\dot{\varepsilon}_{12}] = [\dot{\varepsilon}_{11}, \dot{\varepsilon}_{22}, \dot{\gamma}_{12}] \tag{1.3}$$

in two dimensions with a similar form in three dimensions:

$$\dot{\boldsymbol{\varepsilon}}^{\mathrm{T}} = [\dot{\varepsilon}_{11}, \dot{\varepsilon}_{22}, \dot{\varepsilon}_{33}, 2\dot{\varepsilon}_{12}, 2\dot{\varepsilon}_{23}, 2\dot{\varepsilon}_{31}] \tag{1.4}$$

When such vector forms are used we can write the strain rates in the form

$$\dot{\boldsymbol{\varepsilon}} = \mathbf{S}\mathbf{u} \tag{1.5}$$

where $\mathbf{S}$ is known as the stain operator and $\mathbf{u}$ is the velocity given in Eq. (1.1).

The stress–strain relations for a linear (newtonian) isotropic fluid require the definition of two constants.

The first of these links the *deviatoric stresses* τ_{ij} to the *deviatoric strain rates*:

$$\tau_{ij} \equiv \sigma_{ij} - \delta_{ij}\frac{\sigma_{kk}}{3} = 2\mu\left(\dot{\varepsilon}_{ij} - \delta_{ij}\frac{\dot{\varepsilon}_{kk}}{3}\right) \tag{1.6}$$

In the above equation the quantity in brackets is known as the deviatoric strain, δ_{ij} is the Kronecker delta, and a repeated index means summation; thus

$$\sigma_{ii} \equiv \sigma_{11} + \sigma_{22} + \sigma_{33} \qquad \text{and} \qquad \dot{\varepsilon}_{ii} \equiv \dot{\varepsilon}_{11} + \dot{\varepsilon}_{22} + \dot{\varepsilon}_{33} \tag{1.7}$$

The coefficient μ is known as the dynamic (shear) viscosity or simply viscosity and is analogous to the shear modulus G in linear elasticity.

The second relation is that between the mean stress changes and the volumetric strain rates. This defines the pressure as

$$p = \frac{\sigma_{ii}}{3} = -\kappa\dot{\varepsilon}_{ii} + p_0 \tag{1.8}$$

where κ is a *volumetric viscosity* coefficient analogous to the bulk modulus K in linear elasticity and p_0 is the initial hydrostatic pressure independent of the strain rate (note that p and p_0 are invariably defined as positive when compressive).

We can immediately write the 'constitutive' relation for fluids from Eqs (1.6) and (1.8) as

$$\begin{aligned}\sigma_{ij} &= 2\mu\left(\dot{\varepsilon}_{ij} - \frac{\delta_{ij}\dot{\varepsilon}_{kk}}{3}\right) + \delta_{ij}\kappa\dot{\varepsilon}_{kk} - \delta_{ij}p_0 \\ &= \tau_{ij} - \delta_{ij}p \end{aligned} \tag{1.9a}$$

or

$$\sigma_{ij} = 2\mu\dot{\varepsilon}_{ij} + \delta_{ij}(\kappa - \tfrac{2}{3}\mu)\dot{\varepsilon}_{ii} + \delta_{ij}p_0 \tag{1.9b}$$

Traditionally the Lamé notation is often used, putting

$$\kappa - \tfrac{2}{3}\mu \equiv \lambda \tag{1.10}$$

but this has little to recommend it and the relation (1.9a) is basic. There is little evidence about the existence of volumetric viscosity and we shall take

$$\kappa\dot{\varepsilon}_{ii} \equiv 0 \tag{1.11}$$

in what follows, giving the essential constitutive relation as (now dropping the suffix on p_0)

$$\sigma_{ij} = 2\mu\left(\dot{\varepsilon}_{ij} - \frac{\delta_{ij}\dot{\varepsilon}_{kk}}{3}\right) - \delta_{ij}p \equiv \tau_{ij} - \delta_{ij}p \tag{1.12a}$$

without necessarily implying incompressibility $\dot{\varepsilon}_{ii} = 0$.

In the above,

$$\tau_{ij} = 2\mu\left(\dot{\varepsilon}_{ij} - \frac{\delta_{ij}\dot{\varepsilon}_{kk}}{3}\right) = \mu\left[\left(\frac{\partial u_i}{\partial x_j} + \frac{\partial u_j}{\partial x_i}\right) - \delta_{ij}\frac{2}{3}\frac{\partial u_k}{\partial x_k}\right] \tag{1.12b}$$

All of the above relationships are analogous to those of elasticity, as we shall note again later for incompressible flow. We have also mentioned this in Chapter 12 of Volume 1 where various stabilization procedures are considered for incompressible problems.

Non-linearity of some fluid flows is observed with a coefficient μ depending on strain rates. We shall term such flows 'non-newtonian'.

1.2.2 Mass conservation

If ρ is the fluid density then the balance of mass flow ρu_i entering and leaving an infinitesimal control volume (Fig. 1.1) is equal to the rate of change in density

$$\frac{\partial \rho}{\partial t} + \frac{\partial}{\partial x_i}(\rho u_i) \equiv \frac{\partial \rho}{\partial t} + \boldsymbol{\nabla}^{\mathrm{T}}(\rho \mathbf{u}) = 0 \tag{1.13a}$$

or in traditional cartesian coordinates

$$\frac{\partial \rho}{\partial t} + \frac{\partial}{\partial x}(\rho u) + \frac{\partial}{\partial y}(\rho v) + \frac{\partial}{\partial z}(\rho w) = 0 \tag{1.13b}$$

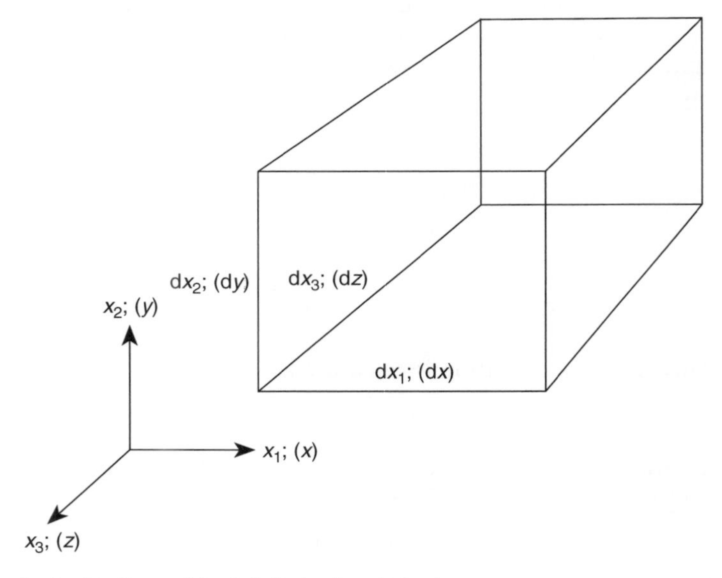

Fig. 1.1 Coordinate direction and the infinitesimal control volume.

1.2.3 Momentum conservation – or dynamic equilibrium

Now the balance of momentum in the jth direction, this is $(\rho u_j)u_i$ leaving and entering a control volume, has to be in equilibrium with the stresses σ_{ij} and body forces ρf_j

giving a typical component equation

$$\frac{\partial(\rho u_j)}{\partial t} + \frac{\partial}{\partial x_i}[(\rho u_j)u_i] - \frac{\partial}{\partial x_i}(\sigma_{ij}) - \rho f_j = 0 \tag{1.14}$$

or using (1.12a),

$$\frac{\partial(\rho u_j)}{\partial t} + \frac{\partial}{\partial x_i}[(\rho u_j)u_i] - \frac{\partial(\tau_{ij})}{\partial x_i} + \frac{\partial p}{\partial x_j} - \rho f_j = 0 \tag{1.15a}$$

with (1.12b) implied.

Once again the above can, of course, be written as three sets of equations in cartesian form:

$$\frac{\partial}{\partial t}(\rho u) + \frac{\partial}{\partial x}(\rho u^2) + \frac{\partial}{\partial y}(\rho uv) + \frac{\partial}{\partial z}(\rho uw) - \frac{\partial \tau_{xx}}{\partial x} - \frac{\partial \tau_{xy}}{\partial y} - \frac{\partial \tau_{xz}}{\partial z} + \frac{\partial p}{\partial x} - \rho f_x = 0 \tag{1.15b}$$

etc.

1.2.4 Energy conservation and equation of state

We note that in the equations of Secs 1.2.2 and 1.2.3 the independent variables are u_i (the velocity), p (the pressure) and ρ (the density). The deviatoric stresses, of course, were defined by Eq. (1.12b) in terms of velocities and hence are not independent.

Obviously, there is one variable too many for this equation system to be capable of solution. However, if the density is assumed constant (as in incompressible fluids) or if a single relationship linking pressure and density can be established (as in isothermal flow with small compressibility) the system becomes complete and is solvable.

More generally, the pressure (p), density (ρ) and absolute temperature (T) are related by an equation of state of the form

$$\rho = \rho(p, T) \tag{1.16}$$

For an ideal gas this takes, for instance, the form

$$\rho = \frac{p}{RT} \tag{1.17}$$

where R is the universal gas constant.

In such a general case, it is necessary to supplement the governing equation system by the equation of *energy conservation*. This equation is indeed of interest even if it is not coupled, as it provides additional information about the behaviour of the system.

Before proceeding with the derivation of the energy conservation equation we must define some further quantities. Thus we introduce e, the *intrinsic energy* per unit mass. This is dependent on the state of the fluid, i.e. its pressure and temperature or

$$e = e(T, p) \tag{1.18}$$

The total energy per unit mass, E, includes of course the kinetic energy per unit mass and thus

$$E = e + \frac{u_i u_i}{2} \tag{1.19}$$

Finally, we can define the *enthalpy* as

$$h = e + \frac{p}{\rho} \qquad \text{or} \qquad H = h + \frac{u_i u_i}{2} = E + \frac{p}{\rho} \tag{1.20}$$

and these variables are found to be convenient.

Energy transfer can take place by convection and by conduction (radiation generally being confined to boundaries). The conductive heat flux q_i is defined as

$$q_i = -k \frac{\partial}{\partial x_i} T \tag{1.21}$$

where k is an isotropic thermal conductivity.

To complete the relationship it is necessary to determine heat source terms. These can be specified per unit volume as q_H due to chemical reaction (if any) and must include the energy dissipation due to internal stresses, i.e. using Eq. (1.12),

$$\frac{\partial}{\partial x_i}(\sigma_{ij} u_j) = \frac{\partial}{\partial x_i}(\tau_{ij} u_j) - \frac{\partial}{\partial x_j}(p u_j) \tag{1.22}$$

The balance of energy in a unit volume can now thus be written as

$$\frac{\partial(\rho E)}{\partial t} + \frac{\partial}{\partial x_i}(\rho u_i E) - \frac{\partial}{\partial x_i}\left(k \frac{\partial T}{\partial x_i}\right) + \frac{\partial}{\partial x_i}(p u_i) - \frac{\partial}{\partial x_i}(\tau_{ij} u_j) - \rho f_i u_i - q_H = 0 \tag{1.23a}$$

or more simply

$$\frac{\partial(\rho E)}{\partial t} + \frac{\partial}{\partial x_i}(\rho u_i H) - \frac{\partial}{\partial x_i}\left(k \frac{\partial T}{\partial x_i}\right) + \frac{\partial}{\partial x_i}(\tau_{ij} u_j) - \rho f_i u_i - q_H = 0 \tag{1.23b}$$

Here, the penultimate term represents the work done by body forces.

1.2.5 Navier–Stokes and Euler equations

The governing equations derived in the preceding sections can be written in the general conservative form

$$\frac{\partial \boldsymbol{\Phi}}{\partial t} + \nabla \mathbf{F} + \nabla \mathbf{G} + \mathbf{Q} = \mathbf{0} \tag{1.24a}$$

or

$$\frac{\partial \boldsymbol{\Phi}}{\partial t} + \frac{\partial \mathbf{F}_i}{\partial x_i} + \frac{\partial \mathbf{G}_i}{\partial x_i} + \mathbf{Q} = \mathbf{0} \tag{1.24b}$$

in which Eqs (1.13), (1.15) or (1.23) provide the particular entries to the vectors.

Thus, the vector of independent unknowns is, using both indicial and cartesian notation,

$$\boldsymbol{\Phi} = \begin{Bmatrix} \rho \\ \rho u_1 \\ \rho u_2 \\ \rho u_3 \\ \rho E \end{Bmatrix} \quad \text{or, in cartesian notation,} \quad \mathbf{U} = \begin{Bmatrix} \rho \\ \rho u \\ \rho v \\ \rho w \\ \rho E \end{Bmatrix} \tag{1.25a}$$

$$\mathbf{F}_i = \begin{Bmatrix} \rho u_i \\ \rho u_1 u_i + p\delta_{1i} \\ \rho u_2 u_i + p\delta_{2i} \\ \rho u_3 u_i + p\delta_{3i} \\ \rho H u_i \end{Bmatrix} \quad \text{or} \quad \mathbf{F}_x = \begin{Bmatrix} \rho u \\ \rho u^2 + p \\ \rho uv \\ \rho uw \\ \rho Hu \end{Bmatrix}, \quad \text{etc.} \tag{1.25b}$$

$$\mathbf{G}_i = \begin{Bmatrix} 0 \\ -\tau_{1i} \\ -\tau_{2i} \\ -\tau_{3i} \\ -(\tau_{ij}u_j) - k\dfrac{\partial T}{\partial x_i} \end{Bmatrix} \quad \text{or}$$

$$\mathbf{G}_x = \begin{Bmatrix} 0 \\ -\tau_{xx} \\ -\tau_{yx} \\ -\tau_{zx} \\ -(\tau_{xx}u + \tau_{xy}v + \tau_{xz}w) - k\dfrac{\partial T}{\partial x} \end{Bmatrix}, \quad \text{etc.} \tag{1.25c}$$

$$\mathbf{Q} = \begin{Bmatrix} 0 \\ -\rho f_1 \\ -\rho f_2 \\ -\rho f_3 \\ -\rho f_i u_i - q_H \end{Bmatrix} \quad \text{or} \quad \mathbf{Q} = \begin{Bmatrix} 0 \\ -\rho f_x \\ -\rho f_y \\ -\rho f_z \\ -\rho(f_x u + f_y v + f_z w) - q_H \end{Bmatrix}, \quad \text{etc.} \tag{1.25d}$$

with

$$\tau_{ij} = \mu\left[\left(\frac{\partial u_i}{\partial x_j} + \frac{\partial u_j}{\partial x_i}\right) - \delta_{ij}\frac{2}{3}\frac{\partial u_k}{\partial x_k}\right]$$

The complete set of (1.24) is known as the *Navier–Stokes equation*. A particular case when viscosity is assumed to be zero and no heat conduction exists is known as the 'Euler equation' ($\tau_{ij} = k = 0$).

The above equations are the basis from which all fluid mechanics studies start and it is not surprising that many alternative forms are given in the literature obtained by combinations of the various equations.[2] The above set is, however, convenient and physically meaningful, defining the conservation of important quantities. It should be noted that only equations written in conservation form will yield the correct, physically meaningful, results in problems where shock discontinuities are present. In Appendix A, we show a particular set of non-conservative equations which are frequently used. There we shall indicate by an example the possibility of obtaining incorrect solutions when a shock exists. The reader is therefore cautioned not to extend the use of non-conservative equations to the problems of high-speed flows.

In many actual situations one or another feature of the flow is predominant. For instance, frequently the viscosity is only of importance close to the boundaries at which velocities are specified, i.e.

$$\Gamma_u \qquad \text{where} \qquad u_i = \bar{u}_i$$

or on which tractions are prescribed:

$$\Gamma_t \qquad \text{where} \qquad n_i \sigma_{ij} = \bar{t}_j$$

In the above n_i are the direction cosines of the outward normal.

In such cases the problem can be considered separately in two parts: one as the *boundary layer* near such boundaries and another as *inviscid flow* outside the boundary layer.

Further, in many cases a steady-state solution is not available with the fluid exhibiting *turbulence*, i.e. a random fluctuation of velocity. Here it is still possible to use the general Navier–Stokes equations now written in terms of the mean flow but with a *Reynolds viscosity* replacing the molecular one. The subject is dealt with elsewhere in detail and in this volume we shall limit ourselves to very brief remarks. The turbulent instability is inherent in the simple Navier–Stokes equations and it is in principle always possible to obtain the transient, turbulent, solution modelling of the flow, providing the mesh size is capable of reproducing the random eddies. Such computations, though possible, are extremely costly and hence the Reynolds averaging is of practical importance.

Two important points have to be made concerning *inviscid flow* (ideal fluid flow as it is sometimes known).

Firstly, the Euler equations are of a purely convective form:

$$\frac{\partial \mathbf{\Phi}}{\partial t} + \frac{\partial \mathbf{F}_i}{\partial x_i} = \mathbf{0} \qquad \mathbf{F}_i = \mathbf{F}_i(\mathbf{U}) \tag{1.26}$$

and hence very special methods for their solutions will be necessary. These methods are applicable and useful mainly in *compressible flow*, as we shall discuss in Chapter 6. Secondly, for incompressible (or nearly incompressible) flows it is of interest to introduce a *potential* that converts the Euler equations to a simple self-adjoint form. We shall mention this potential approximation in Chapter 4. Although potential forms are applicable also to compressible flows we shall not discuss them later as they fail in high-speed supersonic cases.

1.3 Incompressible (or nearly incompressible) flows

We observed earlier that the Navier–Stokes equations are completed by the existence of a state relationship giving [Eq. (1.16)]

$$\rho = \rho(p, T)$$

In (nearly) incompressible relations we shall frequently assume that:

1. The problem is isothermal.

2. The variation of ρ with p is very small, i.e. such that in product terms of velocity and density the latter can be assumed constant.

The first assumption will be relaxed, as we shall see later, allowing some thermal coupling via the dependence of the fluid properties on temperature. In such cases we shall introduce the coupling iteratively. Here the problem of density-induced currents or temperature-dependent viscosity (Chapter 5) will be typical.

If the assumptions introduced above are used we can still allow for small compressibility, noting that density changes are, as a consequence of elastic deformability, related to pressure changes. Thus we can write

$$\mathrm{d}\rho = \frac{\rho}{K}\,\mathrm{d}p \tag{1.27a}$$

where K is the elastic bulk modulus. This can be written as

$$\mathrm{d}\rho = \frac{1}{c^2}\,\mathrm{d}p \tag{1.27b}$$

or

$$\frac{\partial \rho}{\partial t} = \frac{1}{c^2}\frac{\partial p}{\partial t} \tag{1.27c}$$

with $c = \sqrt{K/\rho}$ being the acoustic wave velocity.

Equations (1.24) and (1.25) can now be rewritten omitting the energy transport (and condensing the general form) as

$$\frac{1}{c^2}\frac{\partial p}{\partial t} + \rho\frac{\partial u_i}{\partial x_i} = 0 \tag{1.28a}$$

$$\frac{\partial u_j}{\partial t} + \frac{\partial}{\partial x_i}(u_j u_i) + \frac{1}{\rho}\frac{\partial p}{\partial x_j} - \frac{1}{\rho}\frac{\partial}{\partial x_i}\tau_{ji} - f_j = 0 \tag{1.28b}$$

With $j = 1, 2, 3$ this represents a system of four equations in which the variables are u_j and p.

Written in terms of cartesian coordinates we have, in place of Eq. (1.28a),

$$\frac{1}{c^2}\frac{\partial p}{\partial t} + \rho\frac{\partial u}{\partial x} + \rho\frac{\partial v}{\partial y} + \rho\frac{\partial w}{\partial z} = 0 \tag{1.29a}$$

where the first term is dropped for complete incompressibility ($c = \infty$) and

$$\frac{\partial u}{\partial t} + \frac{\partial}{\partial x}(u^2) + \frac{\partial}{\partial y}(uv) + \frac{\partial}{\partial z}(uw) + \frac{1}{\rho}\frac{\partial p}{\partial x}$$
$$- \frac{1}{\rho}\left(\frac{\partial}{\partial x}\tau_{xx} + \frac{\partial}{\partial y}\tau_{xy} + \frac{\partial}{\partial z}\tau_{xz}\right) - f_x = 0 \tag{1.29b}$$

with similar forms for y and z. In both forms

$$\frac{1}{\rho}\tau_{ij} = \nu\left(\frac{\partial u_i}{\partial x_j} + \frac{\partial u_j}{\partial x_i} - \delta_{ij}\frac{2}{3}\frac{\partial u_k}{\partial x_k}\right)$$

where $\nu = \mu/\rho$ is the kinematic viscosity.

The reader will note that the above equations, with the exception of the convective acceleration terms, are *identical to those governing the problem of incompressible* (*or slightly compressible*) *elasticity*, which we have discussed in Chapter 12 of Volume 1.

1.4 Concluding remarks

We have observed in this chapter that a full set of Navier–Stokes equations can be written incorporating both compressible and incompressible behaviour. At this stage it is worth remarking that

1. More specialized sets of equations such as those which govern shallow-water flow or surface wave behaviour (Chapters 5, 7 and 8) will be of similar forms and need not be repeated here.
2. The essential difference from solid mechanics equations involves the non-self-adjoint convective terms.

Before proceeding with discretization and indeed the finite element solution of the full fluid equations, it is important to discuss in more detail the finite element procedures which are necessary to deal with such convective transport terms.

We shall do this in the next chapter where a standard scalar convective–diffusive–reactive equation is discussed.

References

1. C.K. Batchelor. *An Introduction to Fluid Dynamics*, Cambridge Univ. Press, 1967.
2. H. Lamb. *Hydrodynamics*, 6th ed., Cambridge Univ. Press, 1932.
3. C. Hirsch. *Numerical Computation of Internal and External Flows*, Vol. 1, Wiley, Chichester, 1988.
4. P.J. Roach. *Computational Fluid Mechanics*, Hermosa Press, Albuquerque, New Mexico, 1972.
5. H. Schlichting. *Boundary Layer Theory*, Pergamon Press, London, 1955.
6. L.D. Landau and E.M. Lifshitz. *Fluid Mechanics*, Pergamon Press, London, 1959.
7. R. Temam. *The Navier–Stokes Equation*, North-Holland, 1977.
8. I.G. Currie. *Fundamental Mechanics of Fluids*, McGraw-Hill, 1993.

2

Convection dominated problems – finite element approximations to the convection–diffusion equation

2.1 Introduction

In this chapter we are concerned with the steady-state and transient solutions of equations of the type

$$\frac{\partial \boldsymbol{\Phi}}{\partial t} + \frac{\partial \mathbf{F}_i}{\partial x_i} + \frac{\partial \mathbf{G}_i}{\partial x_i} + \mathbf{Q} = 0 \tag{2.1}$$

where in general $\boldsymbol{\Phi}$ is the basic dependent, vector-valued variable, $\mathbf{Q}$ is a source or reaction term vector and the *flux* matrices $\mathbf{F}$ and $\mathbf{G}$ are such that

$$\mathbf{F}_i = \mathbf{F}_i(\boldsymbol{\Phi}) \tag{2.2a}$$

and in general

$$\mathbf{G}_i = \mathbf{G}_i\left(\frac{\partial \boldsymbol{\Phi}}{\partial x_j}\right) \qquad \mathbf{Q} = \mathbf{Q}(x_i, \boldsymbol{\Phi}) \tag{2.2b}$$

In the above, x_i and i refer in the indicial manner to cartesian coordinates and quantities associated with these.

Equations (2.1) and (2.2) are *conservation laws* arising from a balance of the quantity $\boldsymbol{\Phi}$ with its fluxes $\mathbf{F}$ and $\mathbf{G}$ entering a control volume. Such equations are typical of fluid mechanics which we have discussed in Chapter 1. As such equations may also arise in other physical situations this chapter is devoted to the general discussion of their approximate solution.

The simplest form of Eqs (2.1) and (2.2) is one in which $\boldsymbol{\Phi}$ is a scalar and the fluxes are linear functions. Thus

$$\boldsymbol{\Phi} = \phi \qquad \mathbf{Q} = Q(x_i)$$
$$\mathbf{F}_i = F_i = U_i\phi \qquad \mathbf{G}_i = -k\frac{\partial \phi}{\partial x_i} \tag{2.3}$$

We now have in cartesian coordinates a scalar equation of the form

$$\frac{\partial \phi}{\partial t}+\frac{\partial (U_i\phi)}{\partial x_i}-\frac{\partial}{\partial x_i}\left(k\frac{\partial \phi}{\partial x_i}\right)+Q$$
$$\equiv \frac{\partial U}{\partial t}+\frac{\partial (U_x\phi)}{\partial x}+\frac{\partial (U_y\phi)}{\partial y}-\frac{\partial}{\partial x}\left(k\frac{\partial \phi}{\partial x}\right)-\frac{\partial}{\partial y}\left(k\frac{\partial \phi}{\partial y}\right)+Q=0 \quad (2.4)$$

which will serve as the basic model for most of the present chapter.

In the above equation U_i in general is a known velocity field, ϕ is a quantity being transported by this velocity in a convective manner or by diffusion action, where k is the diffusion coefficient.

In the above the term Q represents any external sources of the quantity ϕ being admitted to the system and also the reaction loss or gain which itself is dependent on the concentration ϕ.

The equation can be rewritten in a slightly modified form in which the convective term has been differentiated as

$$\frac{\partial \phi}{\partial t}+\underline{U_i\frac{\partial \phi}{\partial x_i}}+\phi\frac{\partial U_i}{\partial x_i}-\frac{\partial}{\partial x_i}\left(k\frac{\partial \phi}{\partial x_i}\right)+Q=0 \quad (2.5)$$

We will note that in the above form the problem is self-adjoint with the exception of a convective term which is underlined. The third term disappears if the flow itself is such that its divergence is zero, i.e. if

$$\frac{\partial U_i}{\partial x_i}=0 \quad \text{(summation over } i \text{ implied)} \quad (2.6)$$

In what follows we shall discuss the scalar equation in much more detail as many of the finite element remedies are only applicable to such scalar problems and are not transferable to the vector forms. As in the CBS scheme, which we shall introduce in Chapter 3, the equations of fluid dynamics will be split so that only scalar transport occurs, where this treatment is sufficient.

From Eqs (2.5) and (2.6) we have

$$\frac{\partial \phi}{\partial t}+U_i\frac{\partial \phi}{\partial x_i}-\frac{\partial}{\partial x_i}\left(k\frac{\partial \phi}{\partial x_i}\right)+Q=0 \quad (2.7)$$

We have encountered this equation in Volume 1 [Eq. (3.11), Sec. 3.1] in connection with heat transport, and indeed the general equation (2.1) can be termed the *transport equation* with **F** standing for the *convective* and **G** for *diffusive* flux quantities.

With the variable $\mathbf{\Phi}$ (Eq. 2.1) being approximated in the usual way:

$$\mathbf{\Phi} \approx \hat{\mathbf{\Phi}} = \mathbf{N}\tilde{\mathbf{\Phi}} = \sum \mathbf{N}_k\tilde{\mathbf{\Phi}}_k \quad (2.8)$$

the problem could be presented following the usual (weighted residual) semi-discretization process as

$$\mathbf{M}\dot{\tilde{\mathbf{\Phi}}} + \mathbf{H}\tilde{\mathbf{\Phi}} + \mathbf{f} = \mathbf{0} \quad (2.9)$$

but now even with standard Galerkin (Bubnov) weighting the matrix **H** will not be symmetric. However, this is a relatively minor computational problem compared with inaccuracies and instabilities in the solution which follow the arbitrary use of this weighting function.

This chapter will discuss the manner in which these difficulties can be overcome and the approximation improved.

We shall in the main address the problem of solving Eq. (2.4), i.e. the scalar form, and to simplify matters further we shall often start with the idealized one-dimensional equation:

$$\frac{\partial \phi}{\partial t} + U\frac{\partial \phi}{\partial x} - \frac{\partial}{\partial x}\left(k\frac{\partial \phi}{\partial x}\right) + Q = 0 \tag{2.10}$$

The term $\phi\, \partial U/\partial x$ has been removed here for simplicity. The above reduces in steady state to an ordinary differential equation:

$$U\frac{\mathrm{d}\phi}{\mathrm{d}x} - \frac{\mathrm{d}}{\mathrm{d}x}\left(k\frac{\mathrm{d}\phi}{\mathrm{d}x}\right) + Q = 0 \tag{2.11}$$

in which we shall often assume U, k and Q to be constant. The basic concepts will be evident from the above which will later be extended to multidimensional problems, still treating ϕ as a scalar variable.

Indeed the methodology of dealing with the first space derivatives occurring in differential equations governing a problem, which as shown in Chapter 3 of Volume 1 lead to non-self-adjointness, opens the way for many new physical situations.

The present chapter will be divided into three parts. Part I deals with *steady-state situations* starting from Eq. (2.11), Part II with *transient solutions* starting from Eq. (2.10) and Part III dealing with vector-valued functions. Although the scalar problem will mainly be dealt with here in detail, the discussion of the procedures can indicate the choice of optimal ones which will have much bearing on the solution of the general case of Eq. (2.1). We shall only discuss briefly the extension of some procedures to the vector case in Part III as such extensions are generally heuristic.

Part I: Steady state

2.2 The steady-state problem in one dimension

2.2.1 Some preliminaries

We shall consider the discretization of Eq. (2.11) with

$$\phi \approx \sum N_i \tilde{\phi}_i = \mathbf{N}\tilde{\boldsymbol{\phi}} \tag{2.12}$$

where N_k are shape functions and $\tilde{\boldsymbol{\phi}}$ represents a set of still unknown parameters. Here we shall take these to be the nodal values of ϕ. This gives for a typical internal node i the approximating equation

$$K_{ij}\tilde{\phi}_j + f_i = 0 \tag{2.13}$$

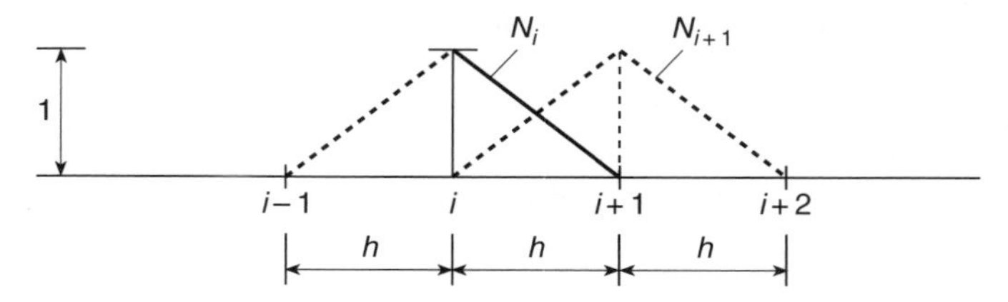

Fig. 2.1 A linear shape function for a one-dimensional problem.

where

$$K_{ij} = \int_0^L W_i U \frac{\mathrm{d}N_j}{\mathrm{d}x}\,\mathrm{d}x + \int_0^L \frac{\mathrm{d}W_i}{\mathrm{d}x} k \frac{\mathrm{d}N_j}{\mathrm{d}x}\,\mathrm{d}x$$
$$f_i = \int_0^L W_i Q\,\mathrm{d}x \tag{2.14}$$

and the domain of the problem is $0 \leqslant x \leqslant L$.

For linear shape functions, Galerkin weighting ($W_i = N_i$) and elements of equal size h, we have for *constant* values of U, k and Q (Fig. 2.1) a typical assembled equation

$$(-Pe - 1)\tilde{\phi}_{i-1} + 2\tilde{\phi}_i + (Pe - 1)\tilde{\phi}_{i+1} + \frac{Qh^2}{k} = 0 \tag{2.15}$$

where

$$Pe = \frac{Uh}{2k} \tag{2.16}$$

is the element *Peclet* number. The above is, incidentally, identical to the usual central finite difference approximation obtained by putting

$$\frac{\mathrm{d}\phi}{\mathrm{d}x} \approx \frac{\tilde{\phi}_{i+1} - \tilde{\phi}_{i-1}}{2h} \tag{2.17a}$$

and

$$\frac{\mathrm{d}^2\phi}{\mathrm{d}x^2} \approx \frac{\tilde{\phi}_{i+1} - 2\tilde{\phi}_i + \tilde{\phi}_{i-1}}{h^2} \tag{2.17b}$$

The algebraic equations are obviously non-symmetric and in addition their accuracy deteriorates as the parameter Pe increases. Indeed as $Pe \to \infty$, i.e. when only convective terms are of importance, the solution is purely oscillatory and bears no relation to the underlying problem, as shown in the simple example where Q is zero of Fig. 2.2 with curves labelled $\alpha = 0$. (Indeed the solution for this problem is now only possible for an odd number of elements and not for even.)

Of course the above is partly a problem of boundary conditions. When diffusion is omitted only a single boundary condition can be imposed and when the diffusion is small we note that the downstream boundary condition ($\phi = 1$) is felt in only a very small region of a *boundary layer* evident from the exact solution[1]

$$\phi = \frac{1 - \mathrm{e}^{Ux/k}}{1 - \mathrm{e}^{UL/k}} \tag{2.18}$$

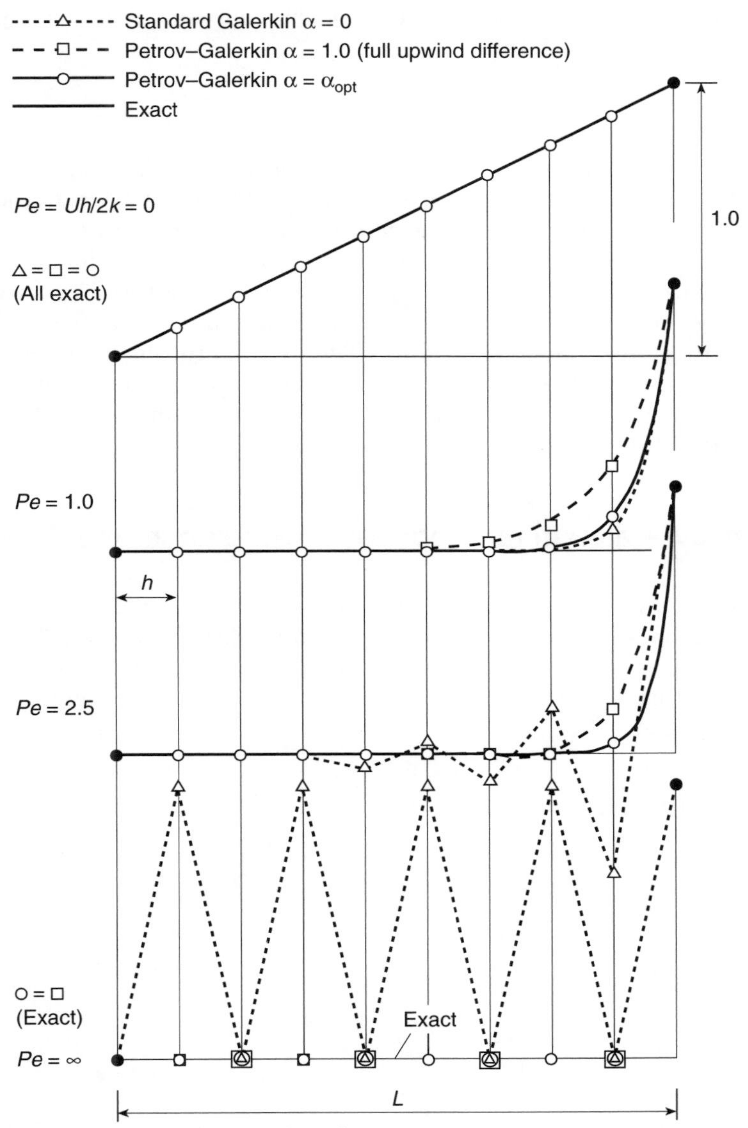

Fig. 2.2 Approximations to $U\,\mathrm{d}\phi/\mathrm{d}x - k\,\mathrm{d}^2\phi/\mathrm{d}x^2 = 0$ for $\phi = 0$, $x = 0$ and $\phi = 1$, $x = L$ for various Peclet numbers.

Motivated by the fact that the propagation of information is in the direction of velocity U, the finite difference practitioners were the first to overcome the bad approximation problem by using *one-sided* finite differences for approximating the first derivative.[2–5] Thus in place of Eq. (2.17a) and with positive U, the approximation was put as

$$\frac{\mathrm{d}\phi}{\mathrm{d}x} \approx \frac{\tilde{\phi}_i - \tilde{\phi}_{i-1}}{h} \tag{2.19}$$

changing the central finite difference form of the approximation to the governing equation as given by Eq. (2.15) to

$$(-2Pe - 1)\tilde{\phi}_{i-1} + (2 + 2Pe)\tilde{\phi}_i - \tilde{\phi}_{i+1} + \frac{Qh^2}{k} = 0 \tag{2.20}$$

With this *upwind* difference approximation, realistic (though not always accurate) solutions can be obtained through the whole range of Peclet numbers of the example of Fig. 2.2 as shown there by curves labelled $\alpha = 1$. However, now exact nodal solutions are only obtained for pure convection ($Pe = \infty$), as shown in Fig. 2.2, in a similar way as the Galerkin finite element form gives exact nodal answers for pure diffusion.

How can such upwind differencing be introduced into the finite element scheme and generalized to more complex situations? This is the problem that we shall now address, and indeed will show that again, as in self-adjoint equations, the finite element solution can result in exact nodal values for the one-dimensional approximation for all Peclet numbers.

2.2.2 Petrov–Galerkin methods for upwinding in one dimension

The first possibility is that of the use of a Petrov–Galerkin type of weighting in which $W_i \neq N_i$.[6–9] Such weightings were first suggested by Zienkiewicz *et al.*[6] in 1975 and used by Christie *et al.*[7] In particular, again for elements with linear shape functions N_i, shown in Fig. 2.1, we shall take, as shown in Fig. 2.3, weighting functions constructed so that

$$W_i = N_i + \alpha W_i^* \tag{2.21}$$

where W_i^* is such that

$$\int_{\Omega_e} W_i^* \,\mathrm{d}x = \pm\frac{h}{2} \tag{2.22}$$

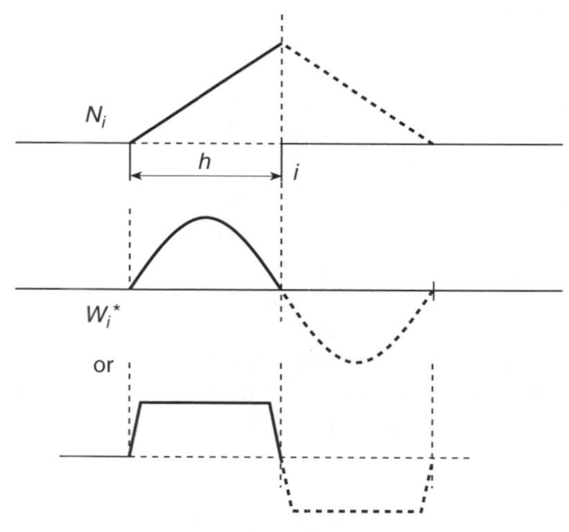

Fig. 2.3 Petrov–Galerkin weight function $W_i = N_i + \alpha W_i^*$. Continuous and discontinuous definitions.

the sign depending on whether U is a velocity directed towards or away from the node.

Various forms of W_i^* are possible, but the most convenient is the following simple definition which is, of course, a discontinuous function (see the note at the end of this section):

$$\alpha W_i^* = \alpha \frac{h}{2} \frac{dN_i}{dx} (\text{sign } U) \tag{2.23}$$

With the above weighting functions the approximation equivalent to that of Eq. (2.15) becomes

$$[-Pe(\alpha + 1) - 1]\tilde{\phi}_{i-1} + [2 + 2\alpha(Pe)]\tilde{\phi}_i + [-Pe(\alpha - 1) - 1]\tilde{\phi}_{i+1} + \frac{Qh^2}{k} = 0 \tag{2.24}$$

Immediately we see that with $\alpha = 0$ the standard Galerkin approximation is recovered [Eq. (2.15)] and that with $\alpha = 1$ the full upwinded discrete equation (2.20) is available, each giving exact nodal values for purely diffusive or purely convective cases respectively.

Now if the value of α is chosen as

$$|\alpha| = \alpha_{\text{opt}} = \coth|Pe| - \frac{1}{|Pe|} \tag{2.25}$$

then exact nodal values will be given *for all values of Pe*. The proof of this is given in reference 7 for the present, one-dimensional, case where it is also shown that if

$$|\alpha| > \alpha_{\text{crit}} = 1 - \frac{1}{|Pe|} \tag{2.26}$$

oscillatory solutions will never arise. The results of Fig. 2.2 show indeed that with $\alpha = 0$, i.e. the Galerkin procedure, oscillations will occur when

$$|Pe| > 1 \tag{2.27}$$

Figure 2.4 shows the variation of α_{opt} and α_{crit} with Pe.*

Although the proof of optimality for the upwinding parameter was given for the case of constant coefficients and constant size elements, nodally exact values will also be given if $\alpha = \alpha_{\text{opt}}$ is chosen for each element individually. We show some typical solutions in Fig. 2.5[10] for a variable source term $Q = Q(x)$, convection coefficients $U = U(x)$ and element sizes. Each of these is compared with a standard Galerkin solution, showing that even when the latter does not result in oscillations the accuracy is improved. Of course in the above examples the Petrov–Galerkin weighting must be applied to all terms of the equation. When this is not done (as in simple finite difference upwinding) totally wrong results will be obtained, as shown in the finite difference results of Fig. 2.6, which was used in reference 11 to discredit upwinding methods. The effect of α on the source term is not apparent in Eq. (2.24) where Q is constant in the whole domain, but its influence is strong when $Q = Q(x)$.

Continuity requirements for weighting functions

The weighting function W_i (or W_i^*) introduced in Fig. 2.3 can of course be discontinuous as far as the contributions to the convective terms are concerned [see Eq. (2.14)],

* Subsequently Pe is interpreted as an absolute value.

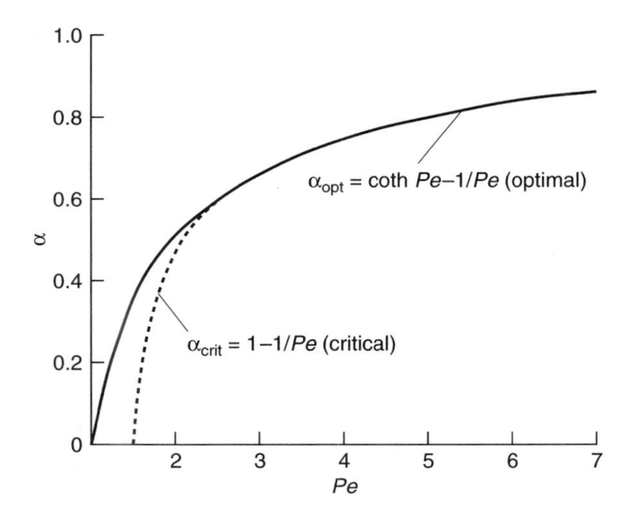

Fig. 2.4 Critical (stable) and optimal values of the 'upwind' parameter α for different values of $Pe = Uh/2k$.

i.e.

$$\int_0^L W_i \frac{dF}{dx}\, dx \qquad \text{or} \qquad \int_0^L W_i U \frac{dN_i}{dx}\, dx$$

Clearly no difficulty arises at the discontinuity in the evaluation of the above integrals. However, when evaluating the diffusion term, we generally introduce integration by parts and evaluate such terms as

$$\int_0^L \frac{dW_i}{dx} k \frac{dN_j}{dx}\, dx$$

in place of the form

$$\int_0^L W_i \frac{d}{dx}\left(k \frac{dN_j}{dx}\right) dx$$

Here a local infinity will occur with discontinuous W_i. To avoid this difficulty we modify the discontinuity of the W_i^* part of the weighting function to occur *within* the element[1] and thus avoid the discontinuity at the node in the manner shown in Fig. 2.3. Now direct integration can be used, showing in the present case zero contributions to the diffusion term, as indeed happens with C_0 continuous functions for W_i^* used in earlier references.

2.2.3 Balancing diffusion in one dimension

The comparison of the nodal equations (2.15) and (2.16) obtained on a uniform mesh and for a constant Q shows that the effect of the Petrov–Galerkin procedure is equivalent to the use of a standard Galerkin process with the addition of a diffusion

$$k_b = \tfrac{1}{2}\alpha U h \tag{2.28}$$

to the original differential equation (2.11).

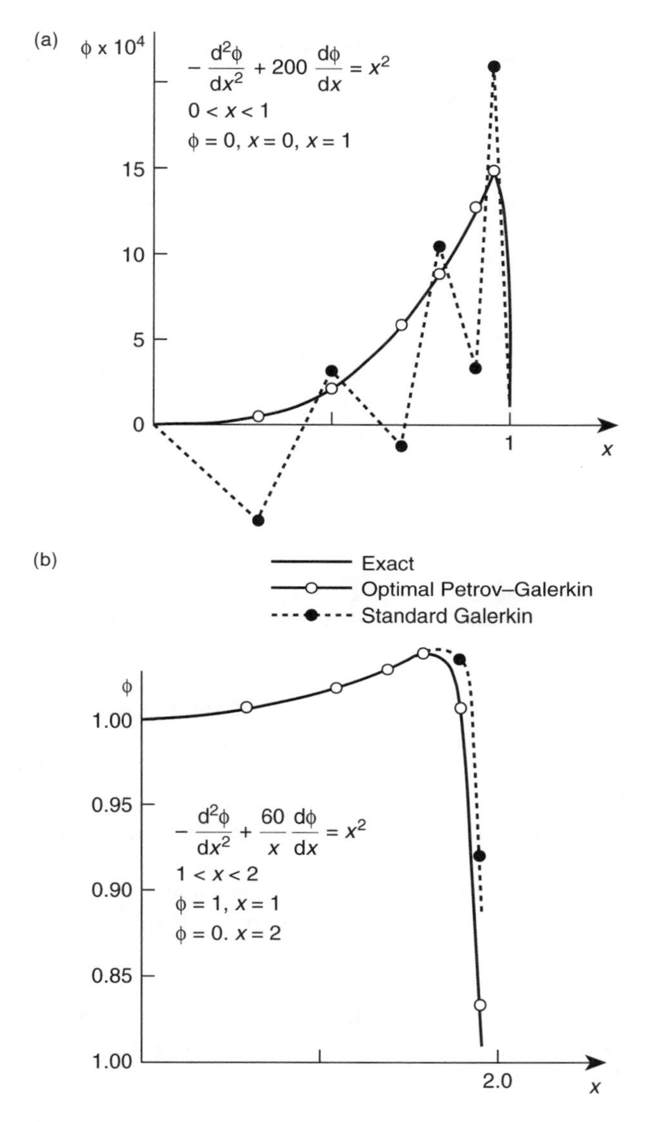

Fig. 2.5 Application of standard Galerkin and Petrov–Galerkin (optimal) approximation: (a) variable source term equation with constants k and h; (b) variable source term with a variable U.

The reader can easily verify that with this substituted into the original equation, thus writing now in place of Eq. (2.11)

$$U\frac{d\phi}{dx} - \frac{d}{dx}\left[(k + k_b)\frac{d\phi}{dx}\right] + Q = 0 \tag{2.29}$$

we obtain an identical expression to that of Eq. (2.24) providing Q is constant and a standard Galerkin procedure is used.

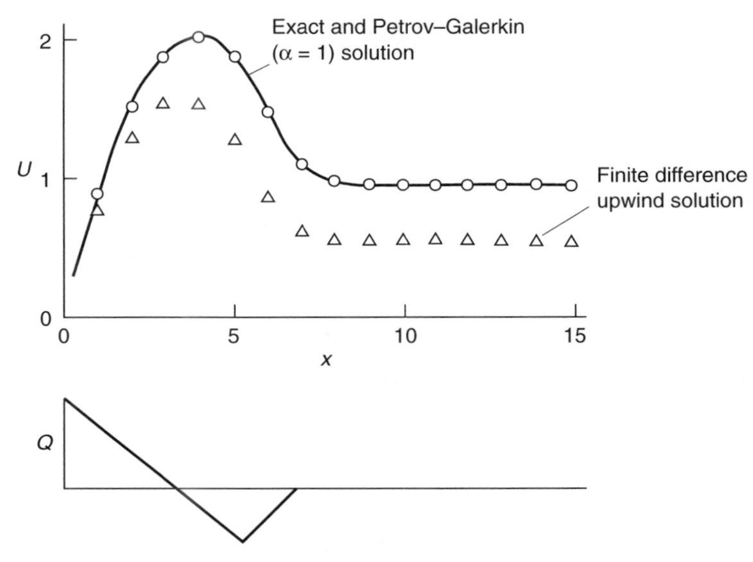

Fig. 2.6 A one-dimensional pure convective problem ($k = 0$) with a variable source term Q and constant U. Petrov–Galerkin procedure results in an exact solution but simple finite difference upwinding gives substantial error.

Such *balancing diffusion* is easier to implement than Petrov–Galerkin weighting, particularly in two or three dimensions, and has some physical merit in the interpretation of the Petrov–Galerkin methods. However, it does not provide the modification of source terms required, and for instance in the example of Fig. 2.6 will give erroneous results identical with a simple finite difference, upwind, approximation.

The concept of *artificial diffusion* introduced frequently in finite difference models suffers of course from the same drawbacks and in addition cannot be logically justified.

It is of interest to observe that a central difference approximation, when applied to the original equations (or the use of the standard Galerkin process), fails by introducing a *negative diffusion* into the equations. This 'negative' diffusion is countered by the present, balancing, one.

2.2.4 A variational principle in one dimension

Equation (2.11), which we are here considering, is not self-adjoint and hence is not directly derivable from any variational principle. However, it was shown by Guymon *et al.*[12] that it is a simple matter to derive a variational principle (or ensure self-adjointness which is equivalent) if the operator is premultiplied by a suitable function p. Thus we write a weak form of Eq. (2.11) as

$$\int_0^L Wp\left[U\frac{\mathrm{d}\phi}{\mathrm{d}x} - \frac{\mathrm{d}}{\mathrm{d}x}\left(k\frac{\mathrm{d}\phi}{\mathrm{d}x}\right) + Q\right]\mathrm{d}x = 0 \tag{2.30}$$

where $p = p(x)$ is as yet undetermined. This gives, on integration by parts,

$$\int_0^L \left[W \frac{\mathrm{d}\phi}{\mathrm{d}x} \left(pU + k\frac{\mathrm{d}p}{\mathrm{d}x} \right) + \frac{\mathrm{d}W}{\mathrm{d}x} (kp) \frac{\mathrm{d}\phi}{\mathrm{d}x} + WpQ \right] \mathrm{d}x + \left| Wpk \frac{\mathrm{d}\phi}{\mathrm{d}x} \right|_0^L = 0 \qquad (2.31)$$

Immediately we see that the operator can be made self-adjoint and a symmetric approximation achieved if the first term in square brackets is made zero (see also Chapter 3 of Volume 1, Sec. 3.11.2, for this derivation). This requires that p be chosen so that

$$pU + k\frac{\mathrm{d}p}{\mathrm{d}x} = 0 \qquad (2.32a)$$

or that

$$p = \text{constant} \times \mathrm{e}^{-Ux/k} = \text{constant} \times \mathrm{e}^{-2(Pe)x/h} \qquad (2.32b)$$

For such a form corresponding to the existence of a variational principle the 'best' approximation is that of the Galerkin method with

$$W = N_i \qquad \phi = \sum N_j \phi_j \qquad (2.33)$$

Indeed, as shown in Volume 1, such a formulation will, in one dimension, yield answers exact at nodes (see Appendix H of Volume 1). It must therefore be equivalent to that obtained earlier by weighting in the Petrov–Galerkin manner. Inserting the approximation of Eq. (2.33) into Eq. (2.31), with Eqs (2.32) defining p using an origin at $x = x_i$, we have for the ith equation of the uniform mesh

$$\int_{-h}^{h} \left[\frac{\mathrm{d}N_i}{\mathrm{d}x} (k\,\mathrm{e}^{-2(Pe)x/h}) \frac{\mathrm{d}N_j}{\mathrm{d}x} \tilde{\phi}_j + N_i \mathrm{e}^{-2(Pe)x/h} Q \right] \mathrm{d}x = 0 \qquad (2.34)$$

with $j = i-1,\ i,\ i+1$. This gives, after some algebra, a typical nodal equation:

$$(1 - \mathrm{e}^{-2(Pe)})\tilde{\phi}_{i+1} + (\mathrm{e}^{-2(Pe)} - \mathrm{e}^{-2(Pe)})\tilde{\phi}_i - (1 - \mathrm{e}^{-2(Pe)})\tilde{\phi}_{i+1}$$
$$- \frac{Qh^2}{2(Pe)k} (\mathrm{e}^{Pe} - \mathrm{e}^{-Pe})^2 = 0 \qquad (2.35)$$

which can be shown to be identical with the expression (2.24) into which $\alpha = \alpha_{\mathrm{opt}}$ given by Eq. (2.25) has been inserted.

Here we have a somewhat more convincing proof of the optimality of the proposed Petrov–Galerkin weighting.[13,14] However, serious drawbacks exist. The numerical evaluation of the integrals is difficult and the equation system, though symmetric overall, is not well conditioned if p is taken as a continuous function of x through the whole domain. The second point is easily overcome by taking p to be *discontinuously* defined, for instance taking the origin of x at point i for *all assemblies* as we did in deriving Eq. (2.35). This is permissible by arguments given in Sec. 2.2 and is equivalent to scaling the full equation system row by row.[13] Now of course the total equation system ceases to be symmetric.

The numerical integration difficulties disappear, of course, if the simple weighting functions previously derived are used. However, the proof of equivalence is important as the problem of determining the optimal weighting is no longer necessary.

2.2.5 Galerkin least square approximation (GLS) in one dimension

In the preceding sections we have shown that several, apparently different, approaches have resulted in identical (or almost identical) approximations. Here yet another procedure is presented which again will produce similar results. In this a combination of the standard Galerkin and least square approximations is made.[15,16]

If Eq. (2.11) is rewritten as

$$L\phi + Q = 0 \qquad \phi \approx \hat{\phi} = \mathbf{N}\tilde{\boldsymbol{\phi}} \tag{2.36a}$$

with

$$L = U\frac{\mathrm{d}}{\mathrm{d}x} - \frac{\mathrm{d}}{\mathrm{d}x}\left(k\frac{\mathrm{d}}{\mathrm{d}x}\right) \tag{2.36b}$$

the standard Galerkin approximation gives for the kth equation

$$\int_0^L N_k L(\mathbf{N})\tilde{\boldsymbol{\phi}}\,\mathrm{d}x + \int_0^L N_k Q\,\mathrm{d}x = 0 \tag{2.37}$$

with boundary conditions omitted for clarity.

Similarly, a least square residual minimization (see Chapter 3 of Volume 1, Sec. 3.14.2) results in

$$R = L\hat{\phi} + Q \qquad \text{and} \qquad \frac{1}{2}\frac{\mathrm{d}}{\mathrm{d}\tilde{\phi}_k}\int_0^L R^2\,\mathrm{d}x = \int_0^L \frac{\mathrm{d}(L\hat{\phi})}{\mathrm{d}\tilde{\phi}_k}(L\hat{\phi} + Q)\,\mathrm{d}x = 0 \tag{2.38}$$

or

$$\int_0^L \left(U\frac{\mathrm{d}N_k}{\mathrm{d}x} - \frac{\mathrm{d}}{\mathrm{d}x}\left(k\frac{\mathrm{d}}{\mathrm{d}x}\right)N_k\right)(L\hat{\phi} + Q) = 0 \tag{2.39}$$

If the final approximation is written as a linear combination of Eqs (2.37) and (2.39), we have

$$\int_0^L \left(N_k + \lambda U\frac{\mathrm{d}N_k}{\mathrm{d}x} - \lambda\frac{\mathrm{d}}{\mathrm{d}x}\left(k\frac{\mathrm{d}}{\mathrm{d}x}\right)N_k\right)(L\hat{\phi} + Q)\,\mathrm{d}x = 0 \tag{2.40}$$

This is of course, the same as the Petrov–Galerkin approximation with an undetermined parameter λ. If the second-order term is omitted (as could be done assuming linear N_k and a curtailment as in Fig. 2.3) and further if we take

$$\lambda = \frac{|\alpha|h}{2|U|} \tag{2.41}$$

the approximation is identical to that of the Petrov–Galerkin method with the weighting given by Eqs (2.21) and (2.22).

Once again we see that a Petrov–Galerkin form written as

$$\int_0^L \left(N_k + \frac{|\alpha|}{2}\frac{Uh}{|U|}\frac{\mathrm{d}N_k}{\mathrm{d}x}\right)\left(U\frac{\mathrm{d}\hat{\phi}}{\mathrm{d}x} - \frac{\mathrm{d}}{\mathrm{d}x}\left(k\frac{\mathrm{d}\hat{\phi}}{\mathrm{d}x}\right) + Q\right)\mathrm{d}x = 0 \tag{2.42}$$

is a result that follows from diverse approaches, though only the variational form of Sec. 2.2.4 explicitly determines the value of α that should optimally be used. In all the other derivations this value is determined by an *a posteriori* analysis.

2.2.6 The finite increment calculus (FIC) for stabilizing the convective–diffusion equation in one dimension

As mentioned in the previous sections, there are many procedures which give identical results to those of the Petrov–Galerkin approximations. We shall also find a number of such procedures arising directly from the transient formulations discussed in Part II of this chapter; however there is one further simple process which can be applied directly to the steady-state equation. This process was suggested by Oñate in 1998[17] and we shall describe its basis below.

We shall start at the stage where the conservation equation of the type given by Eq. (2.5) is derived. Now instead of considering an infinitesimal control volume of length '$\mathrm{d}x$' which is going to zero, we shall consider a finite length δ. Expanding to one higher order by Taylor series (backwards), we obtain instead of Eq. (2.11)

$$-U\frac{\mathrm{d}\phi}{\mathrm{d}x}+\frac{\mathrm{d}}{\mathrm{d}x}\left(k\frac{\mathrm{d}\phi}{\mathrm{d}x}\right)+Q-\frac{\delta}{2}\left[-U\frac{\mathrm{d}\phi}{\mathrm{d}x}+\frac{\mathrm{d}}{\mathrm{d}x}\left(k\frac{\mathrm{d}\phi}{\mathrm{d}x}\right)+Q\right]=0 \qquad (2.43)$$

with δ being the finite distance which is smaller than or equal to that of the element size h. Rearranging terms and substituting $\delta=\alpha h$ we have

$$U\frac{\mathrm{d}\phi}{\mathrm{d}x}-\frac{\mathrm{d}}{\mathrm{d}x}\left[\left(k+\frac{\alpha hU}{2}\right)\frac{\mathrm{d}\phi}{\mathrm{d}x}\right]+Q-\frac{\delta}{2}\frac{\mathrm{d}Q}{\mathrm{d}x}=0 \qquad (2.44)$$

In the above equation we have omitted the higher order expansion for the diffusion term as in the previous section.

From the last equation we see immediately that a stabilizing term has been recovered and the additional term $\alpha hU/2$ is identical to that of the Petrov–Galerkin form (Eq. 2.28).

There is no need to proceed further and we see how simply the finite increment procedure has again yielded exactly the same result by simply modifying the conservation differential equations. In reference 17 it is shown further that arguments can be brought to determine α as being precisely the optimal value we have already obtained by studying the Petrov–Galerkin method.

2.2.7 Higher-order approximations

The derivation of accurate Petrov–Galerkin procedures for the convective diffusion equation is of course possible for any order of finite element expansion. In reference 9 Heinrich and Zienkiewicz show how the procedure of studying exact discrete solutions can yield optimal upwind parameters for quadratic shape functions. However, here the simplest approach involves the procedures of Sec. 2.2.4, which

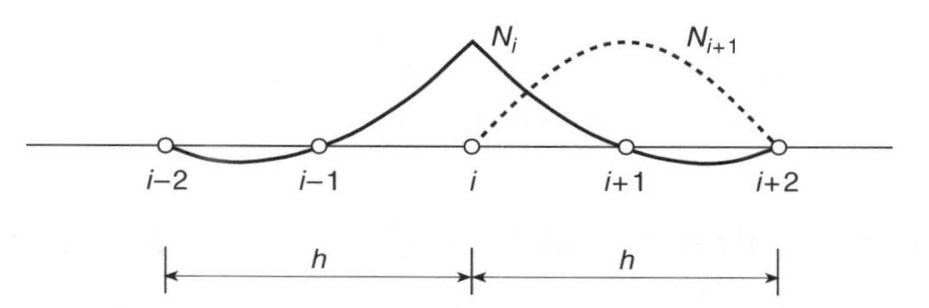

Fig. 2.7 Assembly of one-dimensional quadratic elements.

are available of course for any element expansion and, as shown before, will always give an optimal approximation.

We thus recommend the reader to pursue the example discussed in that section and, by extending Eq. (2.34), to arrive at an appropriate equation linking the two quadratic elements of Fig. 2.7.

For practical purposes for such elements it is possible to extend the Petrov–Galerkin weighting of the type given in Eqs (2.21) to (2.23) now using

$$\alpha_{\text{opt}} = \coth Pe - \frac{1}{Pe} \qquad \text{and} \qquad \alpha W_i^* = \alpha \frac{h}{4} \frac{\mathrm{d}N_i}{\mathrm{d}x} (\text{sign } U) \tag{2.45}$$

This procedure, though not as exact as that for linear elements, is very effective and has been used with success for solution of Navier–Stokes equations.[18]

In recent years, the subject of optimal upwinding for higher-order approximations has been studied further and several references show the developments.[19,20] It is of interest to remark that the procedure known as the *discontinuous Galerkin* method avoids most of the difficulties of dealing with higher-order approximations. This procedure was recently applied to convection–diffusion problems and indeed to other problems of fluid mechanics by Oden and coworkers.[21–23] As the methodology is not available for lowest polynomial order of unity we do not include the details of the method here but for completeness we show its derivation in Appendix B.

2.3 The steady-state problem in two (or three) dimensions

2.3.1 General remarks

It is clear that the application of standard Galerkin discretization to the steady-state scalar convection–diffusion equation in several space dimensions is similar to the problem discussed previously in one dimension and will again yield unsatisfactory answers with high oscillation for local Peclet numbers greater than unity.

The equation now considered is the steady-state version of Eq. (2.7), i.e.

$$U_x \frac{\partial \phi}{\partial x} + U_y \frac{\partial \phi}{\partial y} - \frac{\partial}{\partial x}\left(k \frac{\partial \phi}{\partial x}\right) - \frac{\partial}{\partial y}\left(k \frac{\partial \phi}{\partial y}\right) + Q = 0 \tag{2.46a}$$

in two dimensions or more generally using indicial notation

$$U_i \frac{\partial \phi}{\partial x_i} - \frac{\partial}{\partial x_i}\left(k \frac{\partial \phi}{\partial x_i}\right) + Q = 0 \tag{2.46b}$$

in both two and three dimensions.

Obviously the problem is now of greater practical interest than the one-dimensional case so far discussed, and a satisfactory solution is important. Again, all of the possible approaches we have discussed are applicable.

2.3.2 Streamline (Upwind) Petrov–Galerkin weighting (SUPG)

The most obvious procedure is to use again some form of Petrov–Galerkin method of the type introduced in Sec. 2.2.2 and Eqs (2.21) to (2.25), seeking optimality of α in some heuristic manner. Restricting attention here to two-dimensions, we note immediately that the Peclet parameter

$$\mathbf{Pe} = \frac{\mathbf{U}h}{2k} \qquad \mathbf{U} = \begin{Bmatrix} U_1 \\ U_2 \end{Bmatrix} \tag{2.47}$$

is now a 'vector' quantity and hence that upwinding needs to be 'directional'.

The first reasonably satisfactory attempt to do this consisted of determining the optimal Petrov–Galerkin formulation using αW^* based on components of $\mathbf{U}$ associated to the *sides of elements* and of obtaining the final weight functions by a blending procedure.[8,9]

A better method was soon realized when the analogy between balancing diffusion and upwinding was established, as shown in Sec. 2.2.3. In two (or three) dimensions the convection is only active in the direction of the resultant element velocity $\mathbf{U}$, and hence the corrective, or *balancing, diffusion* introduced by upwinding should be anisotropic with a coefficient different from zero only in the direction of the velocity resultant. This innovation introduced simultaneously by Hughes and Brooks[24,25] and Kelly *et al.*[10] can be readily accomplished by taking the individual weighting functions as

$$\begin{aligned} W_k &= N_k + \alpha W_k^* = N_k + \frac{\alpha h}{2} \frac{U_1(\partial N_k/\partial x_1) + U_2(\partial N_k/\partial x_2)}{|\mathbf{U}|} \\ &\equiv N_k + \frac{\alpha h}{2} \frac{U_i}{|\mathbf{U}|} \frac{\partial N_k}{\partial x_i} \end{aligned} \tag{2.48}$$

where α is determined for each element by the previously found expression (2.22) written as follows:

$$\alpha = \alpha_{\text{opt}} = \coth Pe - \frac{1}{Pe} \tag{2.49}$$

with

$$Pe = \frac{|\mathbf{U}|h}{2k} \tag{2.50a}$$

and

$$|\mathbf{U}| = (U_1^2 + U_2^2)^{1/2} \qquad \text{or} \qquad \sqrt{U_i U_i} \tag{2.50b}$$

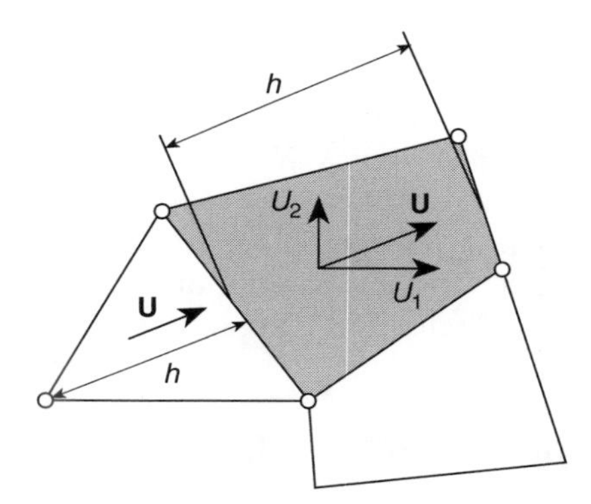

Fig. 2.8 A two-dimensional, streamline assembly. Element size h and streamline directions.

The above expressions presuppose that the velocity components U_1 and U_2 in a particular element are substantially constant and that the element size h can be reasonably defined.

Figure 2.8 shows an assembly of linear triangles and bilinear quadrilaterals for each of which the mean resultant velocity **U** is indicated. Determination of the element size h to use in expression (2.50) is of course somewhat arbitrary. In Fig. 2.8 we show it simply as the maximum size in the direction of the velocity vector.

The form of Eq. (2.48) is such that the 'non-standard' weighting W^* has a zero effect in the direction in which the velocity component is zero. Thus the balancing diffusion is only introduced in the direction of the resultant velocity (convective) vector **U**. This can be verified if Eq. (2.46) is written in tensorial (indicial) notation as

$$U_i \frac{\partial \phi}{\partial x_i} - \frac{\partial}{\partial x_i}\left(k \frac{\partial \phi}{\partial x_i}\right) + Q = 0 \tag{2.51a}$$

In the discretized form the 'balancing diffusion' term [obtained from weighting the first term of the above with W of Eq. (2.48)] becomes

$$\int_\Omega \frac{\partial N}{\partial x_i} \tilde{k}_{ij} \frac{\partial N}{\partial x_j} \, d\Omega \tag{2.51b}$$

with

$$\tilde{k}_{ij} = \frac{\alpha U_i U_j}{|\mathbf{U}|} \frac{h}{2} \tag{2.51c}$$

This indicates a highly anisotropic diffusion with zero coefficients normal to the convective velocity vector directions. It is therefore named the *streamline balancing diffusion*[10,24,25] or streamline upwind Petrov–Galerkin process.

The streamline diffusion should allow discontinuities in the direction normal to the streamline to travel without appreciable distortion. However, with the standard finite element approximations actual discontinuities cannot be modelled and in practice some oscillations may develop when the function exhibits 'shock like' behaviour. For this reason it is necessary to add some smoothing diffusion in the direction normal to the streamlines and some investigators make appropriate suggestions.[26–29]

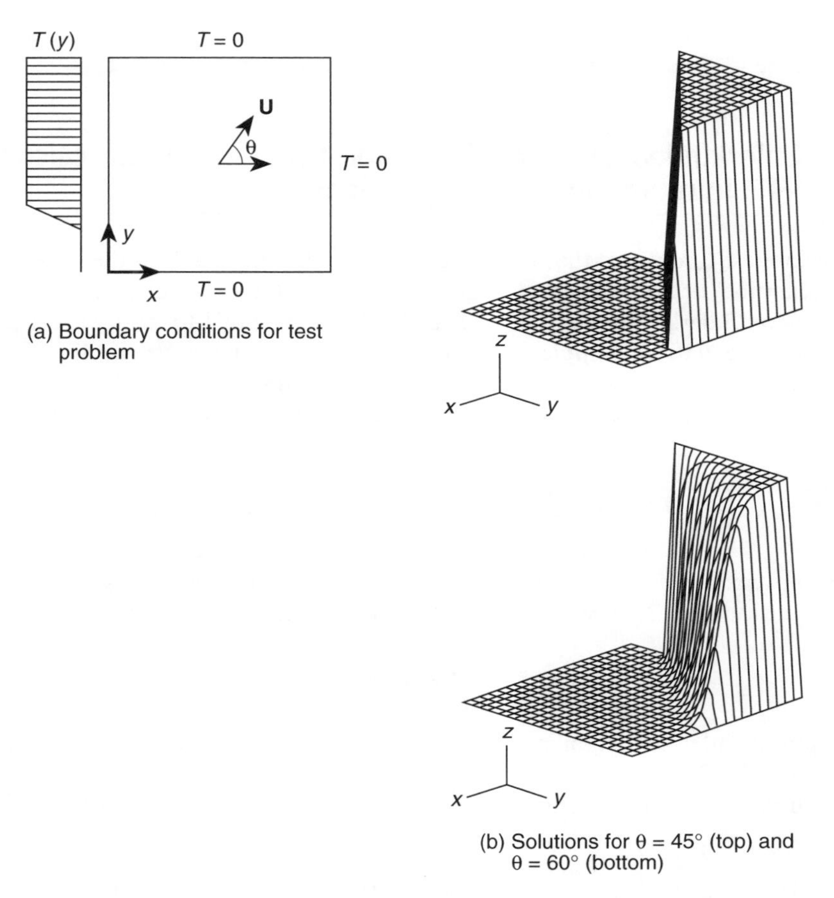

Fig. 2.9 'Streamline' procedures in a two-dimensional problem of pure convection. Bilinear elements.[31]

The mathematical validity of the procedures introduced in this section has been established by Johnson *et al.*[30] for $\alpha = 1$, showing convergence improvement over the standard Galerkin process. However, the proof does not include any optimality in the selection of α values as shown by Eq. (2.49).

Figure 2.9 shows a typical solution of Eq. (2.46), indicating the very small amount of 'cross-wind diffusion', i.e. allowing discontinuities to propagate in the direction of flow without substantial smearing.[31]

A more convincing 'optimality' can be achieved by applying the exponential modifying function, making the problem self-adjoint. This of course follows precisely the procedures of Sec. 2.2.4 and is easily accomplished if the velocities are constant in the element assembly domain. If velocities vary from element to element, again the exponential functions

$$p = e^{-Ux'/k} \tag{2.52}$$

with x' orientated in the velocity direction in each element can be taken. This appears to have been first implemented by Sampaio[31] but problems regarding the origin of coordinates, etc., have once again to be addressed. However, the results are essentially similar here to those achieved by Petrov–Galerkin procedures.

2.3.3 Galerkin least squares (GLS) and finite increment calculus (FIC) in multidimensional problems

It is of interest to observe that the somewhat intuitive approach to the generation of the 'streamline' Petrov–Galerkin weight functions of Eq. (2.48) can be avoided if the least square Galerkin procedures of Sec. 2.2.4 are extended to deal with the multi-dimensional equation. Simple extension of the reasoning given in Eqs (2.36) to (2.42) will immediately yield the weighting of Eq. (2.48).

Extension of the GLS to two or three dimensions gives (again using indicial notation)

$$\int_\Omega \left(N_k + \lambda U_i \frac{\partial N_k}{\partial x_i}\right)\left(U_j \frac{\partial \hat{\phi}}{\partial x_j} - \frac{\partial}{\partial x_j}\left(k \frac{\partial \hat{\phi}}{\partial x_j}\right) + Q\right) \mathrm{d}\Omega = 0 \tag{2.53}$$

In the above equation, higher-order terms are omitted for the sake of simplicity. As in one dimension (Eq. 2.40) we have an additional weighting term. Now assuming

$$\lambda = \frac{\alpha h}{2|\mathbf{U}|} \tag{2.54}$$

we obtain an identical stabilizing term to that of the streamline Petrov–Galerkin procedure (Eq. 2.51).

The finite increment calculus method in multidimensions can be written as[17]

$$U_j \frac{\partial \phi}{\partial x_j} - \frac{\partial}{\partial x_j}\left(k \frac{\partial \phi}{\partial x_j}\right) + Q - \frac{\delta_i}{2} \frac{\partial}{\partial x_i}\left[U_j \frac{\partial \phi}{\partial x_j} - \frac{\partial}{\partial x_j}\left(k \frac{\partial \phi}{\partial x_j}\right) + Q\right] = 0 \tag{2.55}$$

Note that the value of δ_i is now dependent on the coordinate directions. To obtain streamline-oriented stabilization, we simply assume that δ_i is the projection oriented along the streamlines. Now

$$\delta_i = \delta \frac{U_i}{|\mathbf{U}|} \tag{2.56}$$

with $\delta = \alpha h$. Again, omitting the higher order terms in k, the streamline Petrov–Galerkin form of stabilization is obtained (Eq. 2.51). The reader can verify that both the GLS and FIC produce the correct weighting for the source term Q as of course is required by the Petrov–Galerkin method.

2.4 Steady state – concluding remarks

In Secs 2.2 and 2.3 we presented several currently used procedures for dealing with the steady-state convection–diffusion equation with a scalar variable. All of these translate essentially to the use of streamline Petrov–Galerkin discretization, though of course the modification of the basic equations to a self-adjoint form given in Sec. 2.2.4 produces the *full justification* of the special weighting. Which of the procedures is best used in practice is largely a matter of taste, as all can give excellent results. However, we shall see from the second part of this chapter, in which transient problems are dealt with, that other methods can be adopted if time-stepping procedures are used as an iteration to derive steady-state algorithms.

Indeed most of these procedures will again result in the addition of a diffusion term in which the parameter α is now replaced by another one involving the length of the time step Δt. We shall show at the end of the next section a comparison between various procedures for stabilization and will note essentially the same forms in the steady-state situation.

In the last part of this chapter (Part III) we shall address the case in which the unknown ϕ is a vector variable. Here only a limited number of procedures described in the first two parts will be available and even so we do not recommend in general the use of such methods for vector-valued functions.

Before proceeding further it is of interest to consider the original equation with a source term proportional to the variable ϕ, i.e. writing the one-dimensional equation (2.11) as

$$U\frac{\mathrm{d}\phi}{\mathrm{d}x} - \frac{\mathrm{d}}{\mathrm{d}x}\left(k\frac{\mathrm{d}\phi}{\mathrm{d}x}\right) + m\phi + Q = 0 \tag{2.57}$$

Equations of this type will arise of course from the transient Eq. (2.10) if we assume the solution to be decomposed into Fourier components, writing for each component

$$Q = Q^* \,\mathrm{e}^{\mathrm{i}\omega t} \qquad \phi = \phi^* \,\mathrm{e}^{\mathrm{i}\omega t} \tag{2.58}$$

which on substitution gives

$$U\frac{\mathrm{d}\phi^*}{\mathrm{d}x} - \frac{\mathrm{d}}{\mathrm{d}x}\left(k\frac{\mathrm{d}\phi^*}{\mathrm{d}x}\right) + \mathrm{i}\omega\phi^* + Q^* = 0 \tag{2.59}$$

in which ϕ^* can be complex.

The use of Petrov–Galerkin or similar procedures on Eq. (2.57) or (2.59) can again be made. If we pursue the line of approach outlined in Sec. 2.2.4 we note that

(a) the function p required to achieve self-adjointness remains unchanged;

and hence

(b) the weighting applied to achieve optimal results (see Sec. 2.2.3) again remains unaltered – providing of course it is applied to all terms.

Although the above result is encouraging and permits the solution in the frequency domain for transient problems, it does not readily 'transplant' to problems in which time-stepping procedures are required.

Some further points require mentioning at this stage. These are simply that:

1. When pure convection is considered (that is $k = 0$) only one boundary condition – generally that giving the value of ϕ at the inlet – can be specified, and in such a case the violent oscillations observed in Fig. 2.2 with standard Galerkin methods will not occur generally.

2. Specification of no boundary condition at the outlet edge in the case when $k > 0$, which is equivalent to imposing a zero conduction flux there, generally results in quite acceptable solutions with standard Galerkin weighting even for quite high Peclet numbers.

Part II: Transients

2.5 Transients – introductory remarks

2.5.1 Mathematical background

The objective of this section is to develop procedures of general applicability for the solution by direct time-stepping methods of Eq. (2.1) written for scalar values of ϕ, F_i and G_i:

$$\frac{\partial \phi}{\partial t} + \frac{\partial F_i}{\partial x_i} + \frac{\partial G_i}{\partial x_i} + Q = 0 \tag{2.60}$$

though consideration of the procedure for dealing with a vector-valued function will be included in Part III. However, to allow a simple interpretation of the various methods and of behaviour patterns the scalar equation in one dimension [see Eq. (2.10)], i.e.

$$\frac{\partial \phi}{\partial t} + U\frac{\partial \phi}{\partial x} - \frac{\partial}{\partial x}\left(k\frac{\partial \phi}{\partial x}\right) + Q = 0 \tag{2.61a}$$

will be considered. This of course is a particular case of Eq. (2.60) in which $F = F(\phi)$, $U = \partial F/\partial \phi$ and $Q = Q(\phi, x)$ and therefore

$$\frac{\partial F}{\partial x} = \frac{\partial F}{\partial \phi}\frac{\partial \phi}{\partial x} = U\frac{\partial \phi}{\partial x} \tag{2.61b}$$

The problem so defined is non-linear unless U is constant. However, the non-conservative equations (2.61) admit a spatial variation of U and are quite general.

The main behaviour patterns of the above equations can be determined by a change of the independent variable x to x' such that

$$\mathrm{d}x_i' = \mathrm{d}x_i - U_i\,\mathrm{d}t \tag{2.62}$$

Noting that for $\phi = \phi(x_i', t)$ we have

$$\left.\frac{\partial \phi}{\partial t}\right|_{x\,\text{const}} = \frac{\partial \phi}{\partial x_i'}\frac{\partial x_i'}{\partial t} + \left.\frac{\partial \phi}{\partial t}\right|_{x'\,\text{const}} = -U_i\frac{\partial \phi}{\partial x_i'} + \left.\frac{\partial \phi}{\partial t}\right|_{x'\,\text{const}} \tag{2.63}$$

The one-dimensional equation (2.61a) now becomes simply

$$\frac{\partial \phi}{\partial t} - \frac{\partial}{\partial x'}\left(k\frac{\partial \phi}{\partial x'}\right) + Q(x') = 0 \tag{2.64}$$

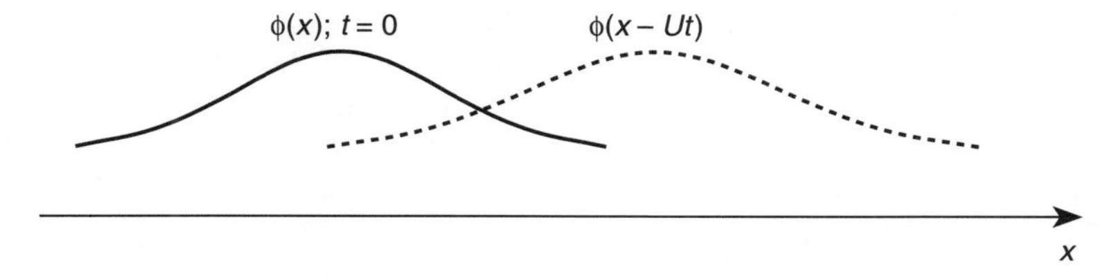

Fig. 2.10 The wave nature of a solution with no conduction. Constant wave velocity U.

and equations of this type can be readily discretized with self-adjoint spatial operators and solved by procedures developed previously in Volume 1.

The coordinate system of Eq. (2.62) describes *characteristic directions* and the moving nature of the coordinates must be noted. A further corollary of the coordinate change is that with no conduction or heat generation terms, i.e. when $k = 0$ and $Q = 0$, we have simply

$$\frac{\partial \phi}{\partial t} = 0$$

or (2.65)

$$\phi(x') = \phi(x - Ut) = \text{constant}$$

along a characteristic [assuming U to be constant, which will be the case if $F = F(\phi)$]. This is a typical equation of a wave propagating with a velocity U in the x direction, as shown in Fig. 2.10. The wave nature is evident in the problem even if the conduction (diffusion) is not zero, and in this case we shall have solutions showing a wave that attenuates with the distance travelled.

2.5.2 Possible discretization procedures

In Part I of this chapter we have concentrated on the essential procedures applicable directly to a steady-state set of equations. These procedures started off from somewhat heuristic considerations. The Petrov–Galerkin method was perhaps the most rational but even here the amount and the nature of the weighting functions were a matter of guess-work which was subsequently justified by consideration of the numerical error at nodal points. The Galerkin least square (GLS) method in the same way provided no absolute necessity for improving the answers though of course the least square method would tend to increase the symmetry of the equations and thus could be proved useful. It was only by results which turned out to be remarkably similar to those obtained by the Petrov–Galerkin methods that we have deemed this method to be a success. The same remark could be directed at the finite increment calculus (FIC) method and indeed to other methods suggested dealing with the problems of steady-state equations.

For the transient solutions the obvious first approach would be to try again the same types of methods used in steady-state calculations and indeed much literature has been devoted to this.[26–44] Petrov–Galerkin methods have been used here quite extensively. However, it is obvious that the application of Petrov–Galerkin methods will lead to non-symmetric mass matrices and these will be difficult to use for any explicit method as lumping is not by any means obvious.

Serious difficulty will also arise with the Galerkin least squares (GLS) procedure even if the temporal variation is generally included by considering space-time finite elements in the whole formulation. This approach to such problems was made by Nguen and Reynen,[32] Carey and Jieng,[33,34] Johnson and coworkers[30,35,36] and others.[37,38] However the use of space-time elements is expensive as explicit procedures are not available.

Which way, therefore, should we proceed? Is there any other obvious approach which has not been mentioned? The answer lies in the wave nature of the equations which indeed not only permits different methods of approach but in many senses is much more direct and fully justifies the numerical procedures which we shall use. We shall therefore concentrate on such methods and we will show that they will lead to artificial diffusions which in form are very similar to those obtained previously by the Petrov–Galerkin method but in a much more direct manner which is consistent with the equations.

The following discussion will therefore be centred on two main directions: (1) the procedures based on the use of the *characteristics* and the wave nature directly leading to so-called characteristic Galerkin methods which we shall discuss in Sec. 2.6; and then (2) we shall proceed to approach the problem through the use of higher-order time approximations called Taylor–Galerkin methods.

Of the two approaches the first one based on the characteristics is in our view more important. However for historical and other reasons we shall discuss both methods which for a scalar variable can be shown to give identical answers.

The solutions of convective scalar equations can be given by both approaches very simply. This will form the basis of our treatment for the solution of fluid mechanics equations in Chapter 3, where both explicit iterative processes as well as implicit methods can be used.

Many of the methods for solving the transient scalar equations of convective diffusion have been applied to the full fluid mechanics equations, i.e. solving the full vector-valued convective–diffusive equations we have given at the beginning of the chapter (Eq. 2.1). This applies in particular to the Taylor–Galerkin method which has proved to be quite successful in the treatment of high-speed compressible gas flow problems. Indeed this particular approach was the first one adopted to solve such problems. However, the simple wave concepts which are evident in the scalar form of the equations do not translate to such multivariant problems and make the procedures largely heuristic. The same can be said of the direct application of the SUPG and GLS methods to multivariant problems. We have shown in Volume 1, Chapter 12 that procedures such as GLS can provide a useful stabilization of difficulties encountered with incompressibility behaviour. This does not justify their widespread use and we therefore recommend the alternatives to be discussed in Chapter 3.

For completeness, however, Part III of this chapter will be added to discuss to some extent the extension of some methods to vector-type variables.

2.6 Characteristic-based methods

2.6.1 Mesh updating and interpolation methods

We have already observed that, if the spatial coordinate is 'convected' in the manner implied by Eq. (2.62), i.e. along the problem *characteristics*, then the convective, first-order, terms disappear and the remaining problem is that of simple diffusion for which standard discretization procedures with the Galerkin spatial approximation are optimal (in the energy norm sense).

The most obvious use of this in the finite element context is to update the position of the mesh points in a lagrangian manner. In Fig. 2.11(a) we show such an update for the one-dimensional problem of Eq. (2.61) occurring in an interval Δt.

For a constant x' coordinate

$$dx = U\,dt \tag{2.66}$$

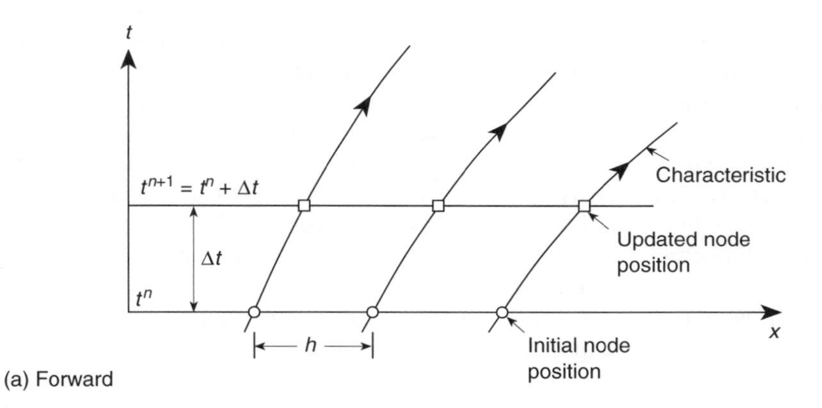

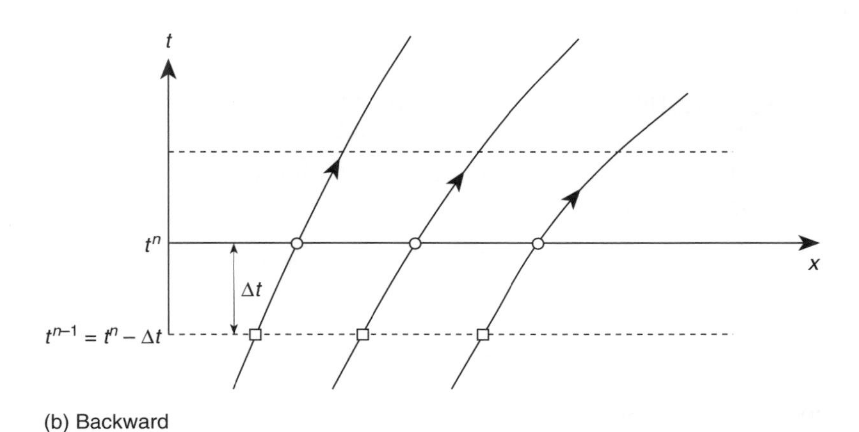

Fig. 2.11 Mesh updating and interpolation: (a) Forward; (b) Backward.

and for a typical nodal point i, we have

$$x_i^{n+1} = x_i^n + \int_{t_n}^{t_{n+1}} U \, \mathrm{d}t \qquad (2.67)$$

where in general the 'velocity' U may be dependent on x. However, if $F = F(\phi)$ and $U = \partial F/\partial\phi = U(\phi)$ then the wave velocity is constant along a characteristic by virtue of Eq. (2.65) and the characteristics are straight lines.

For such a constant U we have simply

$$x_i^{n+1} = x_i^n + U\Delta t \qquad (2.68)$$

for the updated mesh position. This is not always the case and updating generally has to be done with variable U.

On the updated mesh only the time-dependent diffusion problem needs to be solved, using the methods of Volume 1. These we need not discuss in detail here.

The process of continuously updating the mesh and solving the diffusion problem on the new mesh is, of course, impracticable. When applied to two- or three-dimensional configurations very distorted elements would result and difficulties will always arise on the boundaries of the domain. For that reason it seems obvious that after completion of a single step a return to the original mesh should be made by interpolating from the updated values, to the original mesh positions.

This procedure can of course be reversed and characteristic origins traced backwards, as shown in Fig. 2.11(b) using appropriate interpolated starting values.

The method described is somewhat intuitive but has been used with success by Adey and Brebbia[45] and others as early as 1974 for solution of transport equations. The procedure can be formalized and presented more generally and gives the basis of so-called characteristic–Galerkin methods.[46]

The diffusion part of the computation is carried out either on the original or on the final mesh, each representing a certain approximation. Intuitively we imagine in the updating scheme that the *operator is split* with the diffusion changes occurring separately from those of convection. This idea is explained in the procedures of the next section.

2.6.2 Characteristic–Galerkin procedures

We shall consider that the equation of convective diffusion in its one-dimensional form (2.61) is split into two parts such that

$$\phi = \phi^* + \phi^{**} \qquad (2.69)$$

and

$$\frac{\partial \phi^*}{\partial t} + U\frac{\partial \phi}{\partial x} = 0 \qquad (2.70a)$$

is a purely convective system while

$$\frac{\partial \phi^{**}}{\partial t} - \frac{\partial}{\partial x}\left(k\frac{\partial \phi}{\partial x}\right) + Q = 0 \qquad (2.70b)$$

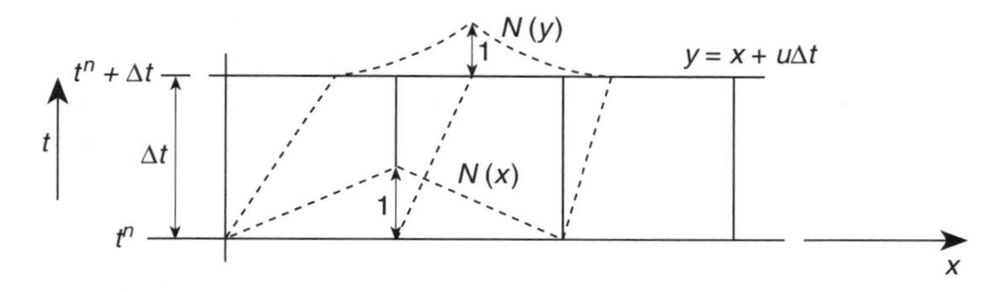

Fig. 2.12 Distortion of convected shape function.

represents the self-adjoint terms [here Q contains the source, reaction and term $(\partial U/\partial x)\phi$].

Both ϕ^* and ϕ^{**} are to be approximated by standard expansions

$$\hat{\phi}^* = \mathbf{N}\tilde{\boldsymbol{\phi}}^* \qquad \hat{\phi}^{**} = \mathbf{N}\tilde{\boldsymbol{\phi}}^{**} \tag{2.71}$$

and in a single time step t^n to $t^n + \Delta t = t^{n+1}$ we shall assume that the initial conditions are

$$t = t^n \qquad \phi^* = 0 \qquad \phi^{**} = \phi^{*n} \tag{2.72}$$

Standard Galerkin discretization of the diffusion equation allows $\tilde{\boldsymbol{\phi}}^{**n+1}$ to be determined on the given fixed mesh by solving an equation of the form

$$\mathbf{M}\Delta\tilde{\boldsymbol{\phi}}^{**n} = \Delta t\mathbf{H}(\tilde{\boldsymbol{\phi}}^n + \theta\Delta\tilde{\boldsymbol{\phi}}^{**n}) + \mathbf{f} \tag{2.73}$$

with

$$\tilde{\boldsymbol{\phi}}^{**n+1} = \tilde{\boldsymbol{\phi}}^{**n} + \Delta\tilde{\boldsymbol{\phi}}^{**n}$$

In solving the convective problem we assume that ϕ^* remains unchanged along the characteristic. However, Fig. 2.12 shows how the initial value of ϕ^{*n} interpolated by standard linear shape functions at time n [see Eq. (2.71)] becomes shifted and distorted. The new value is given by

$$\phi^{*n+1} = \mathbf{N}(y)\tilde{\boldsymbol{\phi}}^{*n} \qquad y = x + U\Delta t \tag{2.74}$$

As we require ϕ^{*n+1} to be approximated by standard shape functions, we shall write a projection for smoothing of these values as

$$\int_\Omega \mathbf{N}^{\mathrm{T}}(\mathbf{N}\tilde{\boldsymbol{\phi}}^{*n+1} - \mathbf{N}(y)\tilde{\boldsymbol{\phi}}^{*n})\,\mathrm{d}x = 0 \tag{2.75}$$

giving

$$\mathbf{M}\tilde{\boldsymbol{\phi}}^{*n+1} = \int_\Omega [\mathbf{N}^{\mathrm{T}}\mathbf{N}(y)\,\mathrm{d}x]\tilde{\boldsymbol{\phi}}^n \tag{2.76a}$$

where $\mathbf{N} = \mathbf{N}(x)$ and $\mathbf{M}$ is

$$\mathbf{M} = \int_\Omega \mathbf{N}^{\mathrm{T}}\mathbf{N}\,\mathrm{d}x \tag{2.76b}$$

The evaluation of the above integrals is of course still complex, especially if the procedure is extended to two or three dimensions. This is generally performed numerically and the stability of the formulation is dependent on the accuracy of such integration.[46] The scheme is stable and indeed exact as far as the convective terms are concerned if the integration is performed exactly (which of course is an unreachable goal). However, stability and indeed accuracy will even then be controlled by the diffusion terms where several approximations have been involved.

2.6.3 A simple explicit characteristic–Galerkin procedure

Many variants of the schemes described in the previous section are possible and were introduced quite early. References 45–56 present some successful versions. However, all methods then proposed are somewhat complex in programming and are time consuming. For this reason a simpler alternative was developed in which the difficulties are avoided at the expense of conditional stability. This method was first published in 1984[57] and is fully described in numerous publications.[58–61] Its derivation involves a local Taylor expansion and we illustrate this in Fig. 2.13.

We can write Eq. (2.61a) along the characteristic as

$$\frac{\partial \phi}{\partial t}(x'(t), t) - \frac{\partial}{\partial x'}\left(k\frac{\partial \phi}{\partial x'}\right) - Q(x') = 0 \tag{2.77}$$

As we can see, in the moving coordinate x', the convective acceleration term disappears and source and diffusion terms are averaged quantities along the characteristic. Now the equation is self-adjoint and the Galerkin spatial approximation is optimal. The time discretization of the above equation along the characteristic (Fig. 2.13) gives

$$\begin{aligned}\frac{1}{\Delta t}(\phi^{n+1} - \phi^n|_{(x-\delta)}) \approx \theta\left[\frac{\partial}{\partial x}\left(k\frac{\partial \phi}{\partial x}\right) - Q\right]^{n+1} \\ + (1-\theta)\left[\frac{\partial}{\partial x}\left(k\frac{\partial \phi}{\partial x}\right) - Q\right]^{n}\Big|_{(x-\delta)}\end{aligned} \tag{2.78}$$

where θ is equal to zero for explicit forms and between zero and unity for semi- and fully implicit forms. As we know, the solution of the above equation in moving coordinates leads to mesh updating and presents difficulties, so we will suggest

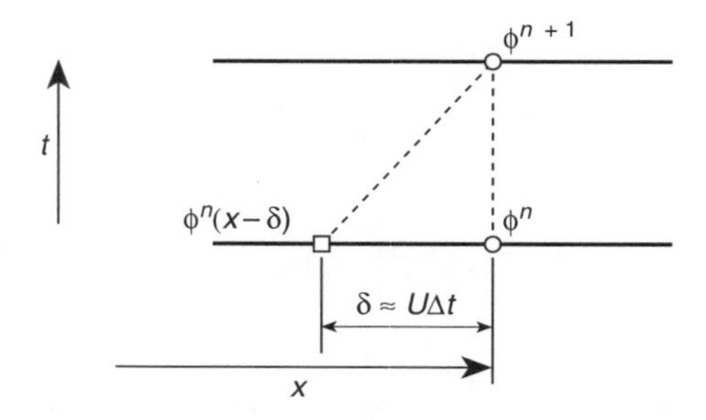

Fig. 2.13. A simple characteristic–Galerkin procedure.

alternatives. From the Taylor expansion we have

$$\phi^n|_{(x-\delta)} \approx \phi^n - \delta \frac{\partial \phi^n}{\partial x} + \frac{\delta^2}{2} \frac{\partial^2 \phi^n}{\partial x^2} + O(\Delta t^3) \tag{2.79}$$

and assuming $\theta = 0.5$

$$\frac{1}{2}\frac{\partial}{\partial x}\left(k\frac{\partial \phi}{\partial x}\right)\Big|_{(x-\delta)} \approx \frac{1}{2}\frac{\partial}{\partial x}\left(k\frac{\partial \phi}{\partial x}\right)^n - \frac{\delta}{2}\frac{\partial}{\partial x}\left[\frac{\partial}{\partial x}\left(k\frac{\partial \phi}{\partial x}\right)^n\right] + O(\Delta t^2) \tag{2.80a}$$

$$\frac{1}{2}Q|_{(x-\delta)} = \frac{Q^n}{2} - \frac{\delta}{2}\frac{\partial Q^n}{\partial x} \tag{2.80b}$$

where δ is the distance travelled by the particle in the x-direction (Fig. 2.13) which is

$$\delta = \bar{U}\Delta t \tag{2.81}$$

where $\bar{U}$ is an average value of U along the characteristic. Different approximations of $\bar{U}$ lead to different stabilizing terms. The following relation is commonly used[62,63]

$$\bar{U} = U^n - U^n \Delta t \frac{\partial U^n}{\partial x} \tag{2.82}$$

Inserting Eqs (2.79)–(2.82) into Eq. (2.78) we have

$$\begin{aligned}\phi^{n+1} - \phi^n = &- \Delta t\left\{U\frac{\partial \phi^n}{\partial x} - \frac{\partial}{\partial x}\left(k\frac{\partial \phi}{\partial x}\right)^{n+1/2} + Q^{n+1/2}\right\} \\ &+ \Delta t\left\{\frac{\Delta t}{2}\frac{\partial}{\partial x}\left[U^2\frac{\partial \phi}{\partial x}\right] - \frac{\Delta t}{2}U\frac{\partial^2}{\partial x^2}\left(k\frac{\partial \phi}{\partial x}\right) + \frac{\Delta t}{2}U\frac{\partial Q}{\partial x}\right\}^n\end{aligned} \tag{2.83a}$$

where

$$\frac{\partial}{\partial x}\left(k\frac{\partial \phi}{\partial x}\right)^{n+1/2} = \frac{1}{2}\frac{\partial}{\partial x}\left(k\frac{\partial \phi}{\partial x}\right)^{n+1} + \frac{1}{2}\frac{\partial}{\partial x}\left(k\frac{\partial \phi}{\partial x}\right)^n \tag{2.83b}$$

and

$$Q^{n+1/2} = \frac{Q^{n+1} + Q^n}{2} \tag{2.83c}$$

In the above equation, higher-order terms (from Eq. 2.80) are neglected. This, as already mentioned, is of an identical form to that resulting from Taylor–Galerkin procedures which will be discussed fully in the next section, and the additional terms add the stabilizing diffusion in the streamline direction. For multidimensional problems, Eq. (2.83a) can be written in indicial notation and approximating $n + 1/2$ terms with n terms (for the fully explicit form)

$$\begin{aligned}\phi^{n+1} - \phi^n = &- \Delta t\left\{U_j\frac{\partial \phi}{\partial x_j} - \frac{\partial}{\partial x_i}\left(k\frac{\partial \phi}{\partial x_i}\right) + Q\right\}^n \\ &+ \Delta t\left\{\frac{\Delta t}{2}\frac{\partial}{\partial x_i}\left[U_iU_j\frac{\partial \phi}{\partial x_j}\right] - \frac{\Delta t}{2}U_k\frac{\partial}{\partial x_k}\left[\frac{\partial}{\partial x_i}\left(k\frac{\partial \phi}{\partial x_i}\right)\right] + \frac{\Delta t}{2}U_i\frac{\partial Q}{\partial x_i}\right\}^n\end{aligned} \tag{2.84}$$

An alternative approximation for $\bar{U}$ recently recommended is[62]

$$\bar{U} = \frac{U^{n+1} + U^n|_{(x-\delta)}}{2} \tag{2.85}$$

Using the Taylor expansion

$$U^n|_{(x-\delta)} \approx U^n - \Delta t U^n \frac{\partial U^n}{\partial x} + O(\Delta t^2) \tag{2.86}$$

from Eqs (2.78)–(2.81) and Eqs (2.85) and (2.86) with θ equal to 0.5 we have

$$\begin{aligned}\frac{1}{\Delta t}(\phi^{n+1} - \phi^n) = &- U^{n+1/2}\frac{\partial \phi^n}{\partial x} + \frac{\Delta t}{2}U^n\frac{\partial U^n}{\partial x}\frac{\partial \phi^n}{\partial x} + \frac{\Delta t}{2}U^{n+1/2}U^{n+1/2}\frac{\partial^2 \phi}{\partial x^2}\\ &+ \frac{\partial}{\partial x}\left(k\frac{\partial \phi}{\partial x}\right)^{n+1/2} - \frac{\Delta t}{2}U^{n+1/2}\frac{\partial}{\partial x}\left[\frac{\partial}{\partial x}\left(k\frac{\partial \phi}{\partial x}\right)^n\right]\\ &- Q + \frac{\Delta t}{2}U^{n+1/2}\frac{\partial Q}{\partial x}\end{aligned} \tag{2.87}$$

where

$$U^{n+1/2} = \frac{U^{n+1} + U^n}{2} \tag{2.88}$$

We can further approximate, as mentioned earlier, $n + 1/2$ terms using n, to get the fully explicit version of the scheme. Thus we have

$$U^{n+1/2} = U^n + O(\Delta t) \tag{2.89}$$

and similarly the diffusion term is approximated. The final form of the explicit characteristic–Galerkin method can be written as

$$\begin{aligned}\Delta\phi = \phi^{n+1} - \phi^n = &- \Delta t\left[U^n\frac{\partial \phi}{\partial x} - \frac{\partial}{\partial x}\left(k\frac{\partial \phi}{\partial x}\right) + Q\right]^n\\ &+ \frac{\Delta t^2}{2}U^n\frac{\partial}{\partial x}\left[U^n\frac{\partial \phi}{\partial x} - \frac{\partial}{\partial x}\left(k\frac{\partial \phi}{\partial x}\right) + Q\right]^n\end{aligned} \tag{2.90}$$

Generalization to multidimensions is direct and can be written in indicial notation for equations of the form Eq. (2.5):

$$\begin{aligned}\Delta\phi = &- \Delta t\left[\frac{\partial(U_j\phi)}{\partial x_j} - \frac{\partial}{\partial x_i}\left(k\frac{\partial \phi}{\partial x_i}\right) + Q\right]^n\\ &+ \frac{\Delta t^2}{2}U_k^n\frac{\partial}{\partial x_k}\left[\frac{\partial(U_j\phi)}{\partial x_j} - \frac{\partial}{\partial x_i}\left(k\frac{\partial \phi}{\partial x_i}\right) + Q\right]^n\end{aligned} \tag{2.91}$$

The reader will notice the difference in the stabilizing terms obtained by two different approximations for $\bar{U}$. However, as we can see the difference between them is small and when U is constant both approximations give identical stabilizing terms. In the rest of the book we shall follow the latter approximation and always use the conservative form of the equations (Eq. 2.91).

As we proved earlier, the Galerkin spatial approximation is justified when the characteristic–Galerkin procedure is used. We can thus write the approximation

$$\phi = \mathbf{N}\tilde{\boldsymbol{\phi}} \tag{2.92}$$

and use the weighting $\mathbf{N}^{\mathrm{T}}$ in the integrated residual expression. Thus we obtain

$$\mathbf{M}(\tilde{\boldsymbol{\phi}}^{n+1} - \tilde{\boldsymbol{\phi}}^{n}) = -\Delta t[(\mathbf{C}\tilde{\boldsymbol{\phi}}^{n} + \mathbf{K}\tilde{\boldsymbol{\phi}}^{n} + \mathbf{f}^{n}) - \Delta t(\mathbf{K}_u\tilde{\boldsymbol{\phi}}^{n} + \mathbf{f}_s^{n})] \tag{2.93}$$

in explicit form without higher-order derivatives and source terms. In the above equation

$$\begin{aligned} \mathbf{M} &= \int_\Omega \mathbf{N}^{\mathrm{T}}\mathbf{N}\,\mathrm{d}\Omega \qquad & \mathbf{C} &= \int_\Omega \mathbf{N}^{\mathrm{T}} \frac{\partial}{\partial x_i}(U_i\mathbf{N})\,\mathrm{d}\Omega \\ \mathbf{K} &= \int_\Omega \frac{\partial \mathbf{N}^{\mathrm{T}}}{\partial x_i} k \frac{\partial \mathbf{N}}{\partial x_i}\,\mathrm{d}\Omega \qquad & \mathbf{f} &= \int_\Omega \mathbf{N}^{\mathrm{T}} Q\,\mathrm{d}\Omega + \text{b.t.} \end{aligned} \tag{2.94}$$

and $\mathbf{K}_u$ and $\mathbf{f}_s^n$ come from the new term introduced by the discretization along the characteristics. After integration by parts, the expression of $\mathbf{K}_u$ and $\mathbf{f}_s$ is

$$\mathbf{K}_u = -\frac{1}{2}\int_\Omega \frac{\partial}{\partial x_i}(U_i\mathbf{N}^{\mathrm{T}})\frac{\partial}{\partial x_i}(U_i\mathbf{N})\,\mathrm{d}\Omega \tag{2.95}$$

$$\mathbf{f}_s = -\frac{1}{2}\int_\Omega \frac{\partial}{\partial x_i}(U_i\mathbf{N}^{\mathrm{T}})Q\,\mathrm{d}\Omega + \text{b.t.} \tag{2.96}$$

where b.t. stands for integrals along region boundaries. Note that the higher-order derivatives are not included in the above equation.

The approximation is valid for any scalar convected quantity even if that is the velocity component U_i itself, as is the case with momentum-conservation equations. For this reason we have elaborated above the full details of the spatial approximation as the matrices will be repeatedly used.

It is of interest that the explicit form of Eq. (2.93) is only conditionally stable. For one-dimensional problems, the stability condition is given as (neglecting the effect of sources)

$$\Delta t \leqslant \Delta t_{\mathrm{crit}} = \frac{h}{|U|} \tag{2.97}$$

for linear elements.

In two-dimensional problems the criteria time step may be computed as[62,63]

$$\Delta t_{\mathrm{crit}} = \frac{\Delta t_\sigma \Delta t_\nu}{\Delta t_\sigma + \Delta t_\nu} \tag{2.98}$$

where Δt_σ is given by Eq. (2.97) and $\Delta t_\nu = h^2/2k$ is the diffusive limit for the critical one-dimensional time step.

Further, with $\Delta t = \Delta t_{\mathrm{crit}}$ the steady-state solution results in an (almost) identical diffusion change to that obtained by using the optimal streamline upwinding procedures discussed in Part I of this chapter. Thus if steady-state solutions are the main objective of the computation such a value of Δt should be used in connection with the $\mathbf{K}_u$ term.

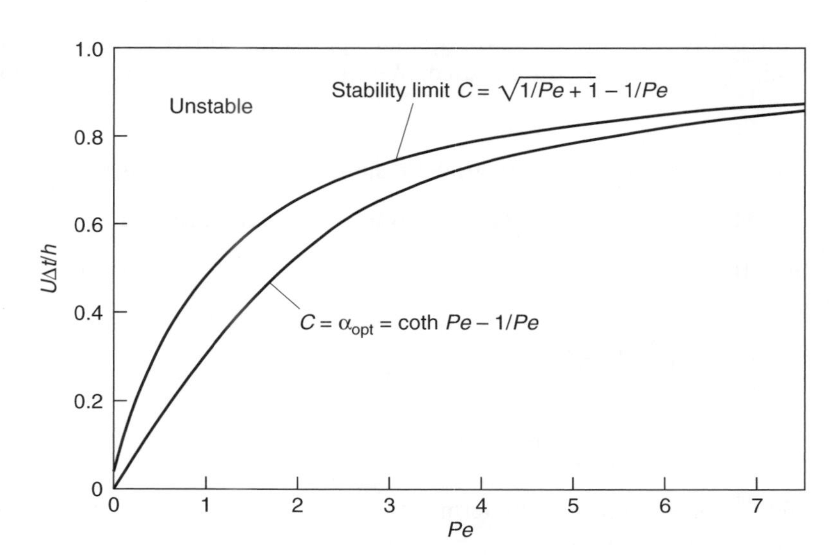

Fig. 2.14 Stability limit for lumped mass approximation and optimal upwind parameter.

A fully implicit form of solution is an expensive one involving unsymmetric matrices. However it is often convenient to apply $\theta \geqslant 1/2$ to the diffusive term only. We call this a *nearly* (or *semi*) *implicit form* and if it is employed we return to the stability condition

$$\Delta t_{\text{crit}} = \frac{h}{|U|} \tag{2.99}$$

which can present an appreciable benefit.

Figure 2.14 shows the stability limit variation prescribed by Eq. (2.97) with a lumped mass matrix.

It is of considerable interest to examine the behaviour of the solution when the steady state is reached – for instance, if we use the time-stepping algorithm of Eq. (2.93) as an iterative process. Now the final solution is given by taking

$$\tilde{\bar{\boldsymbol{\phi}}}^{n+1} = \tilde{\bar{\boldsymbol{\phi}}}^{n} = \tilde{\bar{\boldsymbol{\phi}}}$$

which gives

$$[(\mathbf{C} + \mathbf{K} - \Delta t \mathbf{K}_u]\tilde{\bar{\boldsymbol{\phi}}} + \mathbf{f} - \Delta t \mathbf{f}_s = 0 \tag{2.100}$$

Inspection of Secs 2.2 and 2.3 shows that the above is identical in form with the use of the Petrov–Galerkin approximation. In the latter the matrix $\mathbf{C}$ is identical and the matrix $\mathbf{K}_u$ includes balancing diffusion of the amount given by $\frac{1}{2}\alpha U h$. However, if we take

$$\tfrac{1}{2}\alpha U h = \frac{U^2 \Delta t}{2} \tag{2.101}$$

the identity of the two schemes results. This can be written as a requirement that

$$\alpha = \frac{U \Delta t}{h} = C \tag{2.102}$$

where C is the Courant number.

In Fig. 2.14 we therefore plot the optimal value of α as given in Eq. (2.25) against *Pe. We note immediately that if the time-stepping scheme is operated at or near the critical stability limit of the lumped scheme the steady-state solution reached will be close to that resulting from the optimal Petrov–Galerkin process for the steady state.* However, if smaller time steps than the critical ones are used, the final solution, though stable, will tend towards the standard Galerkin steady-state discretization and may show oscillations if boundary conditions are such that boundary layers are created. Nevertheless, such small time steps result in very accurate transients so we can conclude that it is unlikely that optimality for transients and steady state can be reached simultaneously.

Examination of Eqs (2.93) shows that the characteristic Galerkin algorithm could have been obtained by applying a Petrov–Galerkin weighting

$$\mathbf{N}^{\mathrm{T}} + \frac{\Delta t}{2} U_i \frac{\partial \mathbf{N}^{\mathrm{T}}}{\partial x_i}$$

to the various terms of the governing equation (2.60) excluding the time derivative $\partial \phi / \partial t$ to which the standard Galerkin weighting of $\mathbf{N}^{\mathrm{T}}$ is attached. Comparing the above with the steady-state problem and the weighting given in Eq. (2.48) the connection is obvious.

A two-dimensional application of the characteristic–Galerkin process is illustrated in Fig. 2.15 in which we show pure convection of a disturbance in a circulating flow. It is remarkable to note that almost no dispersion occurs after a complete revolution. The present scheme is here contrasted with the solution obtained by the finite difference scheme of Lax and Wendroff[64] which for a regular one-dimensional mesh gives a scheme identical to the characteristic–Galerkin except for mass matrix, which is lumped in the finite difference scheme.

It seems that here the difference is entirely due to the proper form of the mass matrix $\mathbf{M}$ now used and we note that for transient response the importance of the consistent mass matrix is crucial. However, the numerical convenience of using the lumped form is overwhelming in an explicit scheme. It is easy to recover the performance of the consistent mass matrix by using a simple iteration. In this we write Eq. (2.93) as

$$\mathbf{M}\Delta\tilde{\bar{\boldsymbol{\phi}}}^n = \Delta t \mathbf{S}^n \tag{2.103}$$

with $\mathbf{S}^n$ being the right-hand side of Eq. (2.93) and

$$\tilde{\bar{\boldsymbol{\phi}}}^{n+1} = \tilde{\bar{\boldsymbol{\phi}}}^n + \Delta\tilde{\bar{\boldsymbol{\phi}}}^n$$

Substituting a lumped mass matrix $\mathbf{M}_L$ to ease the solution process we can iterate as follows:

$$(\Delta\tilde{\bar{\boldsymbol{\phi}}})^n_l = \mathbf{M}_L^{-1}[\Delta t \mathbf{S}^n - \mathbf{M}(\Delta\tilde{\bar{\boldsymbol{\phi}}})^n_{l-1}] + (\Delta\tilde{\bar{\boldsymbol{\phi}}})^n_{l-1} \tag{2.104}$$

where l is the iteration number. The process converges very rapidly and in Fig. 2.16 we show the dramatic improvements of results in the solution of a one-dimensional wave propagation with three such iterations done at each time step. At this stage the results are identical to those obtained with the consistent mass matrix.

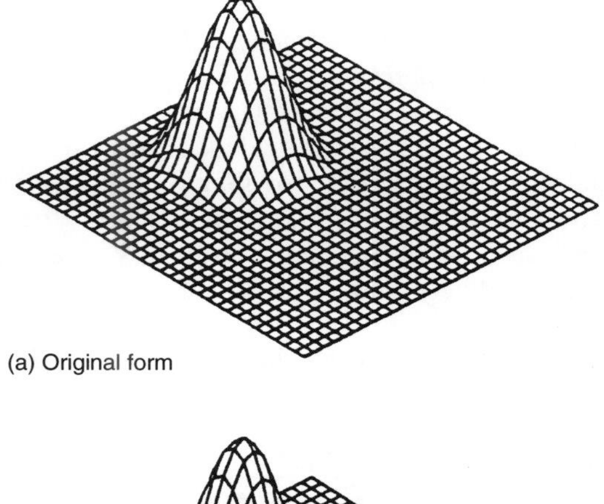

(a) Original form

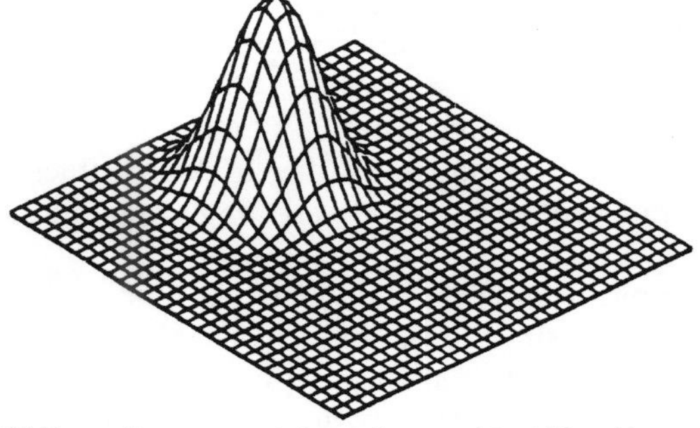

(b) Form after one revolution using consistent M matrix

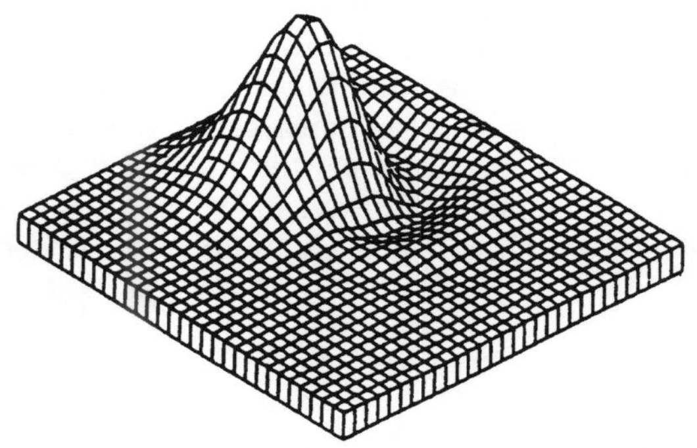

(c) Form after one revolution using lumped mass (Lax–Wendroff)

Fig. 2.15 Advection of a gaussian cone in a rotating fluid by characteristic–Galerkin method: (a) Original form; (b) Form after one revolution using consistent **M** matrix; (c) Form after one revolution using lumped mass (Lax–Wendroff).

2.6.4 Boundary conditions - radiation

As we have already indicated the convection–diffusion problem allows a single boundary condition of the type

$$\phi = \bar{\phi} \text{ on } \Gamma_u \tag{2.105a}$$

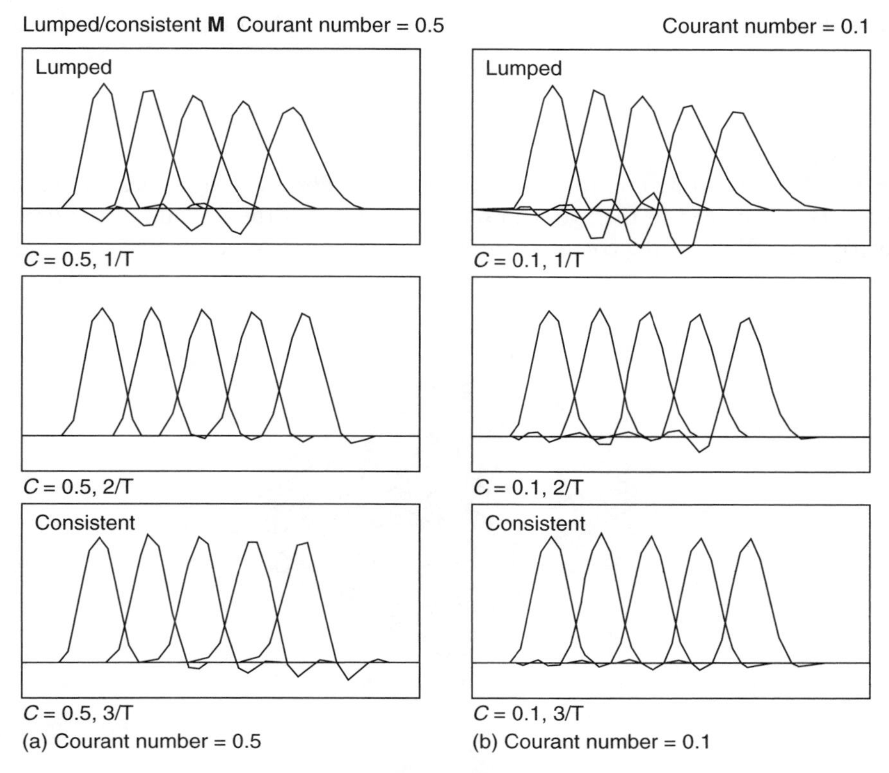

Fig. 2.16 Characteristic–Galerkin method in the solution of a one-dimensional wave progression. Effect of using a lumped mass matrix and of consistent iteration.

and

$$-k\left(\frac{\partial \phi}{\partial n}\right) = \bar{q} \text{ on } \Gamma_q \tag{2.105b}$$

(where $\Gamma = \Gamma_u \cup \Gamma_q$) to be imposed, providing the equation is of second order and diffusion is present.

In the case of pure convection this is no longer the case as the differential equation is of first order. Indeed this was responsible for the difficulty of obtaining a solution in the example of Fig. 2.2 when $Pe \to \infty$ and an exit boundary condition of the type given by Eq. (2.105a) was imposed. In this one-dimensional case for pure convection only the inlet boundary condition can be given; at the exit no boundary condition needs to be prescribed if U, the wave velocity, is positive.

For multidimensional problems of pure convection the same wave specification depends on the value of the normal component of U. Thus if

$$U_i n_i > 0 \tag{2.106}$$

where n_i is the normal direction vector, the wave is leaving the problem and then no boundary condition is specified. If the problem has some diffusion, the same specification of 'no boundary condition' is equivalent to putting

$$-k\left(\frac{\partial \phi}{\partial n}\right) = 0 \tag{2.107}$$

at the exit boundary.

In Fig. 2.17 we illustrate, following the work of Peraire,[65] how cleanly the same wave as that specified in the problem of Fig. 2.15 leaves the domain in the uniform velocity field[61,65] when the correct boundary condition is imposed.

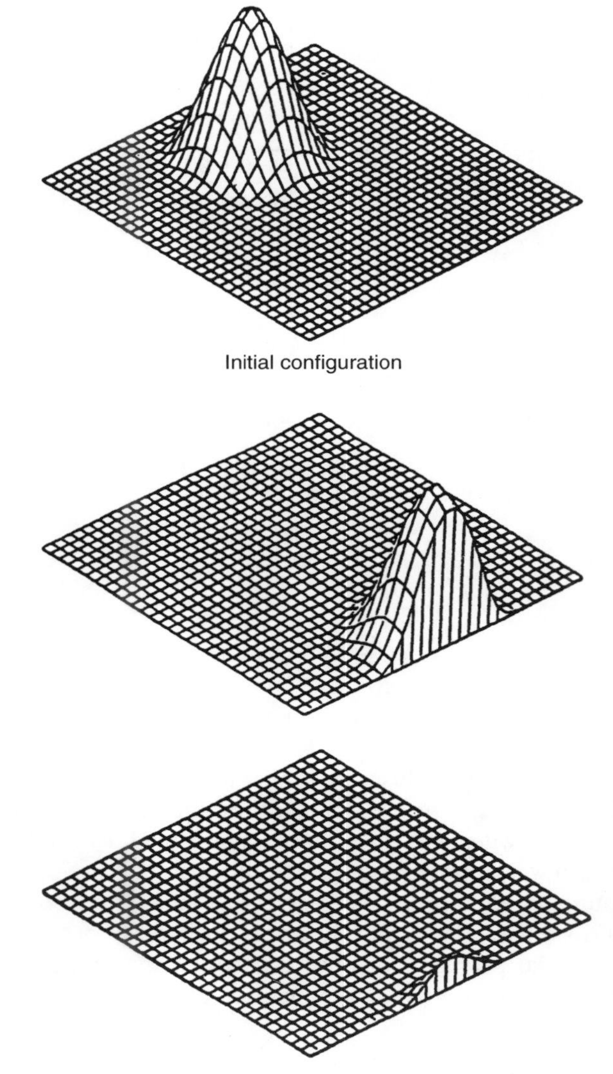

Fig. 2.17 A gaussian distribution advected in a constant velocity field. Boundary condition causes no reflection.

2.7 Taylor–Galerkin procedures for scalar variables

In the Taylor–Galerkin process, the Taylor expansion in time precedes the Galerkin space discretization. Firstly, the scalar variable ϕ is expanded by the Taylor series in time[58,66]

$$\phi^{n+1} = \phi^n + \Delta t \frac{\partial \phi^n}{\partial t} + \frac{\Delta t^2}{2} \frac{\partial^2 \phi^n}{\partial t^2} + O(\Delta t^3) \tag{2.108}$$

From Eq. (2.61a) we have

$$\frac{\partial \phi^n}{\partial t} = \left[-U \frac{\partial \phi}{\partial x} + \frac{\partial}{\partial x}\left(k \frac{\partial \phi}{\partial x} \right) + Q \right]^n \tag{2.109}$$

and

$$\frac{\partial^2 \phi^n}{\partial t^2} = \frac{\partial}{\partial t}\left[-U \frac{\partial \phi}{\partial x} + \frac{\partial}{\partial x}\left(k \frac{\partial \phi}{\partial x} \right) + Q \right]^n \tag{2.110}$$

Substituting Eqs (2.109) and (2.110) into Eq. (2.108) we have

$$\phi^{n+1} - \phi^n = -\Delta t \left[U \frac{\partial \phi}{\partial x} - \frac{\partial}{\partial x}\left(k \frac{\partial \phi}{\partial x} \right) + Q \right]^n - \frac{\Delta t^2}{2} \frac{\partial}{\partial t}\left[U \frac{\partial \phi}{\partial x} - \frac{\partial}{\partial x}\left(k \frac{\partial \phi}{\partial x} \right) + Q \right]^n \tag{2.111}$$

Assuming U and k to be constant we have

$$\phi^{n+1} - \phi^n = -\Delta t \left[U \frac{\partial \phi}{\partial x} - \frac{\partial}{\partial x}\left(k \frac{\partial \phi}{\partial x} \right) + Q \right]^n - \frac{\Delta t^2}{2} \frac{\partial}{\partial x}\left[U \frac{\partial \phi}{\partial t} - \frac{\partial}{\partial x}\left(k \frac{\partial \phi}{\partial t} \right) + Q \right]^n \tag{2.112}$$

Inserting Eq. (2.109) into Eq. (2.112) and neglecting higher-order terms

$$\begin{aligned}\phi^{n+1} - \phi^n = & - \Delta t \left[U \frac{\partial \phi}{\partial x} - \frac{\partial}{\partial x}\left(k \frac{\partial \phi}{\partial x} \right) + Q \right]^n \\ & + \frac{\Delta t^2}{2} \frac{\partial}{\partial x}\left[U^2 \frac{\partial \phi}{\partial x} - U \frac{\partial}{\partial x}\left(k \frac{\partial \phi}{\partial x} \right) + UQ \right]^n + O(\Delta t^3)\end{aligned} \tag{2.113}$$

As we can see the above equation, having assumed constant U and k, is identical to Eq. (2.83a) derived from the characteristic approach. Clearly for scalar variables both characteristic and Taylor–Galerkin procedures give identical stabilizing terms. Thus selection of a method for a scalar variable is a matter of taste. However, the sound mathematical justification of the characteristic–Galerkin method should be mentioned here.

The Taylor–Galerkin procedure for the convection–diffusion equation in multi-dimensions can be written as

$$\begin{aligned}\phi^{n+1} - \phi^n = & - \Delta t \Bigg\{ U_j \frac{\partial \phi}{\partial x_j} - \frac{\partial}{\partial x_i}\left(k \frac{\partial \phi}{\partial x_i} \right) + Q \\ & - \frac{\Delta t}{2} \frac{\partial}{\partial x_i}\left[U_i U_j \frac{\partial \phi}{\partial x_j} - U_i \frac{\partial}{\partial x_j}\left(k \frac{\partial \phi}{\partial x_j} \right) + U_i Q \right] \Bigg\}\end{aligned} \tag{2.114}$$

again showing the complete similarity with the appropriate characteristic–Galerkin form and identity when U_i and k are constant. The Taylor–Galerkin method is the finite element equivalent of the Lax–Wendroff method developed in the finite difference context.[64]

2.8 Steady-state condition

Both the Taylor–Galerkin and characteristic–Galerkin methods give an answer which compares directly with SUPG and GLS giving additional streamline diffusion (higher-order derivatives are omitted)

$$\frac{\Delta t^2}{2}\frac{\partial}{\partial x_i}\left[U_i U_j \frac{\partial \phi}{\partial x_j}\right] \tag{2.115}$$

with Δt replacing the coefficient αh. With the characteristic–Galerkin method being the only method that has a full mathematical justification, we feel that even for steady state problems this should be considered as an appropriate solution technique.

2.9 Non-linear waves and shocks

The procedures developed in the previous sections are in principle of course available for both linear and non-linear problems (with explicit procedures of time stepping being particularly efficient for the latter). Quite generally the convective part of the equation, i.e.

$$\frac{\partial \phi}{\partial t} + \frac{\partial F_i}{\partial x_i} \equiv \frac{\partial \phi}{\partial t} + U_i \frac{\partial \phi}{\partial x_i} = 0 \tag{2.116}$$

will have the vector U_i dependent on ϕ. Thus

$$U_i \equiv \frac{\partial F_i}{\partial \phi} = U_i(\phi) \tag{2.117}$$

In the one-dimensional case with a scalar variable we shall have equations of the type

$$\frac{\partial \phi}{\partial t} + \frac{\partial F}{\partial x} \equiv \frac{\partial \phi}{\partial t} + U(\phi)\frac{\partial \phi}{\partial x} = 0 \tag{2.118}$$

corresponding to waves moving with a non-uniform velocity U. A typical problem in this category is that due to Burger, which is defined by

$$\frac{\partial \phi}{\partial t} + \frac{\partial}{\partial x}\left(\tfrac{1}{2}\phi^2\right) = \frac{\partial \phi}{\partial t} + \phi\frac{\partial \phi}{\partial x} = 0 \tag{2.119}$$

In Fig. 2.18 we illustrate qualitatively how different parts of the wave moving with velocities proportional to their amplitude cause it to steepen and finally develop into a shock form. This behaviour is typical of many non-linear systems and in Chapter 6 we shall see how shocks develop in compressible flow at transonic speeds.

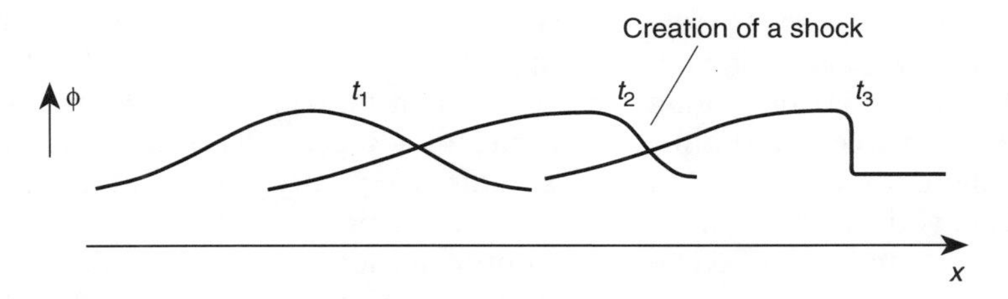

Fig. 2.18 Progression of a wave with velocity $U = \phi$.

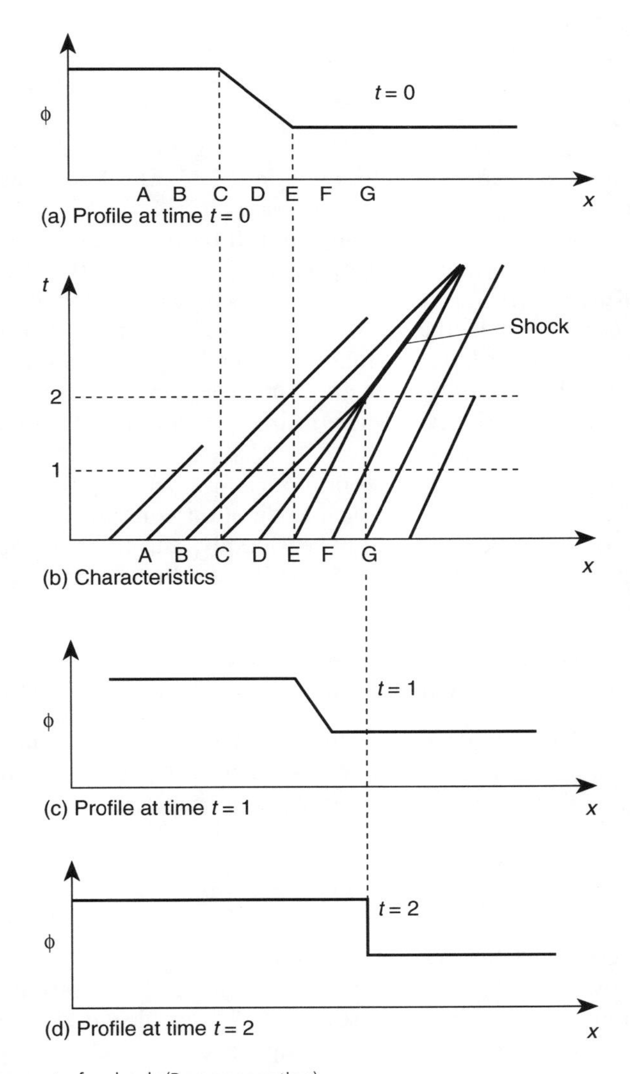

Fig. 2.19 Development of a shock (Burger equation).

To illustrate the necessity for the development of the shock, consider the propagation of a wave with an originally smooth profile illustrated in Fig. 2.19(a). Here as we know the characteristics along which ϕ is constant are straight lines shown in Fig. 2.19(b). These show different propagation speeds intersecting at time $t = 2$ when a discontinuous shock appears. This shock propagates at a finite speed (which here is the average of the two extreme values).

In such a shock the differential equation is no longer valid but the conservation integral is. We can thus write for a small length Δs around the discontinuity

$$\frac{\partial}{\partial t}\int_{\Delta s} \phi \, \mathrm{d}s + F(s + \Delta s) - F(s) = 0 \tag{2.120}$$

or

$$C\Delta\phi - \Delta F = 0 \tag{2.121a}$$

where $C = \lim \Delta s/\Delta t$ is the speed of shock propagation and $\Delta\phi$ and ΔF are the discontinuities in ϕ and F respectively. Equation (2.121a) is known as the Rankine–Hugoniot condition.

We shall find that such shocks develop frequently in the context of compressible flow and shallow-water flow (Chapters 6 and 7) and can often exist even in the presence of diffusive terms in the equation. Indeed, such shocks are not specific to transients but can persist in the steady state. Clearly, approximation of the finite element kind in which we have postulated in general a C_0 continuity to $\hat{\phi}$ can at best *smear* such a discontinuity over an element length, and generally oscillations near such a discontinuity arise even when the best algorithms of the preceding sections are used.

Figure 2.20 illustrates the difficulties of modelling such steep waves occurring even in linear problems in which the physical dissipation contained in the equations is incapable of smoothing the solution out reasonably, and to overcome this problem artificial diffusivity is frequently used. This artificial diffusivity must have the following characteristics:

1. It must vanish as the element size tends to zero.
2. It must not affect substantially the smooth domain of the solution.

A typical diffusivity often used is a finite element version of that introduced by Lapidus[67] for finite differences, but many other forms of local smoothing have been proposed.[68,69] The additional diffusivity is of the form

$$\tilde{k} = C_{\mathrm{Lap}} h^2 \left| \frac{\partial \phi}{\partial x} \right| \tag{2.122}$$

where the last term gives the maximum gradient.

In Fig. 2.21 we show a problem of discontinuous propagation in the Burger equation and how a progressive increase of the C_{Lap} coefficient kills spurious oscillation, but at the expense of rounding of a steep wave.

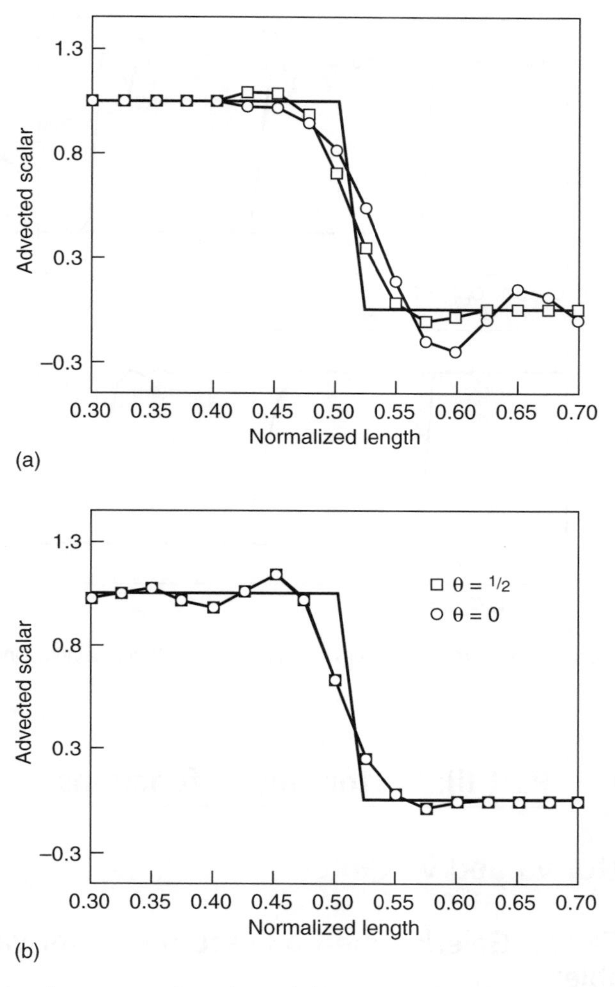

Fig. 2.20 Propagation of a steep wave by Taylor–Galerkin process: (a) Explicit methods $C = 0.5$, step wave at $Pe = 12\,500$; (b) Explicit methods $C = 0.1$, step wave at $Pe = 12\,500$.

For a multidimensional problem with a multidimensional ϕ a degree of anisotropy can be introduced and a possible expression generalizing (2.122) is

$$\tilde{k}_{ij} = C_{\text{Lap}} h^2 \frac{|V_i V_j|}{|\mathbf{V}|} \tag{2.123}$$

where

$$V_i = \frac{\partial \phi}{\partial x_i}$$

Other possibilities are open here and much current work is focused on the subject of 'shock capture'. We shall return to these problems in Chapter 6 where its importance in the high-speed flow of gases is paramount.

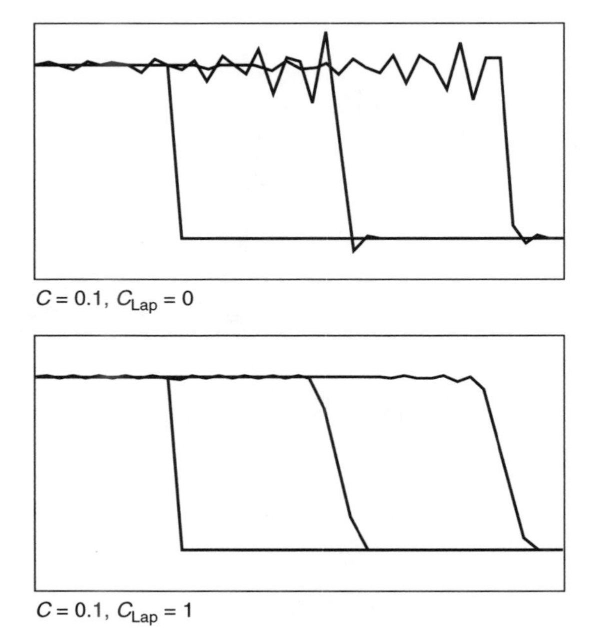

Fig. 2.21 Propagation of a steep front in Burger's equation with solution obtained using different values of Lapidus $C_v = C_{Lap}$.

Part III: Vector-valued functions

2.10 Vector-valued variables

2.10.1 The Taylor–Galerkin method used for vector-valued variables

The only method which adapts itself easily to the treatment of vector variables is that of the Taylor–Galerkin procedure. Here we can repeat the steps of Sec. 2.8 but now addressed to the vector-valued equation with which we started this chapter (Eq. 2.1). Noting that now $\mathbf{\Phi}$ has multiple components, expanding $\mathbf{\Phi}$ by a Taylor series in time we have[66,70]

$$\mathbf{\Phi}^{n+1} = \mathbf{\Phi}^n + \Delta t \left.\frac{\partial \mathbf{\Phi}}{\partial t}\right|_n + \frac{\Delta t^2}{2} \left.\frac{\partial^2 \mathbf{\Phi}}{\partial t^2}\right|_{n+\theta} \tag{2.124}$$

where Θ is a number such that $0 \leqslant \theta \leqslant 1$.

From Eq. (2.1),

$$\left[\frac{\partial \mathbf{\Phi}}{\partial t}\right]_n = -\left[\frac{\partial \mathbf{F}_i}{\partial x_i} + \frac{\partial \mathbf{G}_i}{\partial x_i} + \mathbf{Q}\right]_n \tag{2.125a}$$

and differentiating

$$\left[\frac{\partial^2 \mathbf{\Phi}}{\partial t^2}\right]_{n+\theta} = -\frac{\partial}{\partial t}\left[\frac{\partial \mathbf{F}_i}{\partial x_i} + \frac{\partial \mathbf{G}_i}{\partial x_i} + \mathbf{Q}\right]_{n+\theta} \tag{2.125b}$$

In the above we can write

$$\frac{\partial}{\partial t}\left(\frac{\partial \mathbf{F}_i}{\partial x_i}\right) \equiv \frac{\partial}{\partial x_i}\left(\frac{\partial \mathbf{F}_i}{\partial \mathbf{\Phi}}\frac{\partial \mathbf{\Phi}}{\partial t}\right) = -\frac{\partial}{\partial x_i}\left[\mathbf{A}_i\left(\frac{\partial \mathbf{F}_j}{\partial x_j} + \frac{\partial \mathbf{G}_j}{\partial x_j} + \mathbf{Q}\right)\right] \tag{2.125c}$$

where $\mathbf{A}_i \equiv \partial \mathbf{F}_i/\partial \mathbf{\Phi}$ and if $\mathbf{Q} = \mathbf{Q}(\mathbf{\Phi}, x)$ and $\partial \mathbf{Q}/\partial \mathbf{\Phi} = \mathbf{S}$,

$$\frac{\partial \mathbf{Q}}{\partial t} = \frac{\partial \mathbf{Q}}{\partial \mathbf{\Phi}}\frac{\partial \mathbf{\Phi}}{\partial t} = -\mathbf{S}\left(\frac{\partial \mathbf{F}_i}{\partial x_i} + \frac{\partial \mathbf{G}_i}{\partial x_i} + \mathbf{Q}\right) \tag{2.125d}$$

We can therefore approximate Eq. (2.124) as

$$\begin{aligned}\Delta \mathbf{\Phi}^n &\equiv \mathbf{\Phi}^{n+1} - \mathbf{\Phi}^n \\ &= -\Delta t\left[\frac{\partial \mathbf{F}_i}{\partial x_i} + \frac{\partial \mathbf{G}_i}{\partial x_i} + \mathbf{Q}\right]_n + \frac{\Delta t^2}{2}\left\{\frac{\partial}{\partial x_i}\left[\mathbf{A}_i\left(\frac{\partial \mathbf{F}_j}{\partial x_j} + \frac{\partial \mathbf{G}_j}{\partial x_j} + \mathbf{Q}\right)\right]\right. \\ &\quad \left. + \frac{\partial}{\partial t}\frac{\partial \mathbf{G}_i}{\partial x_i} + \mathbf{S}\left(\frac{\partial \mathbf{F}_j}{\partial x_j} + \frac{\partial \mathbf{G}_j}{\partial x_j} + \mathbf{Q}\right)\right\}_{n+\theta}\end{aligned} \tag{2.126}$$

Omitting the second derivatives of $\mathbf{G}_i$ and interpolating the $n+\theta$ between n and $n+1$ values we have

$$\begin{aligned}\Delta \mathbf{\Phi} &\equiv \mathbf{\Phi}^{n+1} - \mathbf{\Phi}^n \\ &= -\Delta t\left[\frac{\partial \mathbf{F}_i}{\partial x_i} + \mathbf{Q}\right]_n - \Delta t\left(\left[\frac{\partial \mathbf{G}_i}{\partial x_i}\right]_{n+1}\theta + \left[\frac{\partial \mathbf{G}_i}{\partial x_i}\right]_n (1-\theta)\right) \\ &\quad + \frac{\Delta t^2}{2}\left[\frac{\partial}{\partial x_i}\left\{\mathbf{A}_i\left(\frac{\partial \mathbf{F}_j}{\partial x_j} + \mathbf{Q}\right)\right\} + \mathbf{S}\left(\frac{\partial \mathbf{F}_j}{\partial x_j} + \mathbf{Q}\right)\right]_{n+1}\theta \\ &\quad + \frac{\Delta t^2}{2}\left[\frac{\partial}{\partial x_i}\left\{\mathbf{A}_i\left(\frac{\partial \mathbf{F}_j}{\partial x_j} + \mathbf{Q}\right)\right\} + \mathbf{S}\left(\frac{\partial \mathbf{F}_j}{\partial x_j} + \mathbf{Q}\right)\right]_n (1-\theta)\end{aligned} \tag{2.127}$$

At this stage a standard Galerkin approximation is applied which will result in a discrete, implicit, time-stepping scheme that is unconditionally stable if $\theta \geqslant \frac{1}{2}$. As the explicit form is of particular interest we shall only give the details of the discretization process for $\theta = 0$. Writing as usual

$$\mathbf{\Phi} \approx \mathbf{N}\tilde{\mathbf{\Phi}}$$

we have

$$\begin{aligned}\left(\int_\Omega \mathbf{N}^{\mathrm{T}}\mathbf{N}\,\mathrm{d}\Omega\right)\Delta\tilde{\mathbf{\Phi}} = -\Delta t\Bigg[&\int_\Omega \mathbf{N}^{\mathrm{T}}\left(\frac{\partial \mathbf{F}_i}{\partial x_i} + \frac{\partial \mathbf{G}_i}{\partial x_i} + \mathbf{Q}\right)\mathrm{d}\Omega \\ &- \frac{\Delta t}{2}\int_\Omega \mathbf{N}^{\mathrm{T}}\frac{\partial}{\partial x_i}\left\{\mathbf{A}_i\left(\frac{\partial \mathbf{F}_j}{\partial x_j} + \frac{\partial \mathbf{G}_j}{\partial x_j} + \mathbf{Q}\right)\right\}\mathrm{d}\Omega \\ &+ \frac{\Delta t}{2}\int_\Omega \mathbf{N}^{\mathrm{T}}\mathbf{S}\left(\frac{\partial \mathbf{F}_j}{\partial x_j} + \frac{\partial \mathbf{G}_j}{\partial x_j} + \mathbf{Q}\right)\mathrm{d}\Omega\Bigg]_n\end{aligned} \tag{2.128}$$

This can be written in a compact matrix form similar to Eq. (2.93) as

$$\mathbf{M}\Delta\tilde{\mathbf{\Phi}} = -\Delta t[(\mathbf{C} + \mathbf{K}_u + \mathbf{K})\tilde{\mathbf{\Phi}} + \mathbf{f}]^n \tag{2.129a}$$

in which, with

$$\mathbf{G}_i = -k_{ij}\frac{\partial \mathbf{\Phi}}{\partial x_j}$$

we have (on omitting the third derivative terms and the effect of **S**) matrices of the form of Eq. (2.94), i.e.

$$\begin{aligned}
\mathbf{C} &= \int_\Omega \mathbf{N}^{\mathrm{T}}\mathbf{A}_i\frac{\partial \mathbf{N}}{\partial x_i}\,\mathrm{d}\Omega \\
\mathbf{K}_u &= \int_\Omega \frac{\partial \mathbf{N}^{\mathrm{T}}}{\partial x_i}\left(\mathbf{A}_i\mathbf{A}_j\frac{\Delta t}{2}\right)\frac{\partial \mathbf{N}}{\partial x_j}\,\mathrm{d}\Omega \\
\mathbf{K} &= \int_\Omega \frac{\partial \mathbf{N}^{\mathrm{T}}}{\partial x_i}k_{ij}\frac{\partial \mathbf{N}}{\partial x_j}\,\mathrm{d}\Omega \\
\mathbf{f} &= \int_\Omega \left(\mathbf{N}^{\mathrm{T}} + \frac{\Delta t}{2}\mathbf{A}_i\frac{\partial \mathbf{N}^{\mathrm{T}}}{\partial x_i}\right)\mathbf{Q}\,\mathrm{d}\Omega + \text{boundary terms} \\
\mathbf{M} &= \int_\Omega \mathbf{N}^{\mathrm{T}}\mathbf{N}\,\mathrm{d}\Omega
\end{aligned} \tag{2.129b}$$

With $\theta = \frac{1}{3}$ it can be shown that the order of approximation increases and for this scheme a simple iterative solution is possible.[71] We note that with the consistent mass matrix **M** the stability limit for $\theta = \frac{1}{3}$ is increased to $C = 1$.

Use of $\theta = \frac{1}{3}$ apparently requires an implicit solution. However, similar iteration to that used in Eq. (2.104) is rapidly convergent and the scheme can be used quite economically.

2.10.2 Two-step predictor–corrector methods. Two-step Taylor–Galerkin operation

There are of course various alternative procedures for improving the temporal approximation other than the Taylor expansion used in the previous section. Such procedures will be particularly useful if the evaluation of the derivative matrix **A** can be avoided. In this section we shall consider two predictor–corrector schemes (of Runge–Kutta type) that avoid the evaluation of this matrix and are explicit.

The first starts with a standard Galerkin space approximation being applied to the basic equation (2.1). This results in the form

$$\mathbf{M}\frac{\mathrm{d}\tilde{\mathbf{\Phi}}}{\mathrm{d}t} \equiv \mathbf{M}\dot{\tilde{\mathbf{\Phi}}} = \mathbf{P}_C + \mathbf{P}_D + \mathbf{f} = \boldsymbol{\psi} \tag{2.130}$$

where again $\mathbf{M}$ is the standard mass matrix, $\mathbf{f}$ are the prescribed 'forces' and

$$\mathbf{P}_C(\mathbf{\Phi}) = \int_\Omega \mathbf{N}^{\mathrm{T}} \frac{\partial \mathbf{F}_i}{\partial x_i} \, \mathrm{d}\Omega \tag{2.131a}$$

represents the convective 'forces', while

$$\mathbf{P}_D(\mathbf{\Phi}) = \int_\Omega \mathbf{N}^{\mathrm{T}} \frac{\partial \mathbf{G}_i}{\partial x_i} \, \mathrm{d}\Omega \tag{2.131b}$$

are the diffusive ones.

If an explicit time integration scheme is used, i.e.

$$\mathbf{M}\Delta\mathbf{\Phi} \equiv \mathbf{M}(\tilde{\mathbf{\Phi}}^{n+1} - \tilde{\mathbf{\Phi}}^{n}) = \Delta t \boldsymbol{\psi}^n(\tilde{\mathbf{\Phi}}^n) \tag{2.132}$$

the evaluation of the right-hand side does not require the matrix product representation and $\mathbf{A}_i$ does not have to be computed.

Of course the scheme presented is not accurate for the various reasons previously discussed, and indeed becomes *unconditionally unstable* in the absence of diffusion and external force vectors.

The reader can easily verify that in the case of the linear one-dimensional problem the right-hand side is equivalent to a central difference scheme with $\tilde{\mathbf{\Phi}}_{i-1}^n$ and $\tilde{\mathbf{\Phi}}_{i+1}^n$ only being used to find the value of $\mathbf{\Phi}_i^{n+1}$, as shown in Fig. 2.22(a).

The scheme can, however, be recast as a two-step, predictor–corrector operation and conditional stability is regained. Now we proceed as follows:

Step 1. Compute $\tilde{\mathbf{\Phi}}^{n+1/2}$ using an explicit approximation of Eq. (2.132), i.e.

$$\tilde{\mathbf{\Phi}}^{n+1/2} = \tilde{\mathbf{\Phi}}^{n} + \frac{\Delta t}{2}\mathbf{M}^{-1}\boldsymbol{\psi}^n \tag{2.133}$$

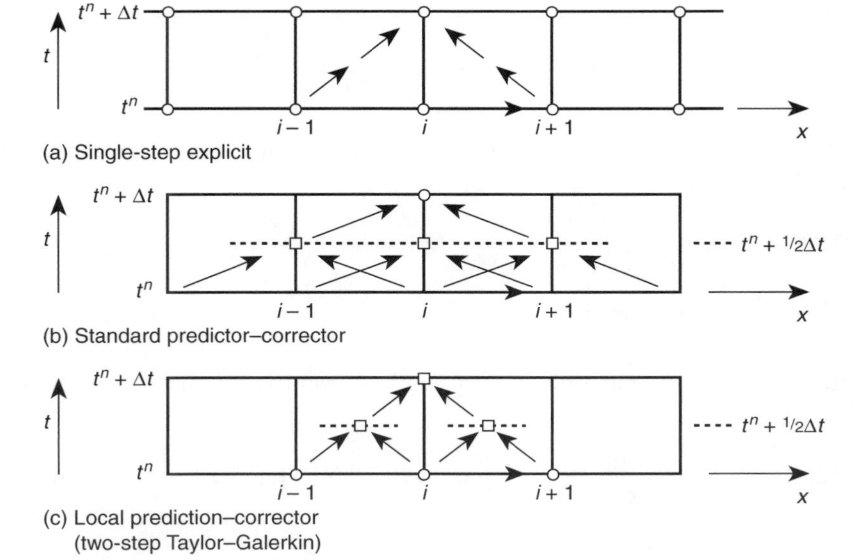

Fig. 2.22 Progression of information in explicit one- and two-step schemes.

and

Step 2. Compute $\tilde{\boldsymbol{\Phi}}^{n+1}$ inserting the improved value of $\tilde{\boldsymbol{\Phi}}^{n+1/2}$ in the right-hand side of Eq. (2.132), giving

$$\tilde{\boldsymbol{\Phi}}^{n+1} = \boldsymbol{\Phi}^n + \Delta t \mathbf{M}^{-1} \boldsymbol{\psi}^{n+1/2} \tag{2.134}$$

This is precisely equivalent to the second-order Runge–Kutta scheme being applied to the ordinary system of differential equations (2.130). Figure 2.22(b) shows in the one-dimensional example how the information 'spreads', i.e. that now $\tilde{\boldsymbol{\Phi}}_i^{n+1}$ will be dependent on values at nodes $i-2, \ldots, i+2$.

It is found that the scheme, though stable, is overdiffusive and numerical results are poor.

An alternative is possible, however, using a two-step Taylor–Galerkin operation. Here we return to the original equation (2.1) and proceed as follows:

Step 1. Find an improved value of $\boldsymbol{\Phi}^{n+1/2}$ *using only the convective and source parts.* Thus

$$\tilde{\boldsymbol{\Phi}}^{n+1/2} = \tilde{\boldsymbol{\Phi}}^n - \frac{\Delta t}{2}\left(\frac{\partial \mathbf{F}_i^n}{\partial x_i} + \mathbf{Q}^n\right) \tag{2.135a}$$

which of course allows the evaluation of $\mathbf{F}_i^{n+1/2}$.

We note, however, that we can also write an approximate expansion as

$$\begin{aligned}\mathbf{F}_i^{n+1/2} &= \mathbf{F}_i^n + \frac{\Delta t}{2}\frac{\partial \mathbf{F}_i^n}{\partial t} = \mathbf{F}_i^n - \frac{\Delta t}{2}\mathbf{A}_i^n \frac{\partial \boldsymbol{\Phi}^n}{\partial t} \\ &= \mathbf{F}_i^n - \frac{\Delta t}{2}\mathbf{A}_i^n\left(\frac{\partial \mathbf{F}_j}{\partial x_j} + \frac{\partial \mathbf{G}_j}{\partial x_j} + \mathbf{Q}\right)^n \end{aligned} \tag{2.135b}$$

This gives

$$\mathbf{A}_i^n\left(\frac{\partial \mathbf{F}_j}{\partial x_j} + \frac{\partial \mathbf{G}_j}{\partial x_j} + \mathbf{Q}\right)^n = -\frac{2}{\Delta t}(\mathbf{F}_i^{n+1/2} - \mathbf{F}_i^n) \tag{2.135c}$$

Step 2. Substituting the above into the Taylor–Galerkin approximation of Eq. (2.128) we have

$$\begin{aligned}\mathbf{M}\Delta\tilde{\boldsymbol{\Phi}} = -\Delta t\bigg[\int_\Omega \mathbf{N}^{\mathrm{T}}\left(\frac{\partial \mathbf{F}_i}{\partial x_i} + \frac{\partial \mathbf{G}_i}{\partial x_i} + \mathbf{Q}\right)^n \mathrm{d}\Omega + \int_\Omega \mathbf{N}^{\mathrm{T}}\frac{\partial}{\partial x_i}(\mathbf{F}_i^{n+1/2} - \mathbf{F}_i^n)\,\mathrm{d}\Omega \\ + \int_\Omega \mathbf{N}^{\mathrm{T}}\mathbf{S}(\mathbf{F}_i^{n+1/2} - \mathbf{F}_i^n)\,\mathrm{d}\Omega\bigg]\end{aligned} \tag{2.135d}$$

and after integration by parts of the terms with respect to the x_i derivatives we obtain simply

$$\begin{aligned}\mathbf{M}\Delta\tilde{\boldsymbol{\Phi}} = -\Delta t\bigg\{\int_\Omega \frac{\partial \mathbf{N}^{\mathrm{T}}}{\partial x_i}(\mathbf{F}_i^{n+1/2} + \mathbf{G}_i^n)\,\mathrm{d}\Omega + \int \mathbf{N}^{\mathrm{T}}[\mathbf{Q} + \mathbf{S}(\mathbf{F}^{n+1/2} - \mathbf{F}^n)]\,\mathrm{d}\Omega \\ + \int_\Gamma \mathbf{N}^{\mathrm{T}}(\mathbf{F}_i^{n+1/2} + \mathbf{G}_i^n)\mathbf{n}_i\,\mathrm{d}\Gamma\bigg\}\end{aligned} \tag{2.136}$$

We note immediately that:

1. The above expression is identical to *using a standard Galerkin approximation on Eq. (2.1)* and an explicit step with $\mathbf{F}_i$ values updated by the simple equation (2.135a).
2. The final form of Eq. (2.136) does not require the evaluation of the matrices $\mathbf{A}_i$ resulting in substantial computation savings as well as yielding essentially the same results. Indeed, some omissions made in deriving Eqs (2.129) did not occur now and presumably the accuracy is improved.

A further practical point must be noted:

3. In non-linear problems it is convenient to interpolate $\mathbf{F}_i$ directly in the finite element manner as

$$\mathbf{F}_i = \mathbf{N}\tilde{\bar{\mathbf{F}}}_i$$

rather than to compute it as $\mathbf{F}_i(\tilde{\bar{\mathbf{\Phi}}})$.

Thus the evaluation of $\mathbf{F}_i^{n+1/2}$ need only be made at the quadrature (integration) points within the element, and the evaluation of $\tilde{\bar{\mathbf{\Phi}}}^{n+1/2}$ by Eq. (2.135a) is only done on such points. For a linear triangle element this reduces to a single evaluation of $\tilde{\mathbf{\Phi}}^{n+1/2}$ and $\mathbf{F}^{n+1/2}$ for each element at its centre, taking of course $\tilde{\mathbf{\Phi}}^{n+1/2}$ and $\mathbf{F}^{n+1/2}$ as the appropriate interpolation average there.

In the simple one-dimensional linear example the information progresses in the manner shown in Fig. 2.22(c). The scheme, which originated at Swansea, can be appropriately called the *Swansea two step*,[57,65,72–80] and has found much use in the direct solution of compressible high-speed gas flow equations. We shall show some of the results obtained by this procedure in Chapter 6. However in Chapter 3 we shall discuss an alternative which is more general and has better performance. It is of interest to remark that the Taylor–Galerkin procedure can be used in contexts other than direct fluid mechanics. The procedure has been used efficiently by Morgan *et al.*[81,82] in solving electromagnetic wave problems.

2.10.3 Multiple wave speeds

When ϕ is a scalar variable, a single wave speed will arise in the manner in which we have already shown at the beginning of Part II. When a vector variable is considered, the situation is very different and in general the number of wave speeds will correspond to the number of variables. If we return to the general equation (2.1), we can write this in the form

$$\frac{\partial \mathbf{\Phi}}{\partial t} + \mathbf{A}_i \frac{\partial \mathbf{\Phi}}{\partial x_i} + \frac{\partial \mathbf{G}_i}{\partial x_i} + \mathbf{Q} = \mathbf{0} \tag{2.137}$$

where $\mathbf{A}_i$ is a matrix of the size corresponding to the variables in the vector $\tilde{\mathbf{\Phi}}$. This is equivalent to the single convective velocity component $A = U$ in a scalar problem and is given as

$$\mathbf{A}_i \equiv \frac{\partial \mathbf{F}_i}{\partial \mathbf{\Phi}} \tag{2.138}$$

This in general may still be a function of $\mathbf{\Phi}$, thus destroying the linearity of the problem.

Before proceeding further, it is of interest to discuss the general behaviour of Eq. (2.1) in the absence of source and diffusion terms. We note that the matrices $\mathbf{A}_i$ can be represented as

$$\mathbf{A}_i = \mathbf{X}_i \mathbf{\Lambda}_i \mathbf{X}_i^{-1} \tag{2.139}$$

by a standard eigenvalue analysis in which $\mathbf{\Lambda}_i$ is a diagonal matrix.

If the matrices $\mathbf{X}_i$ are such that

$$\mathbf{X}_i = \mathbf{X} \tag{2.140}$$

which is always the case in a single dimension, then Eq. (2.137) can be written (in the absence of diffusion or source terms) as

$$\frac{\partial \mathbf{\Phi}}{\partial t} + \mathbf{X} \mathbf{\Lambda}_i \mathbf{X}^{-1} \frac{\partial \mathbf{\Phi}}{\partial x_i} = \mathbf{0} \tag{2.141}$$

Premultiplying by $\mathbf{X}^{-1}$ and introducing new variables (called Riemann invariants) such that

$$\boldsymbol{\phi} = \mathbf{X}^{-1} \mathbf{\Phi} \tag{2.142}$$

we can write the above as a set of decoupled equations in components ϕ of $\boldsymbol{\phi}$ and corresponding Λ of $\mathbf{\Lambda}$:

$$\frac{\partial \phi}{\partial t} + \Lambda_i \frac{\partial \phi}{\partial x_i} = 0 \tag{2.143}$$

each of which represents a wave-type equation of the form that we have previously discussed. A typical example of the above results from a one-dimensional elastic dynamics problem describing stress waves in a bar in terms of stresses (σ) and velocities (v) as

$$\frac{\partial \sigma}{\partial t} - E \frac{\partial v}{\partial x} = 0$$

$$\frac{\partial v}{\partial t} - \frac{1}{\rho} \frac{\partial \sigma}{\partial x} = 0$$

This can be written in the standard form of Eq. (2.1) with

$$\mathbf{\Phi} = \begin{Bmatrix} \sigma \\ v \end{Bmatrix} \qquad F = \begin{Bmatrix} Ev \\ \sigma/\rho \end{Bmatrix}$$

The two variables of Eq. (2.142) become

$$\phi_1 = \sigma - cv \qquad \phi_2 = \sigma + cv$$

where $c = \sqrt{E/\rho}$ and the equations corresponding to (2.143) are

$$\frac{\partial \phi_1}{\partial t} + c \frac{\partial \phi_1}{\partial x} = 0$$

$$\frac{\partial \phi_2}{\partial t} - c \frac{\partial \phi_2}{\partial x} = 0$$

representing respectively two waves moving with velocities $\pm c$.

Unfortunately the condition of Eq. (2.140) seldom pertains and hence the determination of general characteristics and therefore decoupling is not usually possible for more than one space dimension. This is the main reason why the extension of the simple, direct procedures is not generally possible for vector variables. Because of this we shall in Chapter 3 only use the upwinding characteristic-based procedures on scalar systems for which a single wave speed exists and this retains justification of any method proposed.

2.11 Summary and concluding remarks

The reader may well be confused by the variety of apparently unrelated approaches given in this chapter. This may be excused by the fact that optimality guaranteed by the finite element approaches in elliptic, self adjoint problems does not automatically transfer to hyperbolic non-self adjoint ones.

The major part of this chapter is concerned with a scalar variable in the convection–diffusion reaction equation. The several procedures presented for steady-state and transient equations yield identical results. However the characteristic–Galerkin method is optimal for transient problems and gives identical stabilizing terms to that derived by the use of Petrov–Galerkin, GLS and other procedures when the time step used is near the stability limit. For such a problem the optimality is assured simply by splitting the problem into the self-adjoint part where the direct Galerkin approximation is optimal and an advective motion where the unknown variable remains fixed in the characteristic space.

Extension of the various procedures presented to vector variables has been made in the past and we have presented the Taylor–Galerkin method in this context; however its justification is more problematic. For this reason we recommend that when dealing with equations such as those arising in the motion of a fluid an operator split is made in a manner separating several scalar convection–diffusion problems for which the treatment described is used. We shall do so in the next chapter when we introduce the CBS algorithm using the *characteristic-based split*.

References

1. A.N. Brooks and T.J.R. Hughes. Streamline upwind/Petrov–Galerkin formulation for convection dominated flows with particular emphasis on the incompressible Navier Stokes equation. *Comp. Meth. Appl. Mech. Eng.*, **32**, 199–259, 1982.
2. R. Courant, E. Isaacson and M. Rees. On the solution of non-linear hyperbolic differential equations by finite differences. *Comm. Pure Appl. Math.*, **V**, 243–55, 1952.
3. A.K. Runchall and M. Wolfstein. Numerical integration procedure for the steady state Navier–Stokes equations. *J. Mech. Eng. Sci.*, **11**, 445–53, 1969.
4. D.B. Spalding. A novel finite difference formulation for differential equations involving both first and second derivatives. *Int. J. Num. Meth. Eng.*, **4**, 551–9, 1972.
5. K.E. Barrett. The numerical solution of singular perturbation boundary value problem. *Q. J. Mech. Appl. Math.*, **27**, 57–68, 1974.
6. O.C. Zienkiewicz, R.H. Gallagher and P. Hood. Newtonian and non-Newtonian viscous incompressible flow. Temperature induced flows and finite element solutions, in *The*

Mathematics of Finite Elements and Applications (ed. J. Whiteman), Vol. II, Academic Press, London, 1976 (Brunel University, 1975).
7. I. Christie, D.F. Griffiths, A.R. Mitchell and O.C. Zienkiewicz. Finite element methods for second order differential equations with significant first derivatives. *Int. J. Num. Meth. Eng.*, **10**, 1389–96, 1976.
8. O.C. Zienkiewicz, J.C. Heinrich, P.S. Huyakorn and A.R. Mitchell. An upwind finite element scheme for two dimensional convective transport equations. *Int. J. Num. Meth. Eng.*, **11**, 131–44, 1977.
9. J.C. Heinrich and O.C. Zienkiewicz. Quadratic finite element schemes for two dimensional convective–transport problems. *Int. J. Num. Meth. Eng.*, **11**, 1831–44, 1977.
10. D.W. Kelly, S. Nakazawa and O.C. Zienkiewicz. A note on anisotropic balancing dissipation in the finite element method approximation to convective diffusion problems. *Int. J. Num. Meth. Eng.*, **15**, 1705-11, 1980.
11. B.P. Leonard. A survey of finite differences of opinion on numerical muddling of the incomprehensible defective confusion equation, in *Finite Elements for Convection Dominated Flows* (ed. T.J.R. Hughes), AMD Vol. 34, ASME, 1979.
12. G.L. Guymon, V.H. Scott and L.R. Herrmann. A general numerical solution of the two dimensional diffusion–convection equation by the finite element method. *Water Resources Res.*, **6**, 1611–17, 1970.
13. T.J.R. Hughes and J.D. Atkinson. A variational basis of 'upwind' finite elements, in *Variational Methods in the Mechanics of Solids* (ed. S. Nemat-Nasser), pp. 387–91, Pergamon Press, Oxford, 1980.
14. G.F. Carey. Exponential upwinding and integrating factors for symmetrization. *Comm. Appl. Num. Mech.*, **1**, 57–60, 1985.
15. J. Donea, T. Belytschko and P. Smolinski. A generalized Galerkin method for steady state convection–diffusion problems with application to quadratic shape function. *Comp. Meth. Appl. Mech. Eng.*, **48**, 25–43, 1985.
16. T.J.R. Hughes, L.P. Franca, G.M. Hulbert, Z. Johan and F. Sakhib. The Galerkin least square method for advective diffusion equations, in *Recent Developments in Computational Fluid Mechanics* (eds T.E. Tezduyar and T.J.R. Hughes), AMD 95, ASME, 1988.
17. E. Oñate. Derivation of stabilized equations for numerical solution of advective–diffusive transport and fluid flow problems. *Comp. Meth. Appl. Mech. Eng.*, **151**, 233–65, 1998.
18. S. Nakazawa, J.F. Pittman and O.C. Zienkiewicz. Numerical solution of flow and heat transfer in polymer melts, in *Finite Elements in Fluids* (eds R.H. Gallagher *et al.*), Vol. 4, chap. 13, pp. 251–83, Wiley, Chichester, 1982.
19. R. Codina, E. Oñate and M. Cervera. The intrinsic time for the streamline upwind Petrov–Galerkin formulation using quadratic elements. *Comp. Meth. Appl. Mech. Eng.*, **94**, 239–62, 1992.
20. R. Codina. Stability analysis of forward Euler scheme for convection diffusion equation using the SUPG formulation in space. *Int. J. Num. Meth. Eng.*, **36**, 1445–64, 1993.
21. J.T. Oden, I. Babuška and C.E. Baumann. A discontinuous hp finite element method for diffusion problems. *J. Comp. Phys.*, **146**, 491–519, 1998.
22. C.E. Baumann and J.T. Oden. A discontinuous hp finite element method for convection diffusion problems. *Comp. Meth. Appl. Mech. Eng.*, **175**, 311–41, 1999.
23. C.E. Baumann and J.T. Oden. A discontinuous hp finite element method for the Euler and Navier–Stokes equations. *Int. J. Num. Meth. Fluids* **31**, 79–95, 1999.
24. T.J.R. Hughes and A. Brooks. A multi-dimensional upwind scheme with no cross wind diffusion, in *Finite Elements for Convection Dominated Flows* (ed. T.J.R. Hughes), AMD 34, ASME, 1979.

25. T.J.R. Hughes and A.N. Brooks. A theoretical framework for Petrov–Galerkin methods with discontinuous weighting function, in *Finite Elements in Fluids* (eds R.H. Gallagher *et al.*), Vol. 4, pp. 47–65, Wiley, Chichester, 1982.
26. C. Johnson and A. Szepessy. On the convergence of a finite element method for a nonlinear hyperbolic conservation law. *Math. Comput.*, **49**, 427–44, 1987.
27. F. Shakib, T.R.J. Hughes and Z. Johan. A new finite element formulation for computational fluid dynamics: X. The compressible Euler and Navier–Stokes equations. *Comp. Meth. Appl. Mech. Eng.*, **89**, 141–219, 1991.
28. R. Codina. A discontinuity capturing crosswind–dissipation for the finite element solution of convection diffusion equation. *Comp. Meth. Appl. Mech. Eng.*, **110**, 325–42, 1993.
29. P. Nithiarasu, O.C. Zienkiewicz, B.V.K.S. Sai, K. Morgan, R. Codina and M. Vázquez. Shock capturing viscosities for the general fluid mechanics algorithm. *Int. J. Num. Meth. Fluids*, **28**, 1325–53, 1998.
30. C. Johnson, V. Nävert and J. Pitkäranta. Finite element methods for linear, hyperbolic problems. *Comp. Meth. Appl. Mech. Eng.*, **45**, 285–312, 1984.
31. P.A.B. de Sampaio. A modified operator analysis of convection diffusion problems, in *Proc. II National Meeting on Thermal Sciences*, Aguas de Lindoia (Brazil). pp. 180–3, 1988.
32. N. Nguen and J. Reynen. A space–time least square finite element scheme for advection–diffusion equations. *Comp. Meth. Appl. Mech. Eng.*, **42**, 331–42, 1984.
33. G.F. Carey and B.N. Jiang. Least square finite elements for first order hyperbolic systems. *Int. J. Num. Meth. Eng.*, **26**, 81–93, 1988.
34. B.N. Jiang and G.F. Carey. A stable least-square finite element method for non-linear hyperbolic problems. *Int. J. Num. Meth. Fluids*, **8**, 933–42, 1988.
35. C. Johnson. Streamline diffusion elements for problems in fluid mechanics, in *Finite Elements In Fluids* (eds R.H. Gallagher *et al.*), Vol. 6, pp. 251–61, Wiley, Chichester, 1986.
36. C. Johnson. *Numerical Solution of Partial Differential Equations by the Finite Element Method.* Cambridge University Press, Cambridge, 1987.
37. C.C. Yu and J.C. Heinrich. Petrov–Galerkin methods for the time dependent convective transport equation. *Int. J. Num. Meth. Eng.*, **23**, 883–901, 1986.
38. C.C. Yu and J.C. Heinrich. Petrov–Galerkin method for multidimensional, time dependent convective diffusion equation. *Int. J. Num. Meth. Eng.*, **24**, 2201–15, 1987.
39. C.E. Baumann, M.A. Storti and S.R. Idelsohn. A Petrov–Galerkin technique for the solution of transonic and supersonic flows. *Comp. Meth. Appl. Mech. Eng.*, **95**, 49–70, 1992.
40. P.A.B. de Sampaio, P.R.M. Lyra, K. Morgan and N.P. Weatherill. Petrov–Galerkin solutions of the incompressible Navier–Stokes equations in primitive variables with adaptive remeshing. *Comp. Meth. Appl. Mech. Eng.*, **106**, 143–78, 1993.
41. J.A. Cardle. A modification of the Petrov–Galerkin method for the transient convection diffusion equation. *Int. J. Num. Meth. Eng.*, **38**, 171–81, 1995.
42. S.R. Idelsohn, J.C. Heinrich and E. Oñate. Petrov–Galerkin methods for the transient advective–diffusive equation with sharp gradients. *Int. J. Num. Meth. Eng.*, **39**, 1455–73, 1996.
43. R. Codina. Comparison of some finite element methods for solving the diffusion–convection–reaction equation. *Comp. Meth. Appl. Mech. Eng.*, **156**, 185–210, 1998.
44. P.K. Maji and G. Biswas. Analysis of flow in the spiral casing using a streamline upwind Petrov–Galerkin method. *Int. J. Num. Meth. Eng.*, **45**, 147–74, 1999.
45. R.A. Adey and C.A. Brebbia. Finite element solution of effluent dispersion, in *Numerical Methods in Fluid Mechanics* (eds C.A. Brebbia and J.J. Connor). pp. 325–54, Pentech Press, Southampton, 1974.
46. K.W. Morton. Generalised Galerkin methods for hyperbolic problems. *Comp. Meth. Appl. Mech. Eng.*, **52**, 847–71, 1985.

47. R.E. Ewing and T.F. Russell. Multistep Galerkin methods along characteristics for convection–diffusion problems, in *Advances in Computation Methods for PDEs* (eds R. Vichnevetsky and R.S. Stepleman), Vol. IV, IMACS, pp. 28–36, Rutgers University, Brunswick, N.J., 1981.
48. J. Douglas, Jr and T.F. Russell. Numerical methods for convection dominated diffusion problems based on combining the method of characteristics with finite element or finite difference procedures. *SIAM J. Num. Anal.*, **19**, 871–85, 1982.
49. O. Pironneau. On the transport diffusion algorithm and its application to the Navier–Stokes equation. *Num. Math.*, **38**, 309–32, 1982.
50. M. Bercovier, O. Pironneau, Y. Harbani and E. Levine. Characteristics and finite element methods applied to equations of fluids, in *The Mathematics of Finite Elements and Applications* (ed. J.R. Whiteman), Vol. V, pp. 471–8, Academic Press, London, 1982.
51. J. Goussebaile, F. Hecht, C. Labadie and L. Reinhart. Finite element solution of the shallow water equations by a quasi-direct decomposition procedure. *Int. J. Num. Meth. Fluids*, **4**, 1117–36, 1984.
52. M. Bercovier, O. Pironneau and V. Sastri. Finite elements and characteristics for some parabolic–hyperbolic problems. *Appl. Math. Modelling*, **7**, 89–96, 1983.
53. J.P. Benque, J.P. Gregoire, A. Hauguel and M. Maxant. Application des Methodes du decomposition aux calculs numeriques en hydraulique industrielle, in *INRIA*, *6th Coll. Inst. Methodes de Calcul Sci. et Techn.*, Versailles, 12–16 Dec. 1983.
54. A. Bermudez, J. Durany, M. Posse and C. Vazquez. An upwind method for solving transport–diffusion–reaction systems. *Int. J. Num. Meth. Eng.*, **28**, 2021–40, 1984.
55. P.X. Lin, K.W. Morton and E. Suli. Characteristic Galerkin schemes for scalar conservation laws in two and three space dimensions. *SIAM J. Num. Anal.*, **34**, 779–96, 1997.
56. O. Pironneau, J. Liou and T.T.I. Tezduyar. Characteristic Galerkin and Galerkin least squares space-time formulations for the advection–diffusion equation with time dependent domain. *Comp. Meth. Appl. Mech. Eng.*, **100**, 117–41, 1992.
57. O.C. Zienkiewicz, R. Löhner, K. Morgan and S. Nakazawa. Finite elements in fluid mechanics – a decade of progress, in *Finite Elements in Fluids* (eds R.H. Gallagher *et al.*), Vol. 5, chap. 1, pp. 1–26, Wiley, Chichester, 1984.
58. R. Löhner, K. Morgan and O.C. Zienkiewicz. The solution of non-linear hyperbolic equation systems by the finite element method. *Int. J. Num. Meth. Fluids*, **4**, 1043–63, 1984.
59. O.C. Zienkiewicz, R. Löhner, K. Morgan and J. Peraire. High speed compressible flow and other advection dominated problems of fluid mechanics, in *Finite Elements in Fluids* (eds R.H. Gallagher *et al.*), Vol. 6, chap. 2, pp. 41–88, Wiley, Chichester, 1986.
60. R. Löhner, K. Morgan and O.C. Zienkiewicz. An adaptive finite element procedure for compressible high speed flows. *Comp. Meth. Appl. Mech. Eng.*, **51**, 441–65, 1985.
61. O.C. Zienkiewicz, R. Löhner and K. Morgan. High speed inviscid compressive flow by the finite element method, in *The Mathematics of Finite Elements and Applications* (ed. J.R. Whiteman), Vol. VI, pp. 1–25, Academic Press, London, 1985.
62. O.C. Zienkiewicz and R. Codina. A general algorithm for compressible and incompressible flow – Part I. The split, characteristic based scheme. *Int. J. Num. Meth. Fluids*, **20**, 869–85, 1996.
63. O.C. Zienkiewicz, P. Nithiarasu, R. Codina and M. Vázquez. The Characteristic Based Split procedure: An efficient and accurate algorithm for fluid problems. *Int. J. Num. Meth. Fluids*, **31**, 359–92, 1999.
64. P.D. Lax and B. Wendroff. Systems of conservative laws. *Comm. Pure Appl. Math.*, **13**, 217–37, 1960.
65. J. Peraire. A finite method for convection dominated flows. Ph.D. thesis, University of Wales, Swansea, 1986.

66. J. Donea. A Taylor–Galerkin method for convective transport problems. *Int. J. Num. Meth. Eng.*, **20**, 101–19, 1984.
67. A. Lapidus. A detached shock calculation by second order finite differences. *J. Comp. Phys.*, **2**, 154–77, 1967.
68. J.P. Boris and D.L. Brook. Flux corrected transport I Shasta – a fluid transport algorithm that works. *J. Comp. Phys.*, **11**, 38–69, 1973.
69. S.T. Zalesiak. Fully multidimensional flux corrected transport algorithm for fluids. *J. Comp. Phys.*, **31**, 335–62, 1979.
70. V. Selmin, J. Donea and L. Quatrapelle. Finite element method for non-linear advection. *Comp. Meth. Appl. Mech. Eng.*, **52**, 817–45, 1985.
71. L. Bottura and O.C. Zienkiewicz. Experiments on iterative solution of the semi-implicit characteristic Galerkin algorithm. *Comm. Appl. Num. Meth.*, **6**, 387–93, 1990.
72. R. Löhner, K. Morgan, J. Peraire, O.C. Zienkiewicz and L. Kong. Finite element methods for compressible flow, in *Numerical Methods in Fluid Dynamics* (ed. K.W. Morton and M.J. Baines), Vol. II, pp. 27–52, Clarendon Press, Oxford, 1986.
73. J. Peraire, K. Morgan and O.C. Zienkiewicz. Convection dominated problems, in *Numerical Methods for Compressible Flows – Finite Difference, Element and Volume Techniques*, AMD 78, pp. 129–47, ASME, 1987.
74. O.C. Zienkiewicz, J.Z. Zhu, Y.C. Liu, K. Morgan and J. Peraire. Error estimates and adaptivity. From elasticity to high speed compressible flow, in *The Mathematics of Finite Elements and Applications* (ed. J.R. Whiteman), Vol. VII, Academic Press, London, 1988.
75. J. Peraire, J. Peiro, L. Formaggia, K. Morgan and O.C. Zienkiewicz. Finite element Euler computations in three dimensions. *AIAA 26th Aerospace Sciences Metting*, paper AIAA-87-0032, Reno, USA, January 1988.
76. J. Peraire, J. Peiro, L. Formaggia, K. Morgan and O.C. Zienkiewicz. Finite element Euler computations in 3-D. *Int. J. Num. Meth. Eng.*, **26**, 2135–59, 1988.
77. R. Löhner, K. Morgan, J. Peraire and M. Vahdati. Finite element, flux corrected transport (FEM–FCT) for the Euler and Navier–Stokes equations. *Int. J. Num. Meth. Fluids*, **7**, 1093–109, 1987.
78. R. Löhner, K. Morgan and O.C. Zienkiewicz. The use of domain splitting with an explicit hyperbolic solver. *Comp. Meth. Appl. Mech. Eng.*, **45**, 313–29, 1984.
79. R. Löhner and K. Morgan. An unstructured multigrid method for elliptic problems. *Int. J. Num. Meth. Eng.*, **24**, 101–15, 1987.
80. O.C. Zienkiewicz. Explicit (or semiexplicit) general algorithm for compressible and incompressible flows with equal finite element interpolation. Report 90.5, Chalmers Technical University, Gothenborg, 1990.
81. K. Morgan, O. Hassan and J. Peraire. A time domain unstructured grid approach to simulation of electromagnetic scattering in piecewise homogeneous media. *Comp. Meth. Appl. Mech. Eng.*, **134**, 17–36, 1996.
82. K. Morgan, P.J. Brookes, O. Hassan and N.P. Weatherill. Parallel processing for the simulation of problems involving scattering of electromagnetic waves. *Comp. Meth. Appl. Mech. Eng.*, **152**, 157–74, 1998.

3

A general algorithm for compressible and incompressible flows – the characteristic-based split (CBS) algorithm

3.1 Introduction

In the first chapter we have written the fluid mechanics equations in a very general format applicable to both incompressible and compressible flows. The equations included that of energy which for compressible situations is fully coupled with equations for conservation of mass and momentum. However, of course, the equations, with small modifications, are applicable for specialized treatment such as that of incompressible flow where the energy coupling disappears, to the problems of shallow-water equations where the variables describe a somewhat different flow regime. Chapters 4–7 deal with such specialized forms.

The equations have been written in Chapter 1 in fully conservative, standard form [Eq. (1.1)] but all the essential features can be captured by writing the three sets of equations as below.

Mass conservation

$$\frac{\partial \rho}{\partial t} = \frac{1}{c^2}\frac{\partial p}{\partial t} = -\frac{\partial U_i}{\partial x_i} \tag{3.1}$$

where c is the speed of sound and depends on E, p and ρ and assuming constant entropy

$$c^2 = \frac{\partial p}{\partial \rho} = \frac{\gamma p}{\rho} \tag{3.2}$$

where γ is the ratio of specific heats equal to c_p/c_v. For a fluid with a small compressibility

$$c^2 = \frac{K}{\rho} \tag{3.3}$$

where K is the bulk modulus. Depending on the application we use the appropriate relation for c^2.

Momentum conservation

$$\frac{\partial U_i}{\partial t} = -\frac{\partial}{\partial x_j}(u_j U_i) + \frac{\partial \tau_{ij}}{\partial x_j} - \frac{\partial p}{\partial x_i} - \rho g_i \tag{3.4}$$

In the above we define the mass flow fluxes as

$$U_i = \rho u_i \tag{3.5}$$

Energy conservation

$$\frac{\partial(\rho E)}{\partial t} = -\frac{\partial}{\partial x_j}(u_j \rho E) + \frac{\partial}{\partial x_i}\left(k\frac{\partial T}{\partial x_i}\right) - \frac{\partial}{\partial x_j}(u_j p) + \frac{\partial}{\partial x_i}(\tau_{ij} u_j) \tag{3.6}$$

In all of the above u_i are the velocity components; ρ is the density, E is the specific energy, p is the pressure, T is the absolute temperature, ρg_i represents body forces and other source terms, k is the thermal conductivity, and τ_{ij} are the deviatoric stress components given by (Eq. 1.12b)

$$\tau_{ij} = \mu\left(\frac{\partial u_i}{\partial x_j} + \frac{\partial u_j}{\partial x_i} - \frac{2}{3}\delta_{ij}\frac{\partial u_k}{\partial x_k}\right) \tag{3.7}$$

where δ_{ij} is the Kroneker delta $= 1$, if $i = j$ and $= 0$ if $i \neq j$. In general, μ in the above equation is a function of temperature, $\mu(T)$, and appropriate relations will be used. The equations are completed by the universal gas law when the flow is coupled and compressible:

$$p = \rho R T \tag{3.8}$$

where R is the universal gas constant.

The reader will observe that the major difference in the momentum-conservation equations (3.4) and the corresponding ones describing the behaviour of solids (see Volume 1) is the presence of a convective acceleration term. This does not lend itself to the optimal Galerkin approximation as the equations are now non-self-adjoint in nature. However, it will be observed that if a certain operator split is made, the characteristic–Galerkin procedure valid only for scalar variables can be applied to the part of the system which is not self-adjoint but has an identical form to the convection–diffusion equation. We have shown in the previous chapter that the characteristic–Galerkin procedure is optimal for such equations.

It is important to state again here that the equations given above are of the conservation forms. As it is possible for non-conservative equations to yield multiple and/or inaccurate solutions (Appendix A), this fact is very important.

We believe that the algorithm introduced in this chapter is currently the most general one available for fluids, as it can be directly applied to almost all physical situations. We shall show such applications ranging from low Mach number viscous or indeed inviscid flow to the solution of hypersonic flows. In all applications the algorithm proves to be at least as good as other procedures developed and we see no reason to spend much time describing alternatives. We shall note however that the direct use of the Taylor–Galerkin procedures which we have described in the previous chapter (Sec. 2.10) have proved quite effective in compressible gas flows and indeed some of the examples presented will be based on such methods. Further, in problems of very slow viscous flow we find that the treatment can be almost identical to that of incompressible elastic solids and here we shall often find it expedient to use higher-order approximations satisfying the incompressibility conditions (the so-called Babuška–Brezzi restriction) given in Chapter 12 of Volume 1. Indeed on certain occasions the direct use of incompressibility stabilizing processes described in Chapter 12 of Volume 1 can be useful.

The governing equations described above, Eqs (3.1)–(3.8), are often written in non-dimensional form. The scales used to non-dimensionalize these equations vary depending on the nature of the flow. We describe below the scales generally used in compressible flow computations:

$$\bar{t} = \frac{t u_\infty}{L}; \qquad \bar{x}_i = \frac{x_i}{L}; \qquad \bar{\rho} = \frac{\rho}{\rho_\infty}; \qquad \bar{p} = \frac{p}{\rho_\infty u_\infty^2};$$
$$\bar{u}_i = \frac{u_i}{u_\infty}; \qquad \bar{E} = \frac{E}{u_\infty^2}; \qquad \bar{T} = \frac{T c_p}{u_\infty^2}; \qquad \bar{c}^2 = \frac{c^2}{u_\infty^2} \tag{3.9}$$

where an over-bar indicates a non-dimensional quantity, subscript ∞ represents a free stream quantity and L is a reference length. Applying the above scales to the governing equations and rearranging we have the following form:

Conservation of mass

$$\frac{\partial \bar{\rho}}{\partial \bar{t}} = \frac{1}{\bar{c}^2}\frac{\partial \bar{p}}{\partial \bar{t}} = -\frac{\partial \bar{U}_i}{\partial \bar{x}_i} \tag{3.10}$$

Conservation of momentum

$$\frac{\partial \bar{U}_i}{\partial \bar{t}} = -\frac{\partial}{\partial \bar{x}_j}(\bar{u}_j \bar{U}_i) + \frac{1}{Re}\frac{\partial(\bar{\nu}\,\bar{\tau}_{ij})}{\partial \bar{x}_j} - \frac{\partial \bar{p}}{\partial \bar{x}_i} + \bar{\rho}\,\bar{g}_i \tag{3.11}$$

where

$$Re = \frac{u_\infty L}{\nu}; \qquad \bar{g}_i = \frac{g_i L}{u_\infty^2}; \qquad \bar{\nu} = \frac{\nu}{\nu_{\text{ref}}} \tag{3.12}$$

are the Reynolds number, non-dimensional body forces and the viscosity ratio respectively. In the above equation ν is the kinematic viscosity equal to μ/ρ with μ being the dynamic viscosity.

Conservation of energy

$$\frac{\partial(\bar{\rho}\bar{E})}{\partial \bar{t}} = -\frac{\partial}{\partial \bar{x}_j}(\bar{u}_j\,\bar{\rho}\bar{E}) + \frac{1}{RePr}\frac{\partial}{\partial \bar{x}_i}\left(k^*\frac{\partial \bar{T}}{\partial \bar{x}_i}\right) - \frac{\partial}{\partial \bar{x}_i}(\bar{u}_j\,\bar{p}) + \frac{1}{Re}\frac{\partial}{\partial \bar{x}_i}(\bar{\nu}\,\bar{\tau}_{ij}\,\bar{u}_j) \tag{3.13}$$

where Pr is the Prandtl number and k^* is the conductivity ratio given by the relations

$$Pr = \frac{\mu c_p}{k_{\text{ref}}}; \qquad k^* = \frac{k}{k_{\text{ref}}} \tag{3.14}$$

where k_{ref} is a reference thermal conductivity.

Equation of state

$$\bar{p} = \frac{\bar{\rho}R\bar{T}}{c_p} = \bar{\rho}\overline{RT} = \bar{\rho}\frac{(\gamma - 1)}{\gamma}\bar{T} \tag{3.15}$$

In the above equation $R = c_p - c_v$ is used. The following forms of non-dimensional equations are useful to relate the speed of sound, temperature, pressure, energy, etc.

$$\begin{aligned} \bar{E} &= \frac{\bar{T}}{\gamma} + \frac{1}{2}\bar{u_i}\,\bar{u_i} \\ \bar{c}^2 &= (\gamma - 1)\bar{T} \\ \bar{p} &= (\gamma - 1)\left(\bar{\rho}\bar{E} - \frac{1}{2}\frac{\bar{U_i}\,\bar{U_i}}{\bar{\rho}}\right) \end{aligned} \tag{3.16}$$

The above non-dimensional equations are convenient when coding the CBS algorithm. However, the dimensional form will be retained in this and other chapters for clarity.

3.2 Characteristic-based split (CBS) algorithm

3.2.1 The split – general remarks

The split follows the process initially introduced by Chorin[1,2] for incompressible flow problems in the finite difference context. A similar extension of the split to finite element formulation for different applications of incompressible flows have been carried out by many authors.[3–27] However, in this chapter we extend the split to solve the fluid dynamics equations of both compressible and incompressible forms using the characteristic–Galerkin procedure.[28–46] The algorithm in its full form was first introduced in 1995 by Zienkiewicz and Codina[28,29] and followed several years of preliminary research.[47–51]

Although the original Chorin split[1,2] could never be used in a fully explicit code, the new form is applicable for fully compressible flows in both explicit and semi-implicit forms. The split provides a fully explicit algorithm even in the incompressible case for steady-state problems now using an 'artificial' compressibility which does not affect the steady-state solution. When real compressibility exists, such as in gas flows, the computational advantages of the explicit form compare well with other currently used schemes and the additional cost due to splitting the operator is insignificant. Generally for an identical cost, results are considerably improved throughout a large range of aerodynamical problems. However, a further advantage is that both subsonic and supersonic problems can be solved by the same code.

3.2.2 The split – temporal discretization

We can discretize Eq. (3.4) in time using the characteristic–Galerkin process. Except for the pressure term this equation is similar to the convection–diffusion equation

(2.11). This term can however be treated as a known (source type) quantity providing we have an independent way of evaluating the pressure. Before proceeding with the algorithm, we rewrite Eq. (3.4) in the form given below to which the characteristic–Galerkin method can be applied

$$\frac{\partial U_i}{\partial t} = -\frac{\partial}{\partial x_j}(u_j U_i) + \frac{\partial \tau_{ij}}{\partial x_j} + \rho g_i + Q_i^{n+\theta_2} \tag{3.17}$$

with $Q^{n+\theta_2}$ being treated as a known quantity evaluated at $t = t^n + \theta_2 \Delta t$ in a time increment Δt. In the above equation

$$Q_i^{n+\theta_2} = -\frac{\partial p^{n+\theta_2}}{\partial x_i} \tag{3.18}$$

with

$$\frac{\partial p^{n+\theta_2}}{\partial x_i} = \theta_2 \frac{\partial p^{n+1}}{\partial x_i} + (1-\theta_2)\frac{\partial p^n}{\partial x_i} \tag{3.19}$$

or

$$\frac{\partial p^{n+\theta_2}}{\partial x_i} = \frac{\partial p^n}{\partial x_i} + \theta_2 \frac{\partial \Delta p}{\partial x_i} \tag{3.20}$$

In this

$$\Delta p = p^{n+1} - p^n \tag{3.21}$$

Using Eq. (2.91) of the previous chapter and replacing ϕ by U_i, we can write

$$U_i^{n+1} - U_i^n = \Delta t \bigg[-\frac{\partial}{\partial x_j}(u_j U_i)^n + \frac{\partial \tau_{ij}^n}{\partial x_j} + Q_i^{n+\theta_2} - (\rho g_i)^n + \left(\frac{\Delta t}{2} u_k \frac{\partial}{\partial x_k} \left(\frac{\partial}{\partial x_j}(u_j U_i) - Q_i + \rho g_i \right) \right)^n \bigg] \tag{3.22}$$

At this stage we have to introduce the 'split' in which we substitute a suitable approximation for Q which allows the calculation to proceed before p^{n+1} is evaluated. Two alternative approximations are useful and we shall describe these as *Split A* and *Split B* respectively. In the first we remove all the pressure gradient terms from Eq. (3.22); in the second we retain in that equation the pressure gradient corresponding to the beginning of the step, i.e. $\partial p^n / \partial x_i$. Though it appears that the second split might be more accurate, there are other reasons for the success of the first split which we shall refer to later. Indeed Split A is the one which we shall universally recommend.

Split A

In this we introduce an auxiliary variable U_i^* such that

$$\Delta U_i^* = U_i^* - U_i^n = \Delta t \left[-\frac{\partial}{\partial x_j}(u_j U_i) + \frac{\partial \tau_{ij}}{\partial x_j} - \rho g_i + \frac{\Delta t}{2} u_k \frac{\partial}{\partial x_k} \left(\frac{\partial}{\partial x_j}(u_j U_i) + \rho g_i \right) \right]^n \tag{3.23}$$

This equation will be solved subsequently by an explicit time step applied to the discretized form and a complete solution is now possible. The 'correction' given below is available once the pressure increment is evaluated:

$$\Delta U_i = U_i^{n+1} - U_i^n = \Delta U_i^* - \Delta t \frac{\partial p^{n+\theta_2}}{\partial x_i} - \frac{\Delta t^2}{2} u_k \frac{\partial Q_i^n}{\partial x_k} \tag{3.24}$$

From Eq. (3.1) we have

$$\Delta \rho = \left(\frac{1}{c^2}\right)^n \Delta p = -\Delta t \frac{\partial U_i^{n+\theta_1}}{\partial x_i} = -\Delta t \left[\frac{\partial U_i^n}{\partial x_i} + \theta_1 \frac{\partial \Delta U_i}{\partial x_i}\right] \tag{3.25}$$

Replacing U_i^{n+1} by the known intermediate, auxiliary variable U_i^* and rearranging after neglecting higher-order terms we have

$$\Delta \rho = \left(\frac{1}{c^2}\right)^n \Delta p = -\Delta t \left[\frac{\partial U_i^n}{\partial x_i} + \theta_1 \frac{\partial \Delta U_i^*}{\partial x_i} - \Delta t \theta_1 \left(\frac{\partial^2 p^n}{\partial x_i \partial x_i} + \theta_2 \frac{\partial^2 \Delta p}{\partial x_i \partial x_i}\right)\right] \tag{3.26}$$

where the U_i^* and pressure terms in the above equation come from Eq. (3.24).

The above equation is fully self-adjoint in the variable Δp (or $\Delta \rho$) which is the unknown. Now a standard Galerkin-type procedure can be optimally used for spatial approximation. It is clear that the governing equations can be solved after spatial discretization in the following order:

(a) Eq. (3.23) to obtain ΔU_i^*;
(b) Eq. (3.26) to obtain Δp or $\Delta \rho$;
(c) Eq. (3.24) to obtain ΔU_i thus establishing the values at t^{n+1}.

After completing the calculation to establish ΔU_i and Δp (or $\Delta \rho$) the energy equation is dealt with independently and the value of $(\rho E)^{n+1}$ is obtained by the characteristic–Galerkin process applied to Eq. (3.6).

It is important to remark that this sequence allows us to solve the governing equations (3.1), (3.4) and (3.6), in an efficient manner and with adequate numerical damping. Note that these equations are written in conservation form. Therefore, this algorithm is well suited for dealing with supersonic and hypersonic problems, in which the conservation form ensures that shocks will be placed at the right position and a unique solution achieved.

Split B

In this split we also introduce an auxiliary variable U_i^{**} now retaining the known values of $Q_i^n = \partial p^n / \partial x_i$, i.e.

$$\begin{aligned}\Delta U_i^{**} &= U_i^{**} - U_i^n \\ &= \Delta t \left[-\frac{\partial}{\partial x_j}(u_j U_i) + \frac{\partial \tau_{ij}}{\partial x_j} + \frac{\partial p}{\partial x_i} - \rho g_i + \frac{\Delta t}{2} u_k \frac{\partial}{\partial x_k}\left(\frac{\partial}{\partial x_j}(u_j U_i) - Q + \rho g_i\right)\right]^n\end{aligned} \tag{3.27}$$

It would appear that now U_i^{**} is a better approximation of U^{n+1}. We can now write the correction as

$$\Delta U_i = U_i^{n+1} - U_i^n = \Delta U_i^{**} - \theta_2 \Delta t \frac{\partial \Delta p}{\partial x_i} \tag{3.28}$$

i.e. the correction to be applied is smaller than that assuming Split A, Eq. (3.24). Further, if we use the fully explicit form with $\theta_2 = 0$, no mass velocity (U_i) correction is necessary. We proceed to calculate the pressure changes as in Split A as

$$\Delta \rho = \frac{1}{c^2}\Delta p = -\Delta t \left[\frac{\partial U_i^n}{\partial x_i} + \theta_1 \frac{\partial \Delta U_i^{**}}{\partial x_i} - \Delta t \theta_1 \theta_2 \frac{\partial^2 \Delta p}{\partial x_i^2} \right] \tag{3.29}$$

The solution stages follow the same steps as in Split A.

3.2.3 Spatial discretization and solution procedure

Split A

In all of the equations given below the standard Galerkin procedure is used for spatial discretization as this was fully justified for the characteristic–Galerkin procedure in Chapter 2. We now approximate spatially using standard finite element shape functions as

$$\begin{aligned} U_i &= \mathbf{N}_u \tilde{\mathbf{U}}_i \qquad \Delta U_i = \mathbf{N}_u \Delta \tilde{\mathbf{U}}_i \qquad \Delta U_i^* = \mathbf{N}_u \Delta \tilde{\mathbf{U}}_i^* \\ u_i &= \mathbf{N}_u \tilde{\mathbf{u}}_i \qquad p = \mathbf{N}_p \tilde{\mathbf{p}} \qquad \rho = \mathbf{N}_\rho \tilde{\boldsymbol{\rho}} \end{aligned} \tag{3.30}$$

In the above equation

$$\begin{aligned} \tilde{\mathbf{U}}_i &= [U_i^1 \quad U_i^2 \quad \cdots \quad U_i^k \quad \cdots \quad U_i^m]^{\mathrm{T}} \\ \mathbf{N} &= [N^1 \quad N^2 \quad \cdots \quad N^k \quad \cdots \quad N^m] \end{aligned} \tag{3.31}$$

where k is the node (or variable) identifying number (and varies between 1 and m).

Before introducing the above relations, we have the following weak form of Eq. (3.23) for the standard Galerkin approximation (weighting functions are the shape functions)

$$\begin{aligned} &\int_\Omega N_u^k \Delta U_i^* \, \mathrm{d}\Omega \\ &= +\Delta t \left[-\int_\Omega N_u^k \frac{\partial}{\partial x_j}(u_j U_i)\, \mathrm{d}\Omega - \int_\Omega \frac{\partial N_u^k}{\partial x_j} \tau_{ij} \, \mathrm{d}\Omega - \int_\Omega N_u^k (\rho g_i) \, \mathrm{d}\Omega \right]^n \\ &\quad + \frac{\Delta t^2}{2} \left[\int_\Omega \frac{\partial}{\partial x_l}(u_l N_u^k) \left(-\frac{\partial}{\partial x_j}(u_j U_i) + \rho g_i \right) \mathrm{d}\Omega \right]^n \\ &\quad + \Delta t \left[\int_\Gamma N_u^k \tau_{ij} n_j \, \mathrm{d}\Gamma \right]^n \end{aligned} \tag{3.32}$$

It should be noted that in the above equations the weighting functions are the shape functions as the standard Galerkin approximation is used. Also here, the viscous and stabilizing terms are integrated by parts and the last term is the boundary integral arising from integrating by parts the viscous contribution. Since the residual on the boundaries can be neglected, other boundary contributions from the stabilizing terms are negligible. Note from Eq. (2.91) that the whole residual appears in the stabilizing term. However, we have omitted higher-order terms in the above equation for clarity.

As mentioned in Chapter 1, it is convenient to use matrix notation when the finite element formulation is carried out. We start here from Eq. (1.7) of Chapter 1 and we repeat the deviatoric stress and strain relations below

$$\tau_{ij} = 2\mu\left(\dot{\varepsilon}_{ij} - \delta_{ij}\frac{\dot{\varepsilon}_{kk}}{3}\right) \tag{3.33}$$

where the quantity in brackets is the deviatoric strain. In the above

$$\dot{\varepsilon}_{ij} = \frac{1}{2}\left(\frac{\partial u_i}{\partial x_j} + \frac{\partial u_j}{\partial x_i}\right) \tag{3.34}$$

and

$$\dot{\varepsilon}_{ii} = \frac{\partial u_i}{\partial x_i} \tag{3.35}$$

We now define the strain in three dimensions by a six-component vector (or in two dimensions by a three-component vector) as given below (dropping the dot for simplicity)

$$\boldsymbol{\varepsilon} = [\varepsilon_{11} \quad \varepsilon_{22} \quad \varepsilon_{33} \quad 2\varepsilon_{12} \quad 2\varepsilon_{23} \quad 2\varepsilon_{31}]^{\mathrm{T}} = [\varepsilon_x \quad \varepsilon_y \quad \varepsilon_z \quad 2\varepsilon_{xy} \quad 2\varepsilon_{xz} \quad 2\varepsilon_{zx}]^{\mathrm{T}} \tag{3.36}$$

with a matrix $\mathbf{m}$ defined as

$$\mathbf{m} = [1 \quad 1 \quad 1 \quad 0 \quad 0 \quad 0]^{\mathrm{T}} \tag{3.37}$$

We find that the volumetric strain is

$$\varepsilon_v = \varepsilon_{11} + \varepsilon_{22} + \varepsilon_{33} = \varepsilon_x + \varepsilon_y + \varepsilon_z = \mathbf{m}^{\mathrm{T}}\boldsymbol{\varepsilon} \tag{3.38}$$

The deviatoric strain can now be written simply as (see Eq. 3.33)

$$\boldsymbol{\varepsilon}^d = \boldsymbol{\varepsilon} - \tfrac{1}{3}\mathbf{m}\varepsilon_v = \left(\mathbf{I} - \tfrac{1}{3}\mathbf{m}\mathbf{m}^{\mathrm{T}}\right)\boldsymbol{\varepsilon} = \mathbf{I}_d\boldsymbol{\varepsilon} \tag{3.39}$$

where

$$\mathbf{I}_d = \left(\mathbf{I} - \tfrac{1}{3}\mathbf{m}\mathbf{m}^{\mathrm{T}}\right) \tag{3.40}$$

and thus

$$\mathbf{I}_d = \frac{1}{3}\begin{bmatrix} 2 & -1 & -1 & 0 & 0 & 0 \\ -1 & 2 & -1 & 0 & 0 & 0 \\ -1 & -1 & 2 & 0 & 0 & 0 \\ 0 & 0 & 0 & 3 & 0 & 0 \\ 0 & 0 & 0 & 0 & 3 & 0 \\ 0 & 0 & 0 & 0 & 0 & 3 \end{bmatrix} \tag{3.41}$$

If stresses are similarly written in vectorial form as

$$\boldsymbol{\sigma} = [\sigma_{11} \quad \sigma_{22} \quad \sigma_{33} \quad \sigma_{12} \quad \sigma_{23} \quad \sigma_{31}]^{\mathrm{T}} \tag{3.42}$$

where of course σ_{11} is identically equal to σ_x and is also equal to $\tau_{11} - p$ with similar expressions for σ_y and σ_z, while σ_{12} is identical to τ_{12}, etc.

Immediately we can assume that the deviatoric stresses are proportional to the deviatoric strains and write directly from Eq. (3.33)

$$\boldsymbol{\sigma}^d = \mathbf{I}_d \boldsymbol{\sigma} = \mu \mathbf{I}_0 \boldsymbol{\varepsilon}^d = \mu\left(\mathbf{I}_0 - \tfrac{2}{3}\mathbf{m}\mathbf{m}^{\mathrm{T}}\right)\dot{\boldsymbol{\varepsilon}} \tag{3.43}$$

where the diagonal matrix $\mathbf{I}_0$ is

$$\mathbf{I}_0 = \begin{bmatrix} 2 & & & & & \\ & 2 & & & & \\ & & 2 & & & \\ & & & 1 & & \\ & & & & 1 & \\ & & & & & 1 \end{bmatrix} \tag{3.44}$$

To complete the vector derivation the velocities and strains have to be appropriately related and the reader can verify that using the tensorial strain definitions we can write

$$\dot{\boldsymbol{\varepsilon}} = \mathbf{S}\mathbf{u} \tag{3.45}$$

where

$$\mathbf{u} = \begin{bmatrix} u_1 & u_2 & u_3 \end{bmatrix}^{\mathrm{T}} \tag{3.46}$$

and $\mathbf{S}$ is an appropriate strain matrix (operator) defined below

$$\mathbf{S} = \begin{bmatrix} \dfrac{\partial}{\partial x_1} & 0 & 0 \\ 0 & \dfrac{\partial}{\partial x_2} & 0 \\ 0 & 0 & \dfrac{\partial}{\partial x_3} \\ \dfrac{\partial}{\partial x_2} & \dfrac{\partial}{\partial x_1} & 0 \\ 0 & \dfrac{\partial}{\partial x_3} & \dfrac{\partial}{\partial x_2} \\ \dfrac{\partial}{\partial x_3} & 0 & \dfrac{\partial}{\partial x_1} \end{bmatrix} \tag{3.47}$$

where the subscripts 1, 2 and 3 correspond to the x, y and z directions, respectively. Finally the reader will note that the direct link between the strains and velocities will involve a matrix $\mathbf{B}$ defined simply by

$$\mathbf{B} = \mathbf{S}\mathbf{N}_u \tag{3.48}$$

Now from Eqs. (3.30), (3.32) and (3.43), the solution for U_i^* in matrix form is:

Step 1

$$\Delta\tilde{\mathbf{U}}^* = -\mathbf{M}_u^{-1}\Delta t\left[(\mathbf{C}_u\tilde{\mathbf{U}} + \mathbf{K}_\tau\tilde{\mathbf{u}} - \mathbf{f}) - \Delta t(\mathbf{K}_u\tilde{\mathbf{U}} + \mathbf{f}_s)\right]^n \tag{3.49}$$

where the quantities with a ˜ indicate nodal values and all the discretization matrices are similar to those defined in Chapter 2 for convection–diffusion equations (Eqs. 2.94

and 2.95) and are given as

$$\mathbf{M}_u = \int_\Omega \mathbf{N}_u^{\mathrm{T}} \mathbf{N}_u \,\mathrm{d}\Omega \qquad \mathbf{C}_u = \int_\Omega \mathbf{N}_u^{\mathrm{T}} \,(\boldsymbol{\nabla}(\mathbf{u}\mathbf{N}_u))\,\mathrm{d}\Omega$$
$$\mathbf{K}_\tau = \int_\Omega \mathbf{B}^{\mathrm{T}} \mu(\mathbf{I}_0 - \tfrac{2}{3}\mathbf{m}\mathbf{m}^{\mathrm{T}})\mathbf{B}\,\mathrm{d}\Omega \qquad \mathbf{f} = \int_\Omega \mathbf{N}_u^{\mathrm{T}} \rho \mathbf{g}\,\mathrm{d}\Omega + \int_\Gamma \mathbf{N}_u^{\mathrm{T}} \mathbf{t}^d \,\mathrm{d}\Gamma \tag{3.50}$$

where $\mathbf{g}$ is $[g_1\ g_2\ g_3]^{\mathrm{T}}$ and $\mathbf{t}^d$ is the traction corresponding to the deviatoric stress components. The matrix $\mathbf{K}_\tau$ is also defined at several places in Volume 1 (for instance $\mathbf{A}$ in Chapter 12).

In Eq. (3.49) $\mathbf{K}_u$ and $\mathbf{f}_s$ come from the terms introduced by the discretization along the characteristics. After integration by parts, the expressions for $\mathbf{K}_u$ and $\mathbf{f}_s$ are

$$\mathbf{K}_u = -\tfrac{1}{2}\int_\Omega (\boldsymbol{\nabla}^{\mathrm{T}}(\mathbf{u}\mathbf{N}_u))^{\mathrm{T}}(\boldsymbol{\nabla}^{\mathrm{T}}(\mathbf{u}\mathbf{N}_u))\,\mathrm{d}\Omega \tag{3.51}$$

and

$$\mathbf{f}_s = -\tfrac{1}{2}\int_\Omega (\boldsymbol{\nabla}^{\mathrm{T}}(\mathbf{u}\mathbf{N}_u))^{\mathrm{T}} \rho \mathbf{g}\,\mathrm{d}\Omega \tag{3.52}$$

The weak form of the density–pressure equation is

$$\begin{aligned}
\int_\Omega N_p^k \Delta\rho\,\mathrm{d}\Omega &= \int_\Omega N_p^k \frac{1}{c^2}\Delta p\,\mathrm{d}\Omega \\
&= -\Delta t \int_\Omega N_p^k \frac{\partial}{\partial x_i}\left(U_i^n + \theta_1 \Delta U_i^* - \theta_1 \Delta t \frac{\partial p^{n+\theta_2}}{\partial x_i}\right)\mathrm{d}\Omega \\
&= \Delta t \int_\Omega \frac{\partial N_p^k}{\partial x_i}\left[U_i^n + \theta_1\left(\Delta U_i^* - \Delta t \frac{\partial p^{n+\theta_2}}{\partial x_i}\right)\right]\mathrm{d}\Omega \\
&\quad - \Delta t \theta_1 \int_\Gamma N_p^k \left(U_i^n + \Delta U_i^* - \Delta t \frac{\partial p^{n+\theta_2}}{\partial x_i}\right) n_i\,\mathrm{d}\Gamma
\end{aligned} \tag{3.53}$$

In the above, the pressure and ΔU_i^* terms are integrated by parts. Further we shall discretize ρ directly only in problems of compressible gas flows and therefore below we retain p as the main variable. Spatial discretization of the above equation gives

Step 2

$$(\mathbf{M}_p + \Delta t^2 \theta_1 \theta_2 \mathbf{H})\Delta\tilde{\mathbf{p}} = \Delta t[\mathbf{G}\tilde{\mathbf{U}}^n + \theta_1 \mathbf{G}\Delta\tilde{\mathbf{U}}^* - \Delta t \theta_1 \mathbf{H}\tilde{\mathbf{p}}^n - \mathbf{f}_p] \tag{3.54}$$

which can be solved for $\Delta\tilde{\mathbf{p}}$.

The new matrices arising here are

$$\mathbf{H} = \int_\Omega (\nabla\mathbf{N}_p)^{\mathrm{T}} \boldsymbol{\nabla}\mathbf{N}_p\,\mathrm{d}\Omega \qquad \mathbf{M}_p = \int_\Omega \mathbf{N}_p^{\mathrm{T}}\left(\frac{1}{c^2}\right)^n \mathbf{N}_p\,\mathrm{d}\Omega$$
$$\mathbf{G} = \int_\Omega (\boldsymbol{\nabla}\mathbf{N}_p)^{\mathrm{T}} \mathbf{N}_u\,\mathrm{d}\Omega \qquad \mathbf{f}_p = \Delta t \int_\Gamma \mathbf{N}_p^{\mathrm{T}} \mathbf{n}^{\mathrm{T}}[\tilde{\mathbf{U}}^n + \theta_1(\Delta\tilde{\mathbf{U}}^* - \Delta t \boldsymbol{\nabla} p^{n+\theta_2})]\,\mathrm{d}\Gamma \tag{3.55}$$

In the above $\mathbf{f}_p$ contains boundary conditions as shown above as indicated. We shall discuss these forcing terms fully in a later section as this form is vital to the success of the solution process. The weak form of the correction step from Eq. (3.25) is

$$\int_\Omega N_u^k \Delta U_i^{n+1}\,\mathrm{d}\Omega = \int_\Omega N_u^k \Delta U_i^*\,\mathrm{d}\Omega - \Delta t \int_\Omega N_u^k \left(\frac{\partial p^n}{\partial x_i} + \theta_2 \frac{\partial \Delta p}{\partial x_i}\right) \mathrm{d}\Omega$$
$$- \frac{\Delta t^2}{2} \int_\Omega \frac{\partial}{\partial x_j}(u_j N_u^k) \frac{\partial p^n}{\partial x_i}\,\mathrm{d}\Omega \tag{3.56}$$

The final stage of the computation of the mass flow vector U_i^{n+1} is completed by following matrix form

Step 3

$$\Delta\tilde{\mathbf{U}} = \Delta\tilde{\mathbf{U}}^* - \mathbf{M}_u^{-1} \Delta t \left[\mathbf{G}^{\mathrm{T}}(\tilde{\mathbf{p}}^n + \theta_2 \Delta\tilde{\mathbf{p}}) + \frac{\Delta t}{2} \mathbf{P}\tilde{\mathbf{p}}^n\right] \tag{3.57}$$

where

$$\mathbf{P} = \int_\Omega (\boldsymbol{\nabla}(\mathbf{u}\mathbf{N}_u))^{\mathrm{T}} \boldsymbol{\nabla}\mathbf{N}_p\,\mathrm{d}\Omega \tag{3.58}$$

At the completion of this stage the values of $\tilde{\mathbf{U}}^{n+1}$ and $\tilde{\mathbf{p}}^{n+1}$ are fully determined but the computation of the energy $(\rho E)^{n+1}$ is needed so that new values of c^{n+1}, the speed of sound, can be determined.

Once again the energy equation (3.6) is identical in form to that of the scalar problem of convection–diffusion if we observe that p, U_i, etc. are known. The weak form of the energy equation is written using the characteristic–Galerkin approximation of Eq. (2.91) as

$$\int_\Omega N_E^k \Delta(\rho E)^{n+1}\,\mathrm{d}\Omega$$
$$= \Delta t \left[- \int_\Omega N_E^k \frac{\partial}{\partial x_i}(u_i(\rho E + p))\,\mathrm{d}\Omega - \int_\Omega \frac{\partial N_E^k}{\partial x_i}\left(\tau_{ij}u_j + k\frac{\partial T}{\partial x_i}\right)\mathrm{d}\Omega\right]^n$$
$$+ \frac{\Delta t^2}{2}\left[\int_\Omega \frac{\partial}{\partial x_j}(u_j N_E^k)\left[\frac{\partial}{\partial x_i}(-u_i(\rho E + p))\right]\mathrm{d}\Omega\right]^n$$
$$+ \Delta t \left[\int_\Gamma N_E^k \left(\tau_{ij}u_j + k\frac{\partial T}{\partial x_i}\right) n_i\,\mathrm{d}\Gamma\right]^n \tag{3.59}$$

With

$$\rho E = \mathbf{N}_E \tilde{\mathbf{E}} \qquad T = \mathbf{N}_T \tilde{\mathbf{T}} \tag{3.60}$$

we have

Step 4

$$\Delta\tilde{\mathbf{E}} = -\mathbf{M}_E^{-1} \Delta t \left[\mathbf{C}_E \tilde{\mathbf{E}} + \mathbf{C}_p \tilde{\mathbf{p}} + \mathbf{K}_T \tilde{\mathbf{T}} + \mathbf{K}_{\tau E} \tilde{\mathbf{u}} + \mathbf{f}_e - \Delta t(\mathbf{K}_{uE}\tilde{\mathbf{E}} + \mathbf{K}_{up}\tilde{\mathbf{p}} + \mathbf{f}_{es})\right]^n \tag{3.61}$$

where $\tilde{\mathbf{E}}$ contains the nodal values of ρE and again the matrices are similar to those previously obtained (assuming that ρE and T can be suitably scaled in the conduction term).

The matrices and forcing vectors are again similar and given as

$$\mathbf{M}_E = \int_\Omega \mathbf{N}_E^{\mathrm{T}} \mathbf{N}_E \, \mathrm{d}\Omega \qquad \mathbf{C}_E = \int_\Omega \mathbf{N}_E^{\mathrm{T}} \boldsymbol{\nabla}^{\mathrm{T}} (\mathbf{u}\mathbf{N}_E) \, \mathrm{d}\Omega \qquad \mathbf{C}_p = \int_\Omega \mathbf{N}_E^{\mathrm{T}} \boldsymbol{\nabla}^{\mathrm{T}} (\mathbf{u}\mathbf{N}_p) \, \mathrm{d}\Omega$$
$$\mathbf{K}_T = \int_\Omega (\boldsymbol{\nabla}\mathbf{N}_E)^{\mathrm{T}} k \boldsymbol{\nabla}\mathbf{N}_T \, \mathrm{d}\Omega \qquad \mathbf{K}_{\tau E} = \int_\Omega \mathbf{B}^{\mathrm{T}} \mu (\mathbf{I}_0 - \tfrac{2}{3}\mathbf{m}\mathbf{m}^{\mathrm{T}}) \mathbf{B} \, \mathrm{d}\Omega$$
$$\mathbf{K}_{uE} = -\tfrac{1}{2}\int_\Omega (\boldsymbol{\nabla}^{\mathrm{T}}(\mathbf{u}\mathbf{N}_E))^{\mathrm{T}} (\boldsymbol{\nabla}\mathbf{N}_E) \, \mathrm{d}\Omega \qquad \mathbf{f}_e = \int_\Gamma \mathbf{N}_E^{\mathrm{T}} \mathbf{n}^{\mathrm{T}} (\mathbf{t}^d \mathbf{u} + k\boldsymbol{\nabla} T) \, \mathrm{d}\Gamma \tag{3.62}$$
$$\mathbf{K}_{up} = -\tfrac{1}{2}\int_\Omega (\boldsymbol{\nabla}^{\mathrm{T}}(\mathbf{u}\mathbf{N}_E))^{\mathrm{T}} (\boldsymbol{\nabla}\mathbf{N}_p) \, \mathrm{d}\Omega$$

The forcing term $\mathbf{f}_{es}$ contains source terms. If no source terms are available this term is equal to zero.

It is of interest to observe that the process of Step 4 can be extended to include in an identical manner the equations describing the transport of quantities such as turbulence parameters,[36] chemical concentrations, etc., once the first essential Steps 1–3 have been completed.

Split B

With Split B, the discretization and solution procedures have to be modified slightly. Leaving the details of the derivation to the reader and using identical discretization processes, the final steps can be summarized as:

Step 1

$$\Delta\tilde{\mathbf{U}}_i^{**} = -\mathbf{M}_u^{-1}\Delta t\left[(\mathbf{C}_u\tilde{\mathbf{U}} + \mathbf{K}_\tau\tilde{\mathbf{u}} + \mathbf{G}^{\mathrm{T}}\tilde{\mathbf{p}} - \mathbf{f}) - \Delta t\left(\mathbf{K}_u\tilde{\mathbf{U}} + \mathbf{f}_s + \frac{\Delta t}{2}\mathbf{P}\tilde{\mathbf{p}}\right)\right]^n \tag{3.63}$$

where all matrices are the same as in Split A except the forcing term $\mathbf{f}$ which is

$$\mathbf{f} = \int_\Omega \mathbf{N}_u^{\mathrm{T}} \rho \mathbf{g} \, \mathrm{d}\Omega + \int_\Gamma \mathbf{N}_u^{\mathrm{T}} \mathbf{t}^d \, \mathrm{d}\Gamma \tag{3.64}$$

since the pressure term has now been integrated by parts

Step 2

$$(\mathbf{M}_p + \Delta t^2 \theta_1 \theta_2 \mathbf{H})\Delta\tilde{\mathbf{p}} = \Delta t[\mathbf{G}\tilde{\mathbf{U}}^n + \theta_1 \mathbf{G}\Delta\tilde{\mathbf{U}}^{**} - \mathbf{f}_p]^n \tag{3.65}$$

and

Step 3

$$\Delta\tilde{\mathbf{U}} = \Delta\tilde{\mathbf{U}}^{**} - \mathbf{M}_u^{-1}\Delta t\left[\theta_2 \mathbf{G}^{\mathrm{T}}\Delta\tilde{\mathbf{p}}\right] \tag{3.66}$$

Step 4, calculation of the energy, is unchanged. The reader can notice the minor differences in the above equations from those of Split A.

3.3 Explicit, semi-implicit and nearly implicit forms

This algorithm will always contain an explicit portion in the first characteristic–Galerkin step. However the second step, i.e. that of the determination of the pressure increment, can be made either explicit or implicit and various possibilities exist here depending on the choice of θ_2. Now different stability criteria will apply. We refer to schemes as being fully explicit or semi-implicit depending on the choice of the parameter θ_2 as zero or non-zero, respectively.

It is also possible to solve the first step in a partially implicit manner to avoid severe time step restrictions. Now the viscous term is the one for which an implicit solution is sought. We refer to such schemes as quasi- (nearly) implicit schemes. It is necessary to mention that the fully explicit form is only possible for compressible gas flows for which $c \neq \infty$.

3.3.1 Fully explicit form

In fully explicit forms, $\frac{1}{2} \leqslant \theta_1 \leqslant 1$ and $\theta_2 = 0$. In general the time step limitations explained for the convection–diffusion equations are applicable i.e.

$$\Delta t \leqslant \frac{h}{c + |\mathbf{u}|} \tag{3.67}$$

as viscosity effects are generally negligible here.

This particular form is very successful in compressible flow computations and has been widely used by the authors for solving many complex problems. Chapter 6 presents many examples.

3.3.2 Semi-implicit form

In semi-implicit form the following values apply

$$\begin{aligned} \tfrac{1}{2} &\leqslant \theta_1 \leqslant 1 \\ \tfrac{1}{2} &\leqslant \theta_2 \leqslant 1 \end{aligned} \tag{3.68}$$

Again the algorithm is conditionally stable. The permissible time step is governed by the critical step of the characteristic–Galerkin explicit relation solved in Step 1 of the algorithm. This is the standard convection–diffusion problem discussed in Chapter 2 and the same stability limits apply, i.e.

$$\Delta t \leqslant \Delta t_\sigma = \frac{h}{|\mathbf{u}|} \tag{3.69}$$

and/or

$$\Delta t \leqslant \Delta t_\nu = \frac{h^2}{2\nu} \tag{3.70}$$

where ν is the kinematic viscosity. A convenient form incorporating both limits can be written as

$$\Delta t \leqslant \frac{\Delta t_\sigma \Delta t_\nu}{\Delta t_\sigma + \Delta t_\nu} \tag{3.71}$$

The reader can verify that the above relation will give appropriate time step limits with and without the domination of viscosity.

For slightly compressible or incompressible problems in which $\mathbf{M}_p$ is small or zero the semi-implicit form is efficient and it should be noted that the matrix $\mathbf{H}$ of Eqs. (3.54) and (3.65) does not vary during the computation process. Therefore $\mathbf{H}$ can be factored into its triangular parts once leading to an economical direct procedure. As will be seen from the final chapter on computer programming the implicit equation is usually solved by conjugate gradient procedures.

3.3.3 Quasi- (nearly) implicit form

To overcome the severe time step restriction made by the diffusion terms (viscosity, thermal conductivity, etc.), these terms can be treated implicitly. This involves solving separately an implicit form connecting the viscous terms with U_i^* or U_i^{**}. Here, at each step, simultaneous equations need to be solved and this procedure can be of great advantage in certain cases such as high-viscosity flows and low Mach number flows.[13,15,23,40] Now the only time step limitation is $\Delta t \leqslant h/|\mathbf{u}|$ which appears to be a very reasonable and physically meaningful restriction.

3.3.4 Evaluation of time step limit. Local and global time steps

Though they are defined in terms of element sizes the time step limits are best calculated at nodes of the element. In Fig. 3.1 the manner in which the size of the element is easily established at nodes is shown. In such cases, as seen, the element size is not unique for each node. In the calculation, we shall specify, if the scheme is conditionally stable, the time step limit at each node by assigning the minimum

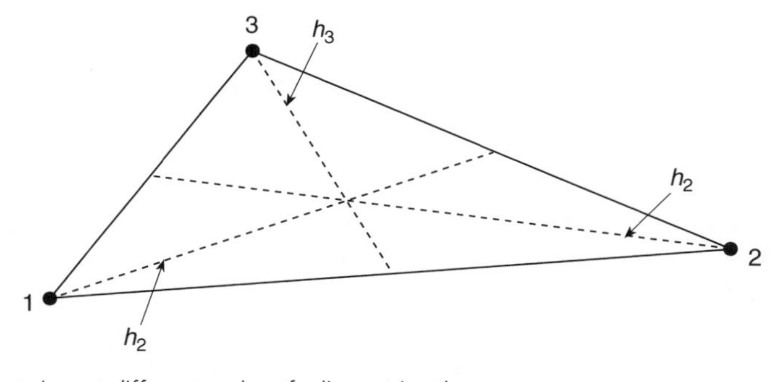

Fig. 3.1 Element sizes at different nodes of a linear triangle.

value for such nodes calculated from all the surrounding elements. When a problem is being solved in true time then obviously the smallest of all nodal values has to be adopted for the solution. In many problems a transient calculation is adopted to find steady-state solutions and *local time stepping* is convenient as it allows more rapid convergence and fewer time steps to be used throughout the problem. Local time stepping can only be applied to problems in which (1) the mass matrix is lumped and (2) the steady-state solution does not itself depend on the mass matrix. Thus with local time stepping we shall use at every node simply the minimum time step found at that node. This of course is equivalent to assuming identical time steps for the whole problem and simply adjusting the lumped masses. Such a problem with adjusted lumped masses is still physically and mathematically meaningful and we know that the convergence will be achieved as it invariably is.

Many steady-state problems have used such localized time stepping in the calculations.

In the context of local and global time stepping it is interesting to note that the stabilizing terms introduced by the characteristic–Galerkin process will not take on the optimal value for any element in which the time step differs from the critical one; that is of course if we use local time stepping we shall automatically achieve this optimal value often throughout all elements at least for steady-state problems. However, on other occasions it may be useful to make sure that (a) in all elements we introduce optimal damping and (b) that the progressive time step for all elements is identical. The latter of course is absolutely necessary if for instance we deal with transient problems where all time steps are real. For such cases it is possible to consider the Δt as being introduced in two stages: (1) as the Δt_{ext} which has of course to preserve stability and must be left at a minimum Δt calculated from any element; and (2) to use in the calculation of each individual element the Δt_{int} which is optimal for an element, as of course exceeding the stability limit does not matter there and we are simply adding better damping characteristics.

This internal–external subdivision is of some importance when incompressibility effects are considered. As shown in the next section, the stabilizing diagonal term occurs in steady-state problems depending on the size of the time step. If the mesh is graded and very small elements dictate the time step over the whole domain we might find that the diagonal term introduced overall is not sufficient to preserve incompressibility. For such problems we recommend the use of internal and external time steps which differ and we introduce these in reference 52.

3.4 'Circumventing' the Babuška–Brezzi (BB) restrictions

In the previous sections we have not restricted the nature of the interpolating shape functions $\mathbf{N}_u$ and $\mathbf{N}_p$. If we choose these interpolations in a manner satisfying the patch test conditions or BB restriction for incompressibility, see Chapter 12, Volume 1 (Chapter 4 of this volume for some permissible interpolations) then of course completely incompressible problems can be dealt with without any special difficulties by both Split A and Split B formulations. However Split A introduces an important bonus which permits us to avoid any restrictions on the nature of the two shape functions used for velocity and pressure. Let us examine here the structure of the equations obtained in steady-state conditions. For simplicity we shall consider here only the Stokes form of the governing equations in which the convective terms disappear. Further we shall take the fluid as incompressible and thus uncoupled from the energy equations. Now the three steps of Eqs. (3.49), (3.54) and (3.57) are written as

$$\begin{aligned}\Delta\tilde{\mathbf{U}}^* &= -\Delta t\mathbf{M}_u^{-1}[\mathbf{K}_\tau\tilde{\mathbf{u}}^n - \mathbf{f}]\\ \Delta\tilde{\mathbf{p}} &= \frac{1}{\Delta t\theta_1\theta_2}\mathbf{H}^{-1}[\mathbf{G}\tilde{\mathbf{U}}^n + \theta_1\mathbf{G}\Delta\tilde{\mathbf{U}}^* - \Delta t\theta_1\mathbf{H}\tilde{\mathbf{p}}^n - \mathbf{f}_p]\\ \Delta\tilde{\mathbf{U}} &= \Delta\tilde{\mathbf{U}}^* - \Delta t\mathbf{M}_u^{-1}\mathbf{G}^{\mathrm{T}}(\tilde{\mathbf{p}}^n + \theta_2\Delta\tilde{\mathbf{p}})\end{aligned} \qquad (3.72)$$

In steady state we have $\Delta\tilde{\mathbf{p}} = \Delta\tilde{\mathbf{U}} = \mathbf{0}$ and eliminating $\Delta\tilde{\mathbf{U}}^*$ we can write (dropping now the superscript n)

$$\mathbf{K}_\tau\tilde{\mathbf{u}} + \mathbf{G}^{\mathrm{T}}\tilde{\mathbf{p}} = \mathbf{f} \qquad (3.73)$$

from the first and third of Eqs. (3.72) and

$$\mathbf{G}\tilde{\mathbf{U}} + \theta_1\Delta t\mathbf{G}\mathbf{M}_u^{-1}\mathbf{G}_{up}^{\mathrm{T}}\tilde{\mathbf{p}} - \Delta t\theta_1\mathbf{H}\tilde{\mathbf{p}} - \mathbf{f}_p = 0 \qquad (3.74)$$

from the second and third of Eqs. (3.72)

We finally have a system which can be written in the form

$$\begin{bmatrix}\mathbf{K}_\tau/\rho & \mathbf{G}^{\mathrm{T}}\\ \mathbf{G} & \Delta t\theta_1[\mathbf{G}\mathbf{M}_u^{-1}\mathbf{G}^{\mathrm{T}} - \mathbf{H}]\end{bmatrix}\begin{Bmatrix}\tilde{\mathbf{U}}\\ \tilde{\mathbf{p}}\end{Bmatrix} = \begin{Bmatrix}\mathbf{f}_1\\ \mathbf{f}_2\end{Bmatrix} \qquad (3.75)$$

here $\mathbf{f}_1$ and $\mathbf{f}_2$ arise from the forcing terms.

The system is now always positive definite and therefore leads to a non-singular solution for *any interpolation functions* $\mathbf{N}_u$, $\mathbf{N}_p$ chosen. In most of the examples discussed in this book and elsewhere equal interpolation is chosen for both the U_i and p variables, i.e. $\mathbf{N}_u = \mathbf{N}_p$. We must however stress that any other interpolation can be used without violating the stability. This is an important reason for the preferred use of the Split A form.

It can be easily verified that if the pressure gradient term is retained as in Eq. (3.27), i.e. if we use Split B the lower diagonal term of Eq. (3.75) is identically zero and the BB conditions in the full scheme cannot be avoided. Now we show this below. From Eqs. (3.63), (3.65) and (3.66), for incompressible Stokes flow we have

$$\begin{aligned}\Delta\tilde{\mathbf{U}}_i^{**} &= -\mathbf{M}_u^{-1}\Delta t\left[\mathbf{K}_\tau\tilde{u} + \mathbf{G}^{\mathrm{T}}\tilde{\mathbf{p}} - \mathbf{f})\right]^n\\ \Delta\tilde{\mathbf{p}} &= \frac{1}{\Delta t\theta_1\theta_2}\mathbf{H}^{-1}\Delta t[\mathbf{G}\tilde{\mathbf{U}}^n + \theta_1\mathbf{G}\Delta\tilde{\mathbf{U}}^{**} - \mathbf{f}_p]^n\\ \Delta\tilde{\mathbf{U}} &= \Delta\tilde{\mathbf{U}}^{**} - \mathbf{M}_u^{-1}\Delta t\left[\theta_2\mathbf{G}^{\mathrm{T}}\Delta\tilde{\mathbf{p}}\right]\end{aligned} \qquad (3.76)$$

At steady state $\Delta\tilde{\mathbf{p}} = \Delta\tilde{\mathbf{U}} = \mathbf{0}$, which gives the following two equations:

$$\mathbf{K}_\tau\tilde{\mathbf{u}} + \mathbf{G}^{\mathrm{T}}\tilde{\mathbf{p}} = \mathbf{f} \qquad (3.77)$$

and

$$\mathbf{G}\tilde{\mathbf{U}} = \mathbf{f}_b \qquad (3.78)$$

Note that $\Delta \mathbf{U}_i^{**}$ is zero from the third of Eq. (3.76). As in Split A we can write the following system

$$\begin{bmatrix} \mathbf{K}_\tau/\rho & \mathbf{G}^{\mathrm{T}} \\ \mathbf{G} & \mathbf{0} \end{bmatrix} \begin{Bmatrix} \tilde{\mathbf{U}} \\ \tilde{\mathbf{p}} \end{Bmatrix} = \begin{Bmatrix} \mathbf{f}_1 \\ \mathbf{f}_2 \end{Bmatrix} \tag{3.79}$$

where $\mathbf{f}_1$ and $\mathbf{f}_2$ arise from the forcing terms as in the Split A form. Clearly here the BB restrictions are not circumvented.

It is interesting to observe that the lower diagonal term which appeared in Eq. (3.75) is equivalent to the difference between the so-called fourth-order and second-order approximations of the laplacian. This justifies the use of similar terms introduced into the computation by some finite difference proponents.[53]

3.5 A single-step version

If the $\Delta \mathbf{U}_i^*$ term in Eq. (3.26) is omitted, the intermediate variable U_i^* need not be determined. Instead we can directly calculate ρ (or p), U_i and ρE. This of course introduces an additional approximation.

The use of the approximation of Eq. (3.1) is not necessary in any expected fully explicit scheme as the density increment is directly obtained if we note that

$$\mathbf{M}_p \Delta \tilde{\mathbf{p}} = \mathbf{M}_u \Delta \tilde{\boldsymbol{\rho}} \tag{3.80}$$

With the above simplifications and Split A we can return to the original equations and using the Galerkin approximation. We can therefore write directly

$$\Delta \tilde{\boldsymbol{\Phi}} = -\mathbf{M}_u^{-1} \Delta t \left[\int_\Omega \mathbf{N}^{\mathrm{T}} \left(\frac{\partial \mathbf{F}_i}{\partial x_i} + \frac{\partial \mathbf{G}_i}{\partial x_i} \right) \mathrm{d}\Omega - \tfrac{1}{2} \Delta t \int_\Omega \mathbf{N}^{\mathrm{T}} \mathbf{D} \, \mathrm{d}\Omega \right]^n \tag{3.81}$$

omitting the source terms for clarity ($\mathbf{F}_i$ and $\mathbf{G}_i$ are explained in Chapter 1, Eq. (1.25)) and noting that now $\tilde{\boldsymbol{\Phi}}$ denotes all the variables. The added stabilizing terms $\mathbf{D}$ are defined below and have to be integrated by parts in the usual manner.

$$\begin{Bmatrix} 2\theta_1 \dfrac{\partial^2}{\partial x_i \partial x_i} p \\ u_i \dfrac{\partial}{\partial x_i} \left[\dfrac{\partial}{\partial x_j} (u_j \rho u_1) + \dfrac{\partial p}{\partial x_1} \right] \\ u_i \dfrac{\partial}{\partial x_i} \left[\dfrac{\partial}{\partial x_j} (u_j \rho u_2) + \dfrac{\partial p}{\partial x_2} \right] \\ u_i \dfrac{\partial}{\partial x_i} \left[\dfrac{\partial}{\partial x_j} (u_j \rho u_3) + \dfrac{\partial p}{\partial x_3} \right] \\ u_i \dfrac{\partial}{\partial x_i} \left[\dfrac{\partial}{\partial x_j} (u_j \rho E + u_j p) \right] \end{Bmatrix} \tag{3.82}$$

The added 'diffusions' are simple and are streamline oriented, and thus do not mask the true effects of viscosity as happens in some schemes (e.g. the Taylor–Galerkin process).

If only steady state results are sought it would appear that Δt multiplying the matrix $\mathbf{D}$ should be set at its optimal value of $\Delta t_{\text{crit}} \approx h/|\mathbf{u}|$ and we generally recommend, providing the viscosity is small, this value for the full scheme.[30]

However the oversimplified scheme of Eq. (3.81) can lose some accuracy and even when steady state is reached will give slightly different results than those obtained using the full sequential updating.[30] However at low Mach numbers the difference is negligible as we shall show later in Sec. 3.7. The small additional cost involved in computing the two-step sequence $\Delta\tilde{\mathbf{U}}^* \rightarrow \Delta\mathbf{p} \rightarrow \Delta\tilde{\mathbf{U}} \rightarrow \Delta\tilde{\mathbf{E}}$ will have to be balanced against the accuracy increase. In general, we have found that the two-step version is preferable.

However it is interesting to consider once again the performance of the single-step scheme in the case of Stokes equations as we did for the other schemes in the previous section. After discretization we have, omitting convective terms, only one additional diffusion term which arises (Eq. 3.82) in the mass conservation equation. After discretization, in steady state

$$\begin{bmatrix} \mathbf{K}_\tau/\rho & \mathbf{G}^{\mathrm{T}} \\ \mathbf{G} & \theta_1 \Delta t \mathbf{H} \end{bmatrix} \begin{Bmatrix} \tilde{\mathbf{U}} \\ \tilde{\mathbf{p}} \end{Bmatrix} = \begin{Bmatrix} \mathbf{f}_1 \\ \mathbf{f}_2 \end{Bmatrix} \tag{3.83}$$

Clearly the single-step algorithm retains the capacity of dealing with full incompressibility without stability problems but of course can only be used for the nearly incompressible range of problems for which $\mathbf{M}_p$ exists. We should remark here that this formulation now achieves precisely the same stabilization as that suggested by Brezzi and Pitkäranta,[54] see Chapter 12, Volume 1.

We shall note the performance of single- and two-step algorithms in Sec. 3.7 of this chapter.

3.6 Boundary conditions

3.6.1 Fictitious boundaries

In a large number of fluid mechanics problems the flow in open domains is considered. A typical open domain describing flow past an aircraft wing is shown in Fig. 3.2. In such problems the boundaries are simply limits of computation and they are therefore fictitious. With suitable values specified at such boundaries, however, accurate solution for the flow inside the isolated domain can be achieved.

Generally as the distance from the object grows, the boundary values tend to those encountered in the free domain flow or the flow at infinity. This is particularly true at the entry and side boundaries shown in Fig. 3.2. At the exit, however, the conditions are different and here the effect of the introduced disturbance can continue for a very large distance denoting the wake of the problem. We shall from time to time discuss problems of this nature but here we shall simply make the following remarks.

1. If the flow is subsonic the specification of all quantities excepting the density can be made on both the side and entry boundaries.

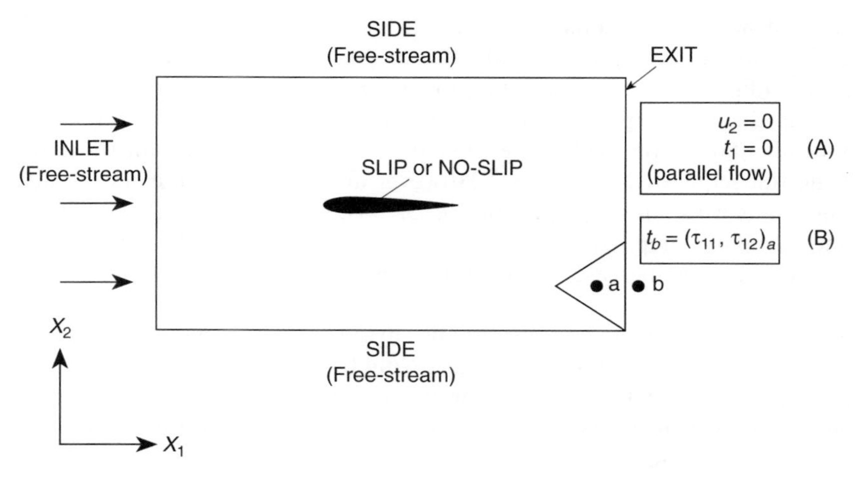

Fig. 3.2 Fictitious and real boundaries.

2. Whereas for supersonic flows all the variables can be prescribed at the inlet, at the exit however no boundary conditions are imposed simply because by definition the disturbances caused by the boundary conditions cannot travel faster than the speed of sound.

With subsonic exit conditions the situation is somewhat more complex and here various possibilities exist. We again illustrate such conditions in Fig. 3.2.

Condition A: Denoting the most obvious assumptions with regard to the traction and velocities.

Condition B: A more sophisticated condition of zero gradient of traction and stresses existing there. Such conditions will of course always apply to the exit domains for incompressible flow. Condition B was first introduced by Zienkiewicz *et al.*[47] and is discussed fully by Papanastasiou *et al.*[55] This condition is of some importance as it gives remarkably good answers.

We shall refer to these open boundary conditions in various classes of problems dealt with later in this book and shall discuss them in detail. In particular the kind of differences that may occur in incompressible flows in conduits under different exit conditions are considered.

Of considerable importance, especially in view of the new schemes, are however conditions which we will encounter on real boundaries.

3.6.2 Real boundaries

By real boundaries we mean limits of fluid domains which are physically defined and here three different possibilities exist.

1. *Solid boundaries with no slip conditions*: On such boundaries the fluid is assumed to stick or attach itself to the boundary and thus all velocity components become zero. Obviously this condition is only possible for viscous flows.
2. *Solid boundaries in inviscid flow* (*slip conditions*): When the flow is inviscid we will always encounter slipping boundary conditions where only the normal velocity component is specified and is in general equal to zero in steady-state motion. Such boundary conditions will invariably be imposed for problems of Euler flow whether it is compressible or incompressible.
3. *Prescribed traction boundary conditions*: The last category is that on which tractions are prescribed. This includes zero traction in the case of free surfaces of fluids or any prescribed tractions such as those caused by wind being imposed on the surface.

These three basic kinds of boundary conditions have to be imposed on the fluid and special consideration has to be given to these when split operator schemes are used.

3.6.3 Application of real boundary conditions in the discretization using the CBS split

We shall first consider the treatment of boundaries described under (1) or (2) of the previous section. On such boundaries

$$u_n = 0, \qquad \text{normal velocity zero} \tag{3.84}$$

and either

$$t_s = 0, \qquad \text{tangential traction zero for inviscid flow}$$

or (3.85)

$$u_s = 0, \qquad \text{tangential velocity zero for viscous flow}$$

In early applications of the CBS algorithm it appeared correct that when computing $\Delta\tilde{U}_i^*$ *no* velocity boundary conditions be imposed and to use instead the value of boundary tractions which corresponds to the deviatoric stresses and pressures computed at time t_n. We note that if the pressure is removed as in Split A these pressures could also be removed from the boundary traction component. However in Split B no such pressure removal is necessary. This requires, in viscous problems, evaluation of the boundary τ_{ij}'s and this point is explained further later.

When computing $\Delta\rho$ or Δ_p we integrate by parts obtaining (Eq. 3.53)

$$\begin{aligned}\int_\Omega N_p^k \frac{1}{c^2} \Delta p \, d\Omega &= -\Delta t \int_\Omega N_p^k \frac{\partial}{\partial x_i}\left(U_i^n + \theta_1 \Delta U_i^* - \theta_1 \Delta t \frac{\partial p^{n+\theta_2}}{\partial x_i}\right) d\Omega \\ &= \Delta t \int_\Omega \frac{\partial N_p^k}{\partial x_i}\left[U_i^n + \theta_1\left(\Delta U_i^* - \Delta t \frac{\partial p^{n+\theta_2}}{\partial x_i}\right)\right] n_i \, d\Gamma \\ &\quad - \Delta t \int_\Gamma N_p^k n_i \left[U_i^n + \theta_1 \left(\Delta U_b^* - \Delta t \frac{\partial p^{n+\theta_2}}{\partial x_i}\right)\right] d\Gamma \end{aligned} \tag{3.86}$$

Here n_i is the outward drawn normal. The last term in the above equation is identically equal to zero from the condition of Eq. (3.24):

$$U_n = n_i U_i = n_i \left[U_i^n + \theta_1 \left(\Delta U_i^* - \Delta t \frac{\partial p^{n+\theta_2}}{\partial x_i} \right) \right] = 0 \tag{3.87}$$

for conditions of Eq. (3.84). For non-zero normal velocity this would simply become the specified normal velocity. This point seems to have baffled some investigators who simply assume

$$\frac{\partial p}{\partial n} = n_i \qquad \frac{\partial p}{\partial x_i} = 0 \tag{3.88}$$

on solid boundaries. Note that this is not exactly true.

Returning to the traction on the boundaries, the traction on the surface can be defined as

$$t_i = \tau_{ij} n_j - p n_i \tag{3.89}$$

Prescribing the above traction using Split A, we replace the stress components in Step 1 (last term in Eq. (3.32)) as follows

$$\int_\Gamma N_u^k \tau_{ij} n_j \, \mathrm{d}\Gamma = \int_{\Gamma - \Gamma_t} N_u^k \tau_{ij} n_j \, \mathrm{d}\Gamma + \int_{\Gamma_t} N_u^k (t_i + p n_i) \, \mathrm{d}\Gamma \tag{3.90}$$

where Γ_t represents the part of the boundary where the traction is prescribed.

The above calculation may involve a substantial error in 'projecting' deviatoric stresses onto the boundary.

The last step requires the solution for the velocity correction terms to obtain finally the $\tilde{U}_i^{n+1}$. Clearly correct velocity boundary values must always be imposed in this step.

Although the above described procedure is theoretically correct and instructive, better results will generally be obtained if the velocity boundary conditions are directly imposed when computing $\tilde{U}_i^*$. Further, the need of calculating any boundary tractions from internal stress is now avoided even if the tangential velocity is taken as zero (no slip condition).

If tractions are specified we shall generally now put the total at Step 1 when computing $\tilde{U}_i^*$, even though at this stage we should subtract the pressure terms as shown in Eq. (3.90). In Step 3, no further modification is needed and, hence, again we avoid the need to compute additional boundary integrals. However, we are still faced with evaluating pressures on such boundaries as these are needed in solving Step 2 [namely Eq. (3.86)]. This still involves the determination of deviatoric stresses on the surface (namely Section 12.7.6 of Volume 1, where we discussed application of the CBS algorithm to solid mechanics problems) and showed boundary pressures computed as $n_i \tau_{ij} n_j + -n_i \bar{t}_i$. Fortunately, on free surfaces of the fluid, the deviatoric stresses are usually negligible and here direct use of the pressure approximated by $-n_i \bar{t}_i$ may be used.

3.7 The performance of two- and single-step algorithms on an inviscid problem

In this section we demonstrate the performance of the single- and two-step algorithms via an inviscid problem of subsonic flow past a NACA0012 aerofoil. The problem domain and finite element mesh used are shown in Fig. 3.3(a) and (b). The discretization near the aerofoil surface is finer than that of other places and a total number of 969 nodes and 1824 elements are used in the mesh.

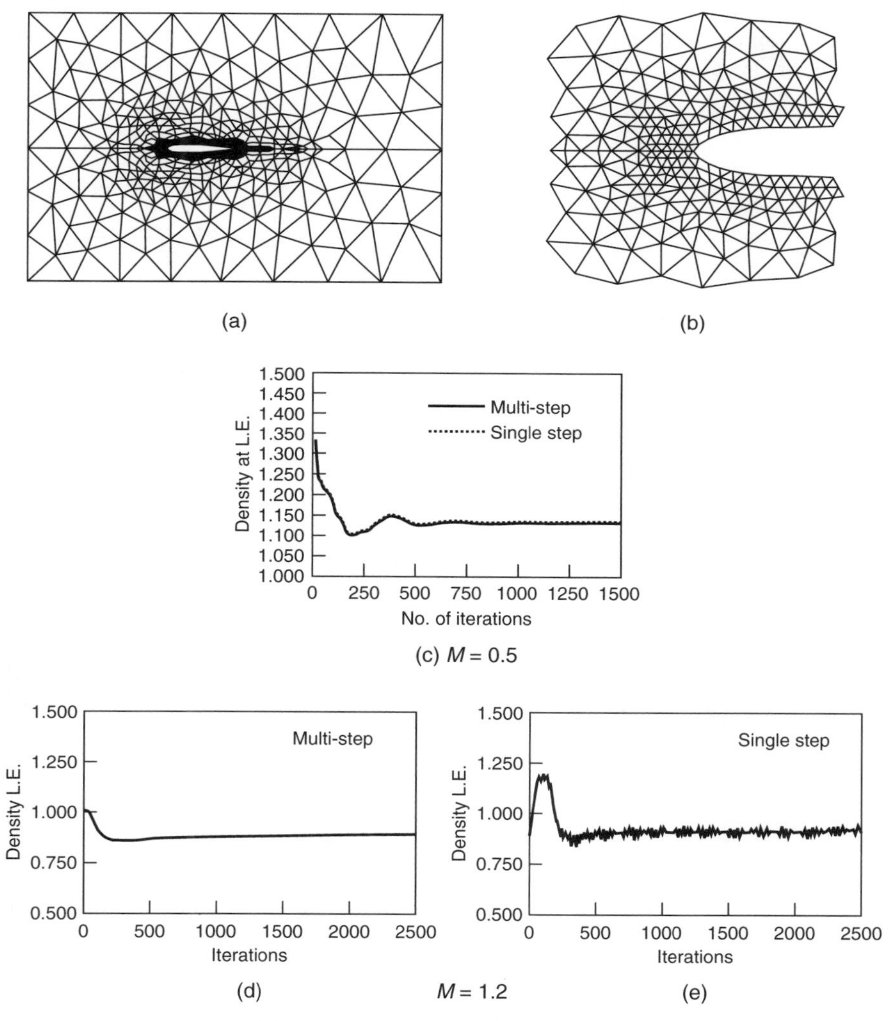

Fig. 3.3 Inviscid flow past a NACA0012 aerofoil $\alpha = 0$: (a) Unstructured mesh 1824 elements and 969 nodes; (b) Details of mesh near stagnation point; (c) Convergence for $M = 0.5$ with two and single step, fully explicit form; (d) Convergence for $M = 1.2$ for two-step scheme; (e) Convergence for $M = 1.2$ for single-step scheme.

The inlet Mach number is assumed to be equal to 0.5 and all variables except the density are prescribed at the inlet. The density is imposed at the exit of the domain. Both the top and bottom sides are assumed to be symmetric with normal component of velocity equal to zero. A slipping boundary is assumed on the surface of the aerofoil. No additional viscosity in any form is used in this problem when we use the CBS algorithm. However other schemes do need additional diffusions to get a reasonable solution.

Figure 3.3(c) shows the comparison of the density evolution at the stagnation point of the aerofoil. It is observed that the difference between the single- and two-step schemes is negligibly small. Further tests on these schemes are carried out at a higher inlet Mach number of 1.2 with the flow being supersonic, and a different mesh with a higher number of nodes (3753) and elements (7351). Here all the variables at the inlet are specified and the exit is free. As we can see from Fig. 3.3(d) and (e), the

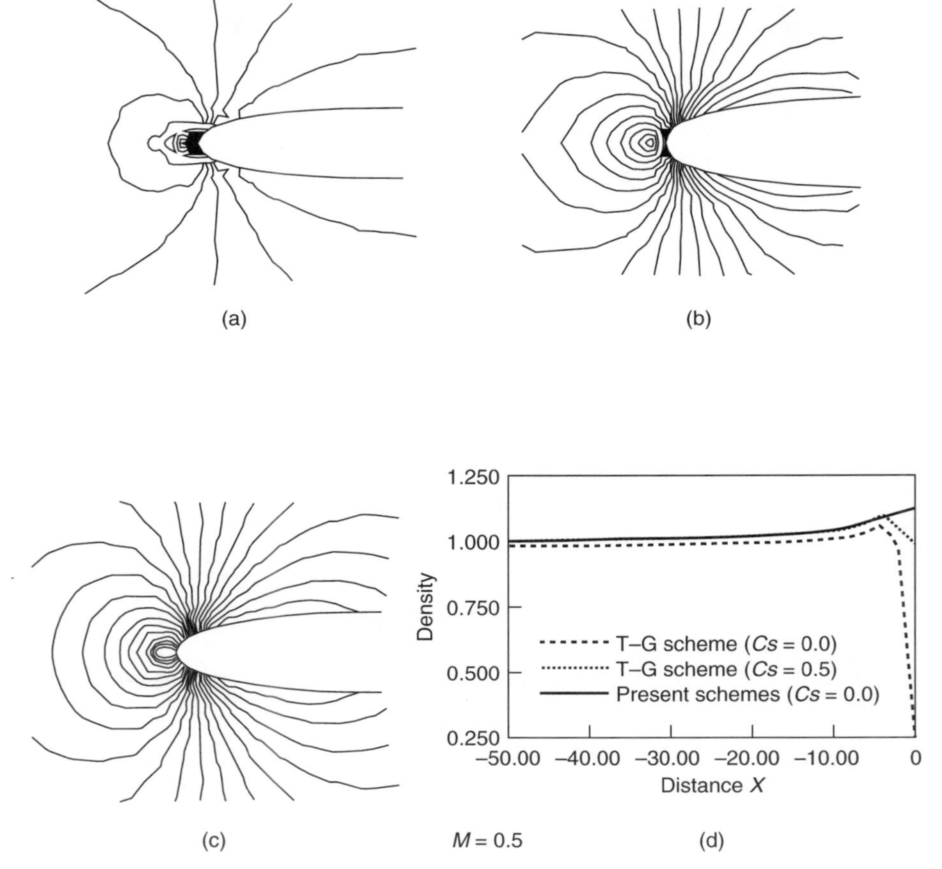

Fig. 3.4 Subsonic inviscid flow past a NACA0012 aerofoil with $\alpha = 0$ and $M = 0.5$: (a) Density contours with TG scheme with no additional viscosity; (b) Density contours with TG scheme with additional viscosity; (c) Density contours with CBS scheme with no additional viscosity; (d) Comparison of density along the stagnation line.

single-step scheme gives spurious oscillations in density values at the stagnation point. Therefore we conclude that here the two-step algorithm is valid for any range of Mach number and the single-step algorithm is limited to low Mach number flows with small compressibility.

In Fig. 3.4 we compare the two-step algorithm results of the subsonic inviscid ($M = 0.5$) results with those obtained by the Taylor–Galerkin scheme for the same mesh. It is observed that the CBS algorithm gives a smooth solution near the stagnation point even though no additional artificial diffusion in any form is introduced. However the Taylor–Galerkin scheme gives spurious solutions and a reasonable solution is obtained from this scheme only with a considerable amount of additional diffusion. Comparison of pressure distribution along the stagnation line shows (Fig. 3.4d) that the Taylor–Galerkin scheme gives an incorrect solution even with additional diffusion. However, the CBS algorithm again gives an accurate solution without the use of any additional artificial diffusion.

3.8 Concluding remarks

The general CBS algorithm is discussed in detail in this chapter for the equations of fluid dynamics in their conservation form. Comparison between the single- and two-step algorithms in the last section shows that the latter scheme is valid for all ranges of flow. In later chapters, we generally apply the two-step algorithm for different flow applications. Another important conclusion made from this chapter is about the accuracy of the present scheme. As observed in the last section, the present CBS algorithm gives excellent performance when the flow is slightly compressible compared to the Taylor–Galerkin algorithm. In the following chapters we show further tests on the algorithm for a variety of problems including general compressible and incompressible flow problems, shallow-water problems, etc.

References

1. A.J. Chorin. Numerical solution of Navier–Stokes equations. *Math. Comput.*, **22**, 745–62, 1968.
2. A.J. Chorin. On the convergence of discrete approximation to the Navier–Stokes equations. *Math. Comput.*, **23**, 341–53, 1969.
3. G. Comini and S. Del Guidice. Finite element solution of incompressible Navier–Stokes equations. *Num. Heat Transfer, Part A*, **5**, 463–78, 1972.
4. G.E. Schneider, G.D. Raithby and M.M. Yovanovich. Finite element analysis of incompressible fluid flow incorporating equal order pressure and velocity interpolation, in C. Taylor, K. Morgan and C.A. Brebbia (eds.), *Numerical Methods in Laminar and Turbulent Flows*, Pentech Press, Plymouth, 1978.
5. J. Donea, S. Giuliani, H. Laval and L. Quartapelle. Finite element solution of unsteady Navier–Stokes equations by a fractional step method. *Comp. Meth. Appl. Mech. Eng.*, **33**, 53–73, 1982.
6. P.M. Gresho, S.T. Chan, R.L. Lee and C.D. Upson. A modified finite element method for solving incompressible Navier–Stokes equations. Part I theory. *Int. J. Num. Meth. Fluids*, **4**, 557–98, 1984.

7. M. Kawahara and K. Ohmiya. Finite element analysis of density flow using the velocity correction method. *Int. J. Num. Meth. Fluids*, **5**, 981–93, 1985.
8. J.G. Rice and R.J. Schnipke. An equal-order velocity–pressure formulation that does not exhibit spurious pressure modes. *Comp. Meth. Appl. Mech. Eng.*, **58**, 135–49, 1986.
9. B. Ramaswamy, M. Kawahara and T. Nakayama. Lagrangian finite element method for the analysis of two dimensional sloshing problems. *Int. J. Num. Meth. Fluids*, **6**, 659–70, 1986.
10. B. Ramaswamy. Finite element solution for advection and natural convection flows. *Comp. Fluids*, **16**, 349–88, 1988.
11. M. Shimura and M. Kawahara. Two dimensional finite element flow analysis using velocity correction procedure. *Struct. Eng./Earthquake Eng.*, **5**, 255–63, 1988.
12. S.G.R. Kumar, P.A.A. Narayana, K.N. Seetharamu and B. Ramaswamy. Laminar flow and heat transfer over a two dimensional triangular step. *Int. J. Num. Meth. Fluids*, **9**, 1165–77, 1989.
13. B. Ramaswamy, T.C. Jue and J.E. Akin. Semi-implicit and explicit finite element schemes for coupled fluid thermal problems. *Int. J. Num. Meth. Eng.*, **34**, 675–96, 1992.
14. R. Rannacher. On Chorin projection method for the incompressible Navier–Stokes equations. *Lecture Notes in Mathematics*, **1530**, 167–83, 1993.
15. B. Ramaswamy. Theory and implementation of a semi-implicit finite element method for viscous incompressible flows. *Comp. Fluids*, **22**, 725–47, 1993.
16. C.B. Yiang and M. Kawahara. A three step finite element method for unsteady incompressible flows. *Comput. Mech.*, **11**, 355–70, 1993.
17. G. Ren and T. Utnes. A finite element solution of the time dependent incompressible Navier–Stokes equations using a modified velocity correction method. *Int. J. Num. Meth. Fluids*, **17**, 349–64, 1993.
18. B.V.K.S. Sai, K.N. Seetharamu and P.A.A. Narayana. Solution of transient laminar natural convection in a square cavity by an explicit finite element scheme. *Num. Heat Transfer, Part A*, **25**, 593–609, 1994.
19. M. Srinivas, M.S. Ravisanker, K.N. Seetharamu and P.A. Aswathanarayana. Finite element analysis of internal flows with heat transfer. *Sadhana – Academy Proc. Eng.*, **19**, 785–816, 1994.
20. P.M. Gresho, S.T. Chan, M.A. Christon and A.C. Hindmarsh. A little more on stabilized Q(1)Q(1) for transient viscous incompressible flow. *Int. J. Num. Meth Fluids*, **21**, 837–56, 1995.
21. Y.T.K. Gowda, P.A.A. Narayana and K.N. Seetharamu. Mixed convection heat transfer past in-line cylinders in a vertical duct. *Num. Heat Transfer, Part A*, **31**, 551–62, 1996.
22. A.R. Chaudhuri, K.N. Seetharamu and T. Sundararajan. Modelling of steam surface condenser using finite element methods. *Comm. Num. Meth. Eng.*, **13**, 909–21, 1997.
23. P. Nithiarasu, T. Sundararajan and K.N. Seetharamu. Finite element analysis of transient natural convection in an odd-shaped enclosure. *Int. J. Num. Meth. Heat and Fluid Flow*, **8**, 199–220, 1998.
24. P.K. Maji and G. Biswas. Three-dimensional analysis of flow in the spiral casing of a reaction turbine using a differently weighted Petrov Galerkin method. *Comp. Meth. Appl. Mech. Eng.*, **167**, 167–90, 1998.
25. P.D. Minev and P.M. Gresho. A remark on pressure correction schemes for transient viscous incompressible flow. *Comm. Num. Meth. Eng.*, **14**, 335–46, 1998.
26. J. Blasco, R. Codina and A. Huerta. A fractional-step method for the incompressible Navier–Stokes equations related to a predictor–multicorrector algorithm. *Int. J. Num. Meth. Fluids*, **28**, 1391–419, 1998.
27. B.S.V.P. Patnaik, P.A.A. Narayana and K.N. Seetharamu. Numerical simulation of vortex shedding past a cylinder under the influence of buoyancy. *Int. J. Heat Mass Transfer*, **42**, 3495–507, 1999.

28. O.C. Zienkiewicz and R. Codina. Search for a general fluid mechanics algorithm. *Frontiers of Computational Fluid Dynamics*, Eds. D.A. Caughey and M.M. Hafez. J. Wiley, New York, 101–13, 1995.
29. O.C. Zienkiewicz and R. Codina. A general algorithm for compressible and incompressible flow, Part I. The split characteristic based scheme. *Int. J. Num. Meth. Fluids*, **20**, 869–85, 1995.
30. O.C. Zienkiewicz, B.V.K.S. Sai, K. Morgan, R. Codina and M. Vázquez. A general algorithm for compressible and incompressible flow – Part II. Tests on the explicit form. *Int. J. Num. Meth. Fluids*, **20**, 887–913, 1995.
31. O.C. Zienkiewicz and P. Ortiz. A split characteristic based finite element model for shallow water equations. *Int. J. Num. Meth. Fluids*, **20**, 1061–80, 1995.
32. P. Ortiz and O.C. Zienkiewicz. Modelizacin por elementos finitos en hidrulica e hidrodinmica costera. *Centro de Estudios y Experimentation de Obras Pblicas*, CEDEX, Madrid, 1995.
33. O.C. Zienkiewicz. A new algorithm for fluid mechanics. Compressible and incompressible behaviour. *Proc. 9th Int. Conf. Finite Elements Fluids – New Trends and Applications*, 49–55, Venezia, Oct. 1995.
34. R. Codina, M. Vázquez and O.C. Zienkiewicz. A fractional step method for compressible flows: Boundary conditions and incompressible limit. *Proc. 9th Int. Conf. Finite Elements Fluids – New Trends and Applications*, 409–18, Venezia, Oct. 1995.
35. P. Ortiz and O.C. Zienkiewicz. Tide and bore propagation in the Severn Estuary by a new algorithm. *Proc. 9th Int. Conf. Finite Elements Fluids – New Trends and Applications*, 1543–52, Venezia, Oct. 1995.
36. O.C. Zienkiewicz, B.V.K.S. Sai, K. Morgan and R. Codina. Split characteristic based semi-implicit algorithm for laminar/turbulent incompressible flows. *Int. J. Num. Meth. Fluids*, **23**, 1–23, 1996.
37. P. Ortiz and O.C. Zienkiewicz. An improved finite element model for shallow water problems, in G.F. Garey (Ed.). *Finite Element Modelling of Environmental Problems*, Wiley, New York, 61–84, 1996.
38. R. Codina, M. Vázquez and O.C. Zienkiewicz. General algorithm for compressible and incompressible flows, Part III – A semi-implicit form. *Int. J. Num. Meth. Fluids*, **27**, 13–32, 1998.
39. B.V.K.S. Sai, O.C. Zienkiewicz, M.T. Manzari, P.R.M. Lyra and K. Morgan. General purpose vs special algorithms for high speed flows with shocks. *Int. J. Num. Meth. Fluids*, **27**, 57–80, 1998.
40. R. Codina, M. Vázquez and O.C. Zienkiewicz. A fractional step method for the solution of compressible Navier–Stokes equations. Eds. M. Hafez and K. Oshima. *Computational Fluid Dynamics Review 1998*, **Vol 1**, 331–47, World Scientific Publishing, 1998.
41. N. Massarotti, P. Nithiarasu and O.C. Zienkiewicz. Characteristic-based-split (CBS) algorithm for incompressible flow problems with heat transfer. *Int. J. Num. Meth. Fluids*, **8**, 969–90, 1998.
42. O.C. Zienkiewicz, P. Nithiarasu, R. Codina, M. Vázquez and P. Ortiz. Characteristic-based-split algorithm, Part I; The theory and general discussion. *Invited Lecture and Proc. of ECCOMAS CFD 1998*, **V2**, Ed. K.D. Papailiou *et al.*, pp. 4–16, Athens, Greece.
43. N. Massarotti, P. Nithiarasu and O.C. Zienkiewicz. Characteristic-based-split algorithm, Part II; Incompressible flow problems with heat transfer. *Invited Lecture and Proc. of ECCOMAS CFD 1998*, **V2**, Ed. K.D. Papailiou *et al.*, pp. 17–21, Athens, Greece.
44. O.C. Zienkiewicz and P. Ortiz. The CBS (characteristic-based-split) algorithm in hydraulic and shallow water flow. *Second Int. Sym. Rever Sedimentation and Env. Hydraulics*, University of Hong Kong, 3–12 Dec. 1998.
45. O.C. Zienkiewicz, J. Rojek, R.L. Taylor and M. Pastor. Triangles and tetrahedra in explicit dynamic codes for solids. *Int. J. Num. Meth. Eng.*, **43**, 565–83, 1999.

46. O.C. Zienkiewicz, P. Nithiarasu , R. Codina and M. Vázquez. The characteristic based split procedure: An efficient and accurate algorithm for fluid problems. *Int. J. Num. Meth. Fluids*, **31**, 359–92, 1999.
47. O.C. Zienkiewicz, J. Szmelter and J. Peraire. Compressible and incompressible flow: an algorithm for all seasons. *Comp. Meth. Appl. Mech. Eng.*, **78**, 105–21, 1990.
48. O.C. Zienkiewicz. Explicit or semi-explicit general algorithm for compressible and incompressible flows with equal finite element interpolation. *Report 90/5*, Chalmers University of Technology, 1990.
49. O.C. Zienkiewicz. Finite elements and computational fluid mechanics, Metodos Numericos en Ingenieria. *SEMNI Congress*, 56–61, Gran Canaria, 1991.
50. O.C. Zienkiewicz and J. Wu. Incompressibility without tears! How to avoid restrictions of mixed formulation. *Int. J. Num. Meth. Eng.*, **32**, 1184–203, 1991.
51. O.C. Zienkiewicz and J. Wu. A general explicit or semi-explicit algorithm for compressible and incompressible flows. *Int. J. Num. Meth. Eng.*, **35**, 457–79, 1992.
52. P. Nithiarasu and O.C. Zienkiewicz. On stabilization of the CBS algorithm. Internal and external time steps. *Int. J. Num. Meth. Eng.*, (in press).
53. A. Jameson and D.J. Mavripolis. Finite volume solution of the two dimensional Euler equations on a regular triangular mesh. *AIAA-85-0435*, 1985.
54. F. Brezzi and J. Pitkäranta. On the stabilization of finite element approximations of the Stokes problem. *Efficient Solution of Elliptic Problems, Notes on Numerical Mechanics*, Ed. W. Hackbusch, **10**, Vieweg Publishers, Wiesbaden, 1984.
55. T.C. Papanastasiou, N. Malamataris and K. Ellwood. A new outflow boundary condition. *Int. J. Num. Meth. Fluids*, **14**, 587–608, 1992.

4

Incompressible laminar flow – newtonian and non-newtonian fluids

4.1 Introduction and the basic equations

The problems of incompressible flows dominate a large part of the fluid mechanics scene. For this reason, they are given special attention in this book and we devote two chapters to this subject. In the present chapter we deal with various steady-state and transient situations in which the flow is forced by appropriate pressure gradients and boundary forces. In the next chapter we shall consider free surface flows in which gravity establishes appropriate wave patterns as well as the so-called buoyancy force in which the only driving forces are density changes caused by temperature variations. At this stage we shall also discuss briefly the important subject of turbulence.

We have already mentioned in Volume 1 the difficulties that are encountered generally with incompressibility when this is present in the equations of solid mechanics. We shall find that exactly the same problems arise again in fluids especially with very slow flows where the acceleration can be neglected and viscosity is dominant (so-called Stokes flow). Complete identity with solids is found here (namely Chapter 12, Volume 1).

The essential difference in the governing equations for incompressible flows from those of compressible flows is that the coupling between the equations of energy and the other equations is very weak and thus frequently the energy equations can be considered either completely independently or as an iterative step in solving the incompressible flow equations.

To proceed further we return to the original equations of fluid dynamics which have been given in Chapters 1 and 3; we repeat these below for problems of small compressibility.

Conservation of mass

$$\frac{\partial \rho}{\partial t} = \frac{1}{c^2}\frac{\partial p}{\partial t} = -\frac{\partial U_i}{\partial x_i} \tag{4.1}$$

and $c^2 = K/\rho$ where K is the bulk modulus. Here in the incompressible limit, the density ρ is assumed to be constant and in this situation the term on the left-hand side is simply zero.

Conservation of momentum

$$\frac{\partial U_i}{\partial t} = -\frac{\partial}{\partial x_j}(u_j U_i) + \frac{\partial \tau_{ij}}{\partial x_j} - \frac{\partial p}{\partial x_i} - \rho g_i \tag{4.2}$$

In the above we define the mass flow fluxes as

$$U_i = \rho u_i \tag{4.3}$$

Conservation of energy is now uncoupled and can be solved independently:

$$\frac{\partial(\rho E)}{\partial t} = -\frac{\partial}{\partial x_j}(u_j \rho E) + \frac{\partial}{\partial x_i}\left(k\frac{\partial T}{\partial x_i}\right) - \frac{\partial}{\partial x_j}(u_j p) + \frac{\partial}{\partial x_i}(\tau_{ij}u_j) \tag{4.4}$$

In the above u_i are the velocity components; E is the specific energy ($c_v T$), p is the pressure, T is the absolute temperature, ρg_i represents the body force and other source terms, and τ_{ij} are the deviatoric stress components given by (Eq. 1.12b).

$$\tau_{ij} = \mu\left(\frac{\partial u_i}{\partial x_j} + \frac{\partial u_j}{\partial x_i} - \frac{2}{3}\delta_{ij}\frac{\partial u_k}{\partial x_k}\right) \tag{4.5}$$

With the substitution made for density changes we note that the essential variables in the first two equations become those of pressure and velocity. In exactly the same way as these, we can specify the variables linking displacements and pressure in the case of incompressible solids. It is thus possible to solve these equations in one of many ways described in Chapter 12 of Volume 1 though, of course, the use of the CBS algorithm is obvious.

Unless the viscosity and in fact the bulk modulus have a strong dependence on temperature the problem is very weakly linked with the energy equation which can be solved independently.

The energy equation for incompressible materials is best written in terms of the absolute temperature T avoiding the specific energy. The equation now becomes simply

$$c_v\rho\left[\frac{\partial T}{\partial t} + u_j\frac{\partial T}{\partial x_j}\right] = \frac{\partial}{\partial x_i}\left(k\frac{\partial T}{\partial x_i}\right) + \frac{\partial}{\partial x_i}(\tau_{ij}u_j) - \frac{\partial}{\partial x_j}(u_j p) \tag{4.6}$$

and we note that this is now a scalar convection–diffusion equation of the type we have already encountered in Chapter 2, written in terms of the variable temperature as the unknown. In the above equation, the last two work dissipation terms are often neglected for fully incompressible flows. Note that the above equation is derived assuming the density and c_v (specific heat at constant volume) to be constants.

In this chapter we shall in general deal with problems for which the coupling is weak and the temperature equations do not present any difficulties. However in Chapter 5 we shall deal with buoyancy effects causing atmospheric or general circulation induced by small density changes induced by temperature differences.

If viscosity is a function of temperature, it is very often best to proceed simply by iterating over a cycle in which the velocity and pressure are solved with the assumption of known viscosity and that is followed by the solution of temperature. Many practical problems have been so solved very satisfactorily. We shall show some of these applications in the field of material forming later on in this chapter.

In the main part of this chapter we shall consider the solution of viscous, newtonian or non-newtonian fluids and we shall in the main use the CBS algorithm described in Chapter 3, though on occasion we shall depart from this due to the similarity with the equations of solid mechanics and use a more direct approach either by satisfying the BB stability conditions of Chapter 12 in Volume 1 for the velocity and pressure variables, or by using reduced integration in the context of a pure velocity formulation with a penalty parameter.

However, before proceeding further, it is of interest to note that the very special case of zero viscosity can be solved in a very much simpler manner and in the next section we shall do so. Here we introduce the idea of potential flow with irrotational constraints and with such a formulation the convective acceleration disappears and the final equations become self-adjoint. For such problems the Galerkin approximation can be used directly. We have already discussed this in Chapter 7 of Volume 1.

4.2 Inviscid, incompressible flow (potential flow)

In the absence of viscosity and compressibility equations, Eqs (4.1) and (4.2) can be written as

$$\frac{\partial u_i}{\partial x_i} = 0 \tag{4.7}$$

and

$$\frac{\partial u_i}{\partial t} + \frac{\partial}{\partial x_j}(u_j u_i) + \frac{1}{\rho}\frac{\partial p}{\partial x_i} - g_i = 0 \tag{4.8}$$

These Euler equations are not convenient for numerical solution, and it is of interest to introduce a potential, ϕ, defining velocities as

$$u_1 = -\frac{\partial \phi}{\partial x_1} \qquad u_2 = -\frac{\partial \phi}{\partial x_2} \qquad u_3 = -\frac{\partial \phi}{\partial x_3} \tag{4.9}$$

or

$$\mathbf{u} = -\boldsymbol{\nabla}\phi \qquad \text{or} \qquad u_i = -\frac{\partial \phi}{\partial x_i}$$

If such a potential exists then insertion of (4.9) into (4.7) gives a single governing equation

$$\frac{\partial^2 \phi}{\partial x_i \partial x_i} \equiv \boldsymbol{\nabla}^2 \phi = 0 \tag{4.10}$$

which, with appropriate boundary conditions, can be readily solved in the manner described in Chapter 7 of Volume 1. For contained flow we can of course impose the normal velocity u_n on the boundaries:

$$u_n = -\frac{\partial \phi}{\partial n} \tag{4.11}$$

and, as we know from the discussions in Volume 1, this provides a *natural* boundary condition.

Indeed, at this stage it is not necessary to discuss the application of finite elements to this particular equation, which was considered at length in Volume 1 and for which many solutions are available.[1] In Fig. 4.1 an example of a typical potential solution is given.

Of course we must be assured that the potential function ϕ exists, and indeed determine what conditions are necessary for its existence. Here we observe that so far we have not used in the definition of the problem the important momentum-conservation equations (4.8), to which we shall now return. However, we first note that a single-valued potential function implies that

$$\frac{\partial^2 \phi}{\partial x_j \partial x_i} = \frac{\partial^2 \phi}{\partial x_i \partial x_j} \tag{4.12}$$

and hence that, using the definition (4.9),

$$\omega_1 = \frac{\partial u_1}{\partial x_2} - \frac{\partial u_2}{\partial x_1} = 0 \qquad \omega_2 = \frac{\partial u_2}{\partial x_3} - \frac{\partial u_3}{\partial x_2} = 0 \qquad \omega_3 = \frac{\partial u_3}{\partial x_1} - \frac{\partial u_1}{\partial x_3} = 0 \tag{4.13}$$

This is a statement of the *irrotationality* of the flow which we see is implied by the existence of the potential.

Inserting the definition of potential into the first term of Eq. (4.8) and using Eqs (4.7) and (4.13) we can rewrite this equation as

$$-\frac{\partial}{\partial x_i}\left(\frac{\partial \phi}{\partial t}\right) + \frac{\partial}{\partial x_i}\left[\frac{1}{2}u_j u_j + \frac{p}{\rho} + P\right] = 0 \tag{4.14}$$

in which P is the potential of the body forces giving these as

$$g_i = -\frac{\partial P}{\partial x_i} \tag{4.15}$$

This is alternatively written as

$$\nabla\left(-\frac{\partial \phi}{\partial t} + H + P\right) = 0 \tag{4.16}$$

where H is the *enthalpy*, given as $H = \frac{1}{2}u_i u_i + p/\rho$.

If isothermal conditions pertain, the specific energy is constant and the above implies that

$$-\frac{\partial \phi}{\partial t} + \frac{1}{2}u_i u_i + \frac{p}{\rho} + P = \text{constant} \tag{4.17}$$

for the whole domain. This can be taken as a corollary of the existence of the potential and indeed is a condition for its existence. In steady-state flows it provides the well-known Bernoulli equation that allows the pressures to be determined throughout the whole potential field when the value of the constant is established.

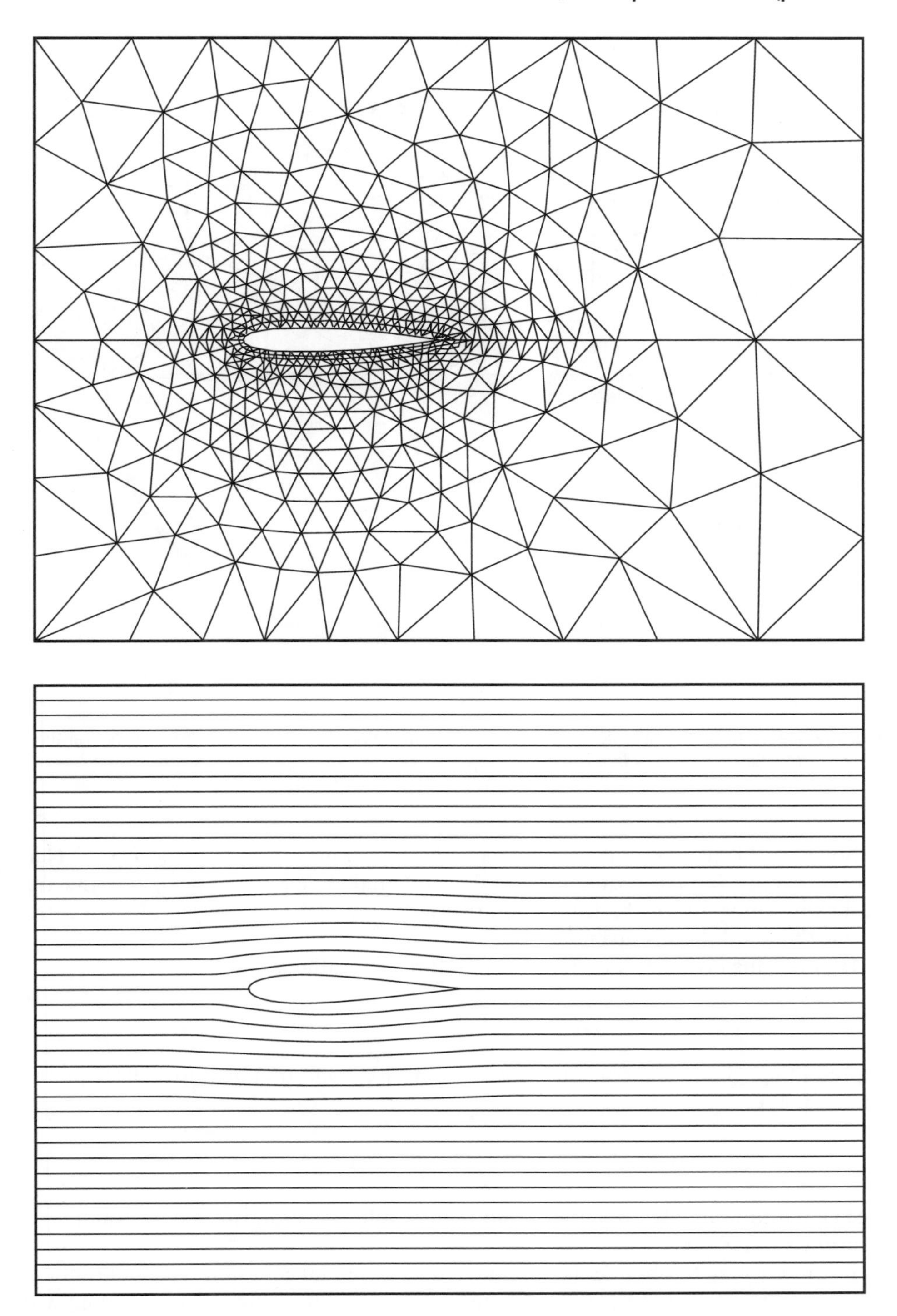

Fig. 4.1 Potential flow solution around an aerofoil. Mesh and streamline plots.

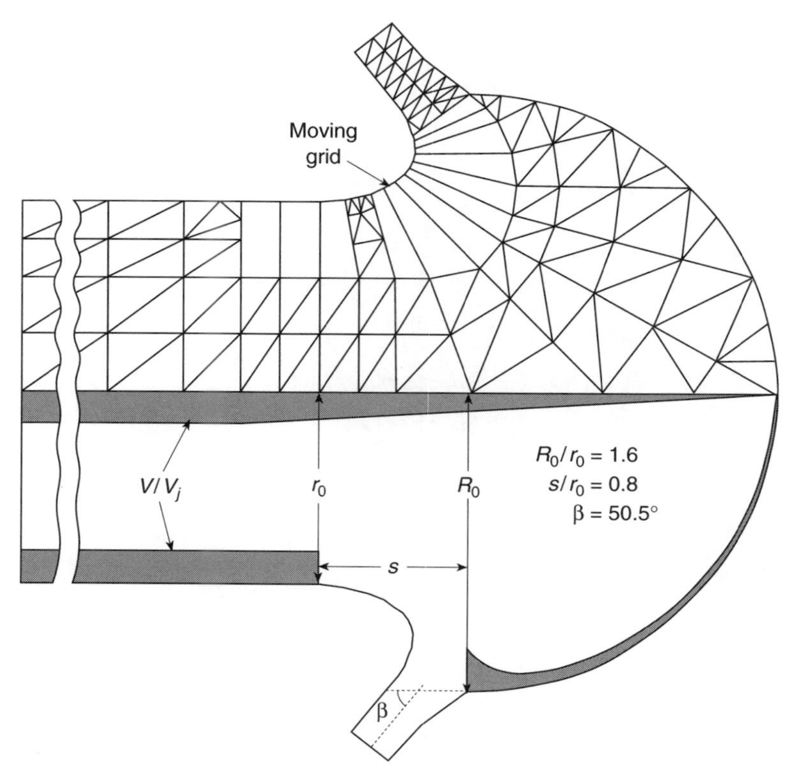

Fig. 4.2 Free surface potential flow, illustrating an axisymmetric jet impinging on a hemispherical thrust reverser (from Sarpkaya and Hiriart[3]).

Some problems of specific interest are those of flow with a free surface.[2–4] Here the governing Laplace equation for the potential remains identical, but the free surface position has to be found iteratively. In Fig. 4.2 an example of such a free surface flow solution is given.[3]

In problems involving gravity the body force potential is simply

$$P = gx_3$$

representing gravity forces, and the free surface condition requires that (in two dimensions)

$$\tfrac{1}{2}(u_1^2 + u_3^2) - gx_3 = 0$$

Such conditions involve an iterative, non-linear solution, as illustrated by examples of overflows in reference 2.

It is interesting to observe that the governing potential equation is self-adjoint and that the introduction of the potential has side-stepped the difficulties of dealing with convective terms.

4.3 Use of the CBS algorithm for incompressible or nearly incompressible flows

4.3.1 The semi-implicit form

For problems of incompressibility with K being equal to infinity or indeed when K is very large, we have no choice of using the fully explicit procedure and we must therefore proceed with the CBS algorithm in its semi-implicit form (Chapter 3, Sec. 3.3.2). This of course will use an explicit solution for the momentum equation followed by an implicit solution of the pressure laplacian form (the Poisson equation).

The solution which has to be obtained implicitly involves only the pressure variable and we will further notice that, from the contents of Chapter 3, at each step the basic equation remains unchanged and therefore the solution can be repeated simply with different right-hand side vectors.

The convergence rate of course depends on the time step used and here we have the time step limitation given by the Courant number

$$\Delta t_1 \leqslant \Delta t_{\text{crit}} = \frac{h}{|\mathbf{u}|} \tag{4.18}$$

for inviscid problems and for viscous problems

$$\Delta t_2 \leqslant \Delta t_{\text{crit}} = \frac{h^2}{2\nu} \tag{4.19}$$

is an additional limitation. Here we note immediately that the viscosity lowers the limit quite substantially and therefore convergence may not be exceedingly rapid. The examples which we shall show nevertheless indicate its good performance and on each of the figures we give the number of iterations used to arrive at final solutions.

The classical problem on which we would like to judge the performance is that of the closed cavity driven by the motion of a lid.[5–7] There are various ways of assuming the boundary conditions but the most common is one in which the velocity along the top surface increases from the corner node to the driven value in the length of one element (so-called ramp conditions).†

The solution was obtained for different values of Reynolds number thus testing the performance of the viscous formulation.

The problem has been studied by many investigators and probably the most detailed investigation was that of Ghia *et al.*,[5] in which they quote many solutions and data for different Reynolds numbers. We shall use those results for comparison.

In the first figure, Fig. 4.3, we show the geometry, boundary conditions and finite element mesh. The mesh is somewhat graded near the walls using a geometrical progression.

† Some investigators use the leaking lid formulation in which the velocity along the top surface is constant and varies to zero within an element in the sides. It is preferable however to use the formulation where velocity is zero on all nodes of the vertical sides.

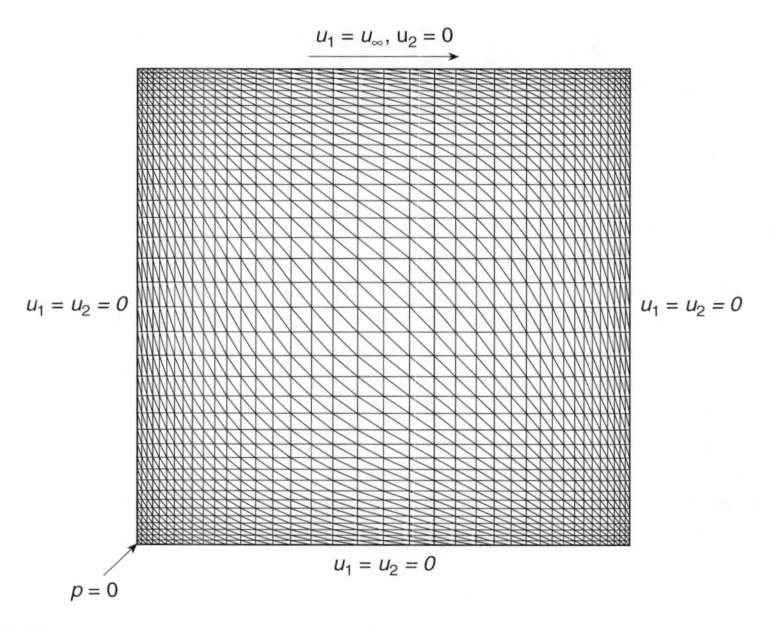

Fig. 4.3 Lid-driven cavity. Geometry, boundary conditions and mesh.

The velocity distribution along the centre-line for four different Reynolds numbers ranging from 0 when we have pure Stokes flow to the Reynolds number of 5000 is shown in Fig. 4.4. Similarly, the pressure distribution along the central horizontal line is given in Fig. 4.5 for different Reynolds numbers.

In Fig. 4.6 we show the contours of pressure and stream function again for the same Reynolds numbers.

In Fig. 4.7 we compare the pressure distribution at the mid-height of the cavity for different meshes at Reynolds number equal to zero (Stokes flow).

The reader will observe how closely the results obtained by the CBS algorithm follow those of Ghia *et al.*[5] calculated using finite differences on a much finer mesh (121×121).

4.3.2 Quasi-implicit solution

We have already remarked in Chapter 3 that the reduction of the explicit time step due to viscosity can be very inconvenient and may require a larger number of iterations as shown in previous figures. The example of the cavity is precisely in that category and at higher Reynolds numbers the reader will certainly note a very large number of iterations which have to be performed before results become reasonably steady. Here the time step is governed only by the relation given in Eq. (4.18). We have rerun the problems with a Reynolds number of 5000 using the quasi-implicit solution[8] which is explicit as far as the convective terms are concerned. The solution obtained is shown in Fig. 4.8. The reader will observe that only a much

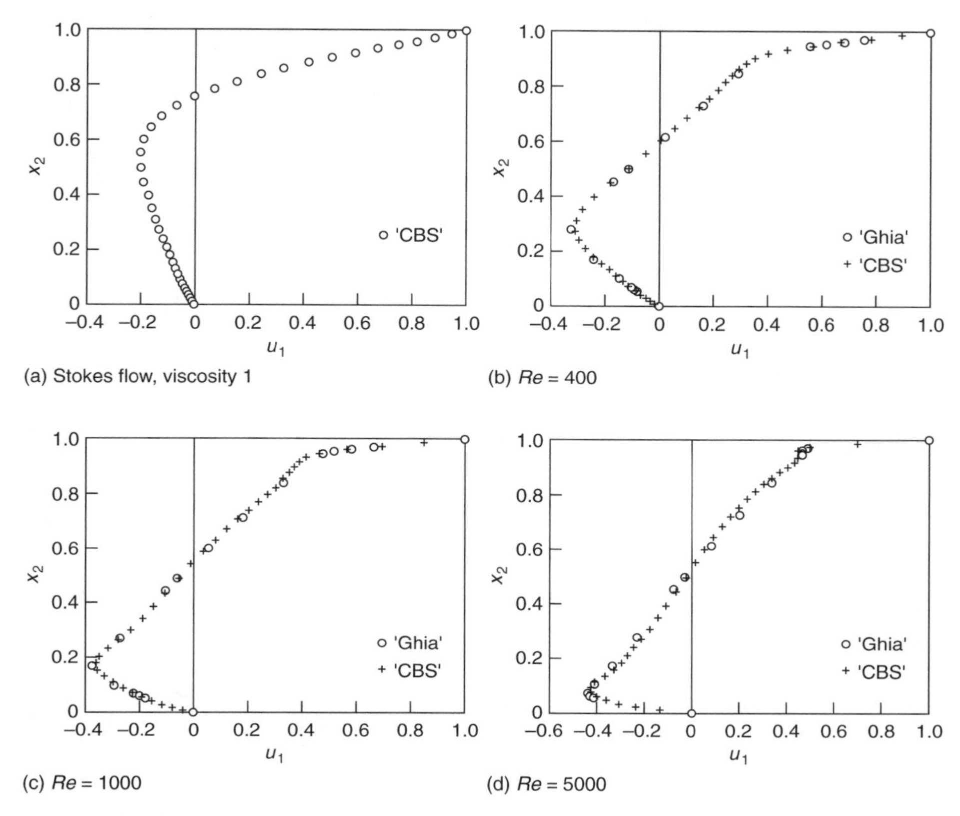

Fig. 4.4 Lid-driven cavity. u_1: velocity distribution along vertical centre-line for different Reynolds numbers (semi-implicit form).

smaller number of iterations is required to reach steady state and gives an accurate solution even at the higher Reynolds numbers. Here a solution for a Reynolds number of 10 000 is given in Fig. 4.9.

4.3.3 Fully explicit mode and artificial compressibility

It is of course impossible to model fully incompressible problems explicitly as the length of the stable time step is simply zero. However, the reader will observe that for steady-state solutions the first term of the continuity equation, i.e.

$$\frac{1}{c^2}\frac{\partial p}{\partial t} \tag{4.20}$$

does not enter the steady-state calculations and we could thus use any reasonably large value of c^2 instead of infinity. This artifice has been used with some success and the solution for a cavity is reported in reference 9 so we do not repeat the results here.

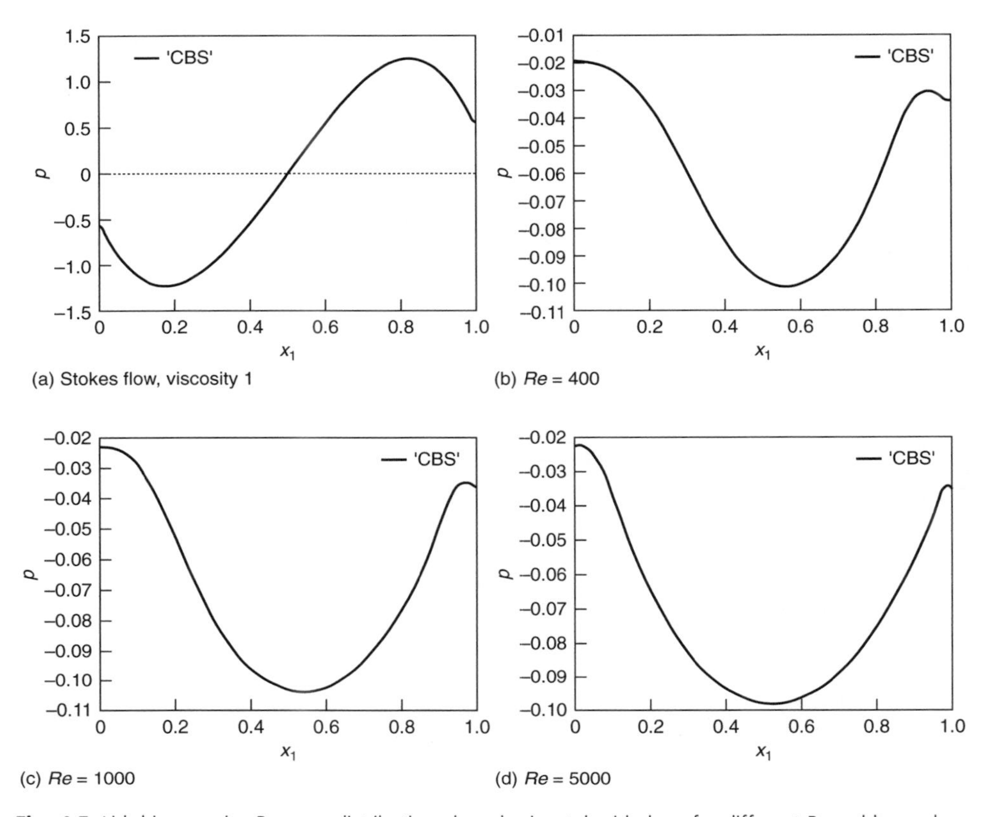

Fig. 4.5 Lid-driven cavity. Pressure distribution along horizontal mid-plane for different Reynolds numbers (semi-implicit form).

As we have mentioned in Chapter 3, it is important when using explicit procedures to make sure that the damping introduced is sufficient for ensuring that an oscillation-free solution can be obtained. With the explicit algorithm the time steps will inevitably be small as they are governed by the compressible wave velocity. It is convenient here, and indeed sometimes essential, to introduce the internal Δt_{int} which is different from the external Δt_{ext}. This matter is discussed by Nithiarasu *et al.* in reference 10 where several examples are shown proving the effectiveness of this process.

4.4 Boundary-exit conditions

The exit boundary conditions described in the previous chapter (Chapter 3, Sec. 3.6) are tested here for flow past a backward facing step. The geometry and boundary conditions are shown in Fig. 4.10(a). Figures 4.10 and 4.11 show the results obtained using the exit boundary conditions discussed in Chapter 3. For the sake of

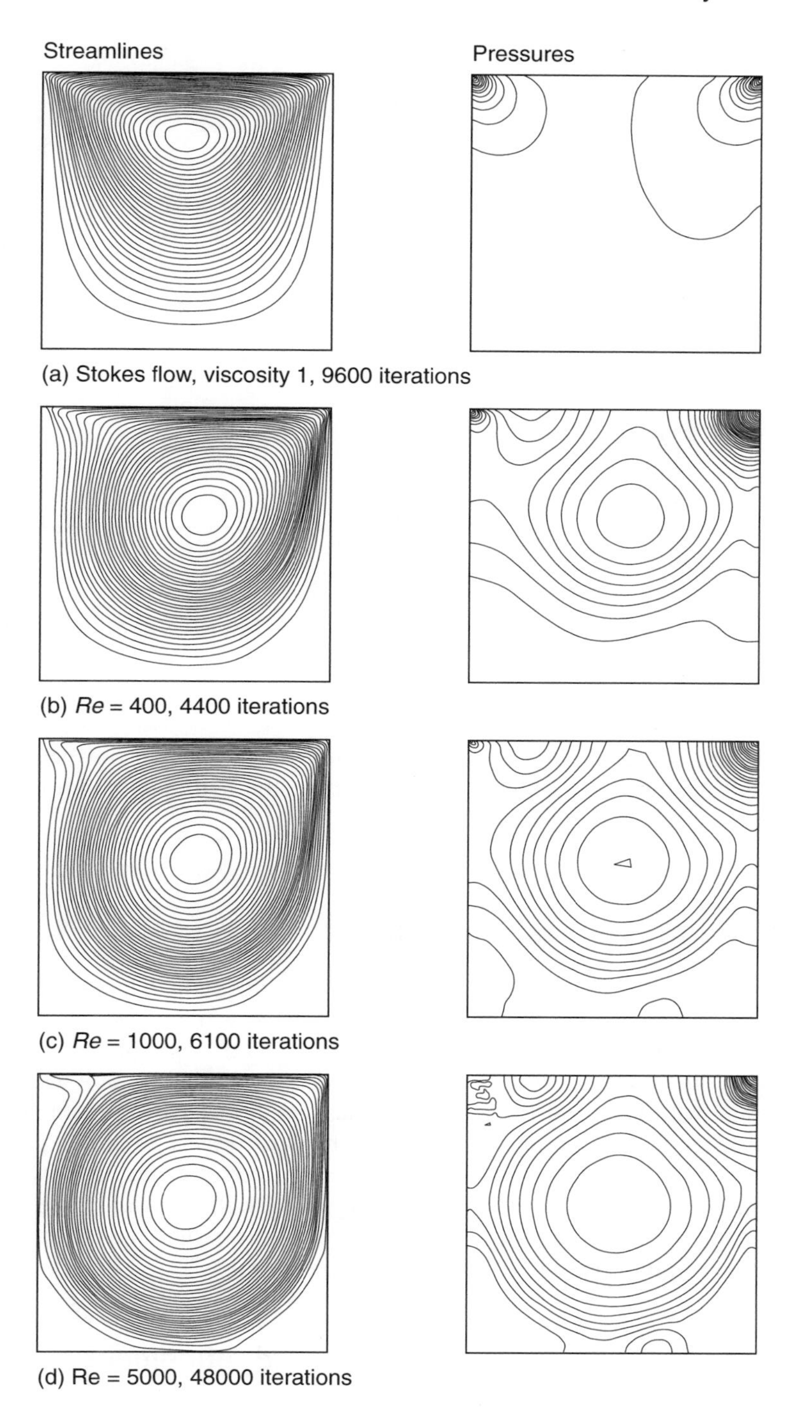

Fig. 4.6 Lid-driven cavity. Streamlines and pressure contours for different Reynolds numbers (semi-implicit form).

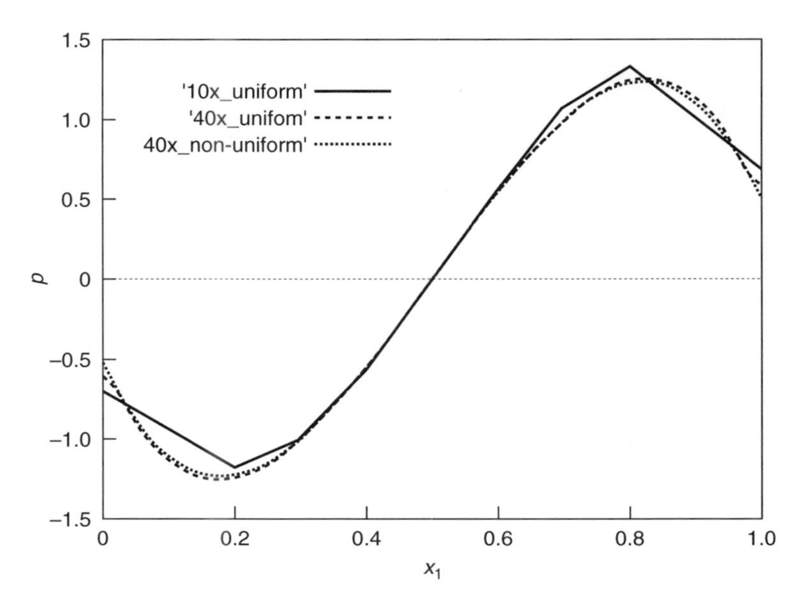

Fig. 4.7 Pressure distribution at mid-horizontal plane for Stokes flow in a cavity for different meshes (semi-implicit form) and different boundary conditions.

comparison, the results predicted by taking a longer domain downstream are also presented. As shown, the results predicted using the boundary conditions explained in Chapter 3, Sec. 3.6 are very accurate.

4.5 Adaptive mesh refinement

We have discussed the matter of adaptive refinement in Chapters 14 and 15 of Volume 1 in some detail. In that volume we have generally strived to obtain the energy norm error to be equal within all elements. The same procedures concerning the energy norm error can be extended of course to viscous flow especially when this is relatively slow and the problem is nearly elliptic. However, the energy norm has little significance at high speeds and here we revert to other considerations which simply give an *error indicator* rather than an error estimator. Two procedures are available and will be used in this chapter as well as a later one dealing with compressible flows. References 11–69 list some of the earlier and latest contributions to the field of adaptive procedures in fluid mechanics.

4.5.1 Second gradient (curvature) based refinement

Here the meaning of error analysis is somewhat different from that of the energy norm and we follow an approach where the error value is constant in each element. In what

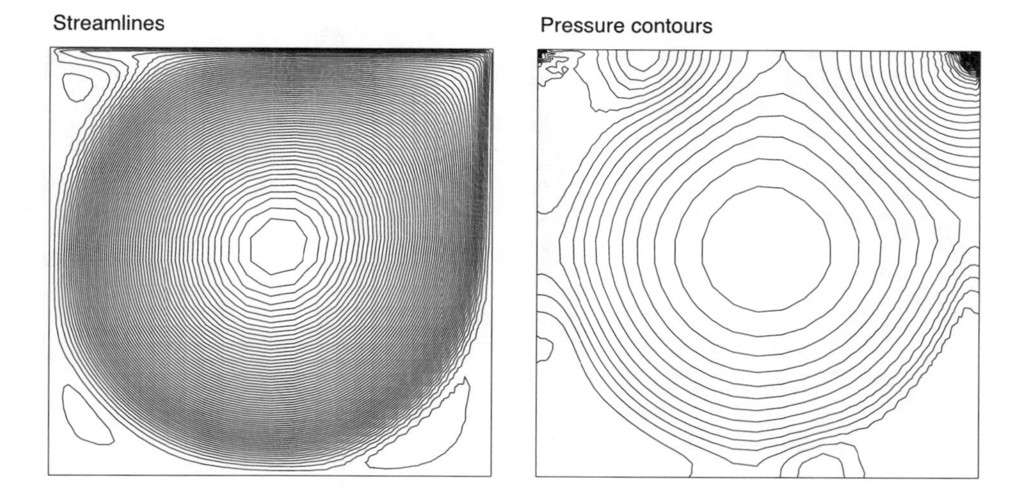

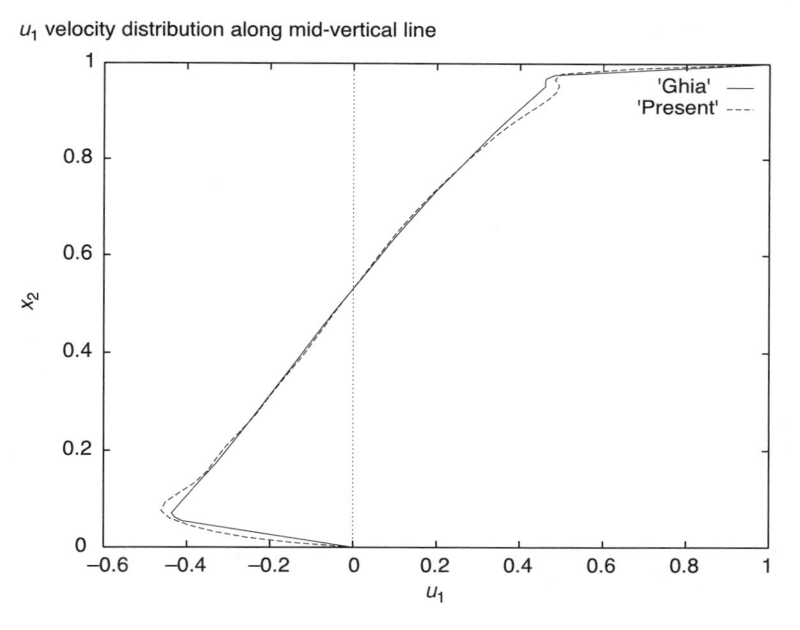

Fig. 4.8 Lid-driven cavity. Quasi-implicit solution for a Reynolds number of 5000.

follows we shall consider first-order (linear) elements and the so-called h refinement process in which increased accuracy is achieved by variation of element size. The p refinement in which the order of the element polynomial expression is changed is of course possible. Many studies are available on hp refinements where both h and p refinements are carried out simultaneously. This has been widely studied by Oden *et al.*[32,33,51,52] but we believe that such refinements impose many limitations on

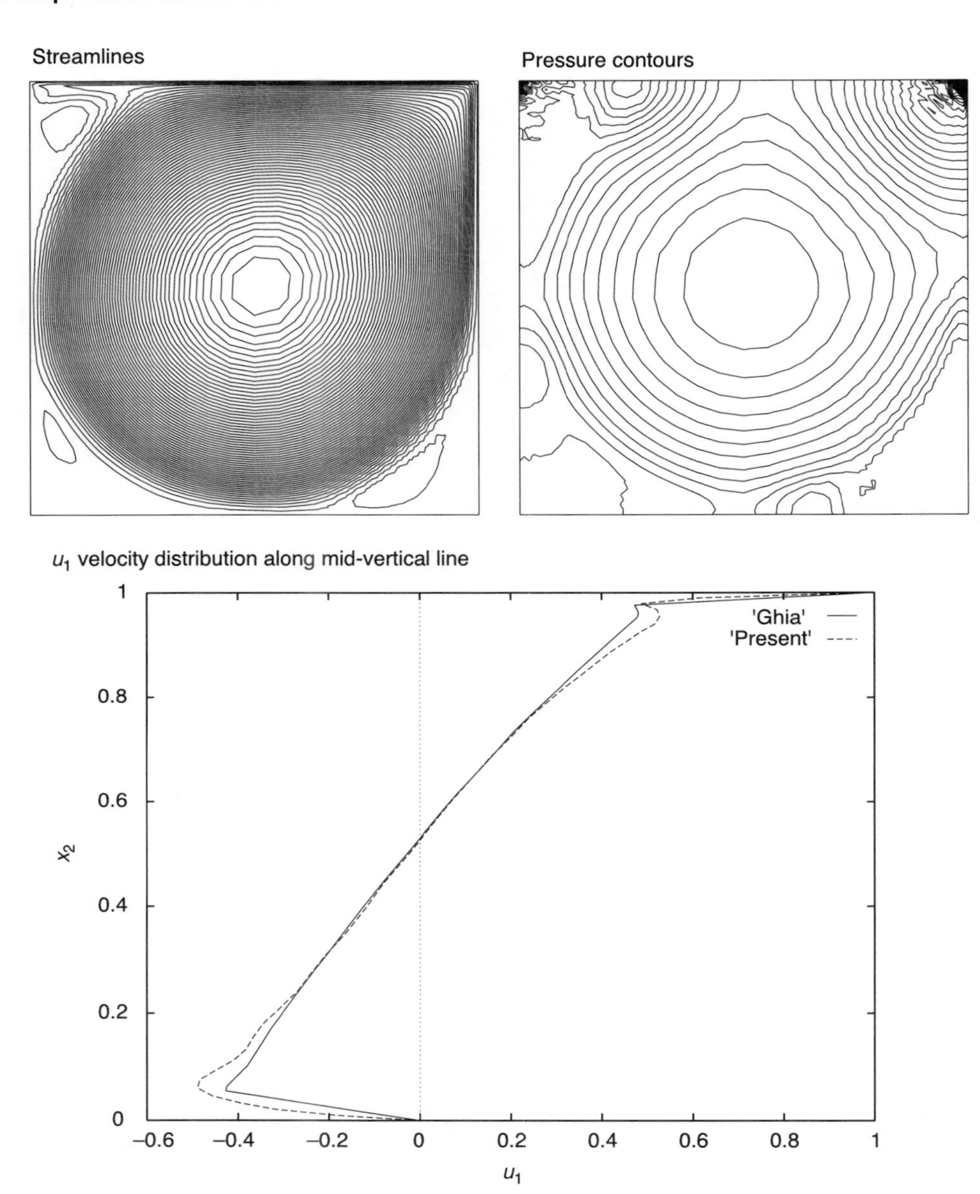

Fig. 4.9 Lid-driven cavity. Quasi-implicit solution for a Reynolds number of 10 000.

mesh generation and solution procedures and as most fluid mechanics problems involve an explicit time marching algorithm, the higher-order elements are not popular.

The determination of error indicators in linear elements is achieved by consideration of the so-called *interpolation error*. Thus if we take a one-dimensional element of length h and a scalar function ϕ, it is clear that the error in ϕ is of order $O(h^2)$ and that

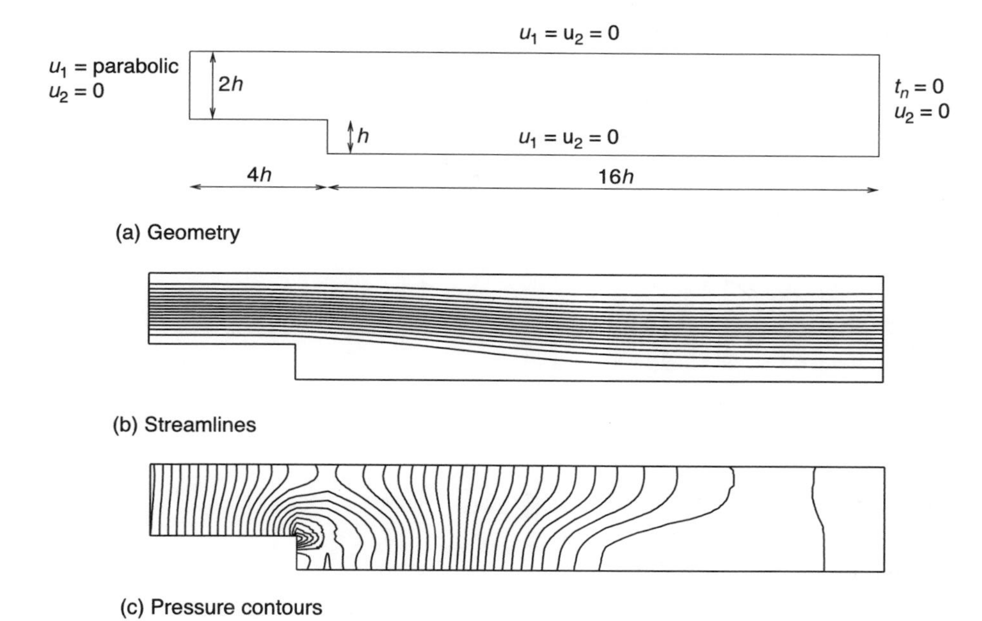

Fig. 4.10 Flow past a backward facing step. Exercise on exit boundary conditions, Re = 100.

it can be written as (see reference 22 for details)

$$e = \phi - \phi^h = ch^2 \frac{d^2\phi}{dx^2} \approx ch^2 \frac{d^2\phi^h}{dx^2} \tag{4.21}$$

where ϕ^h is the finite element solution and c is a constant.

If, for instance, we further assume that $\phi = \phi^h$ at the nodes, i.e. that the nodal error is zero, then e represents the values on a parabola with a curvature of $d^2\phi^h/dx^2$. This allows c, the unknown constant, to be determined, giving for instance the maximum

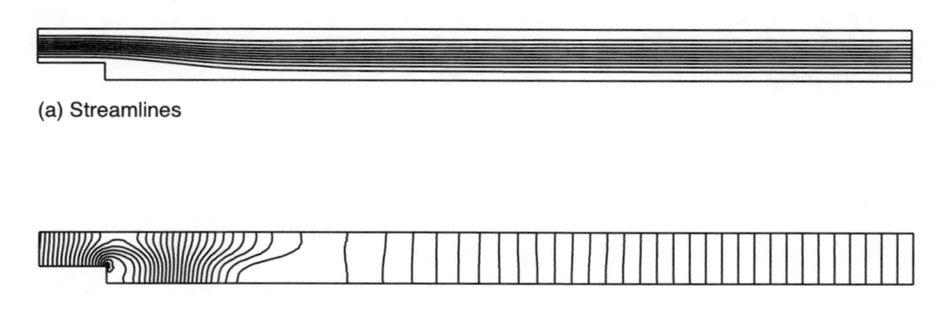

Fig. 4.11 Flow past a backward facing step. Solution with a longer domain, Re = 100.

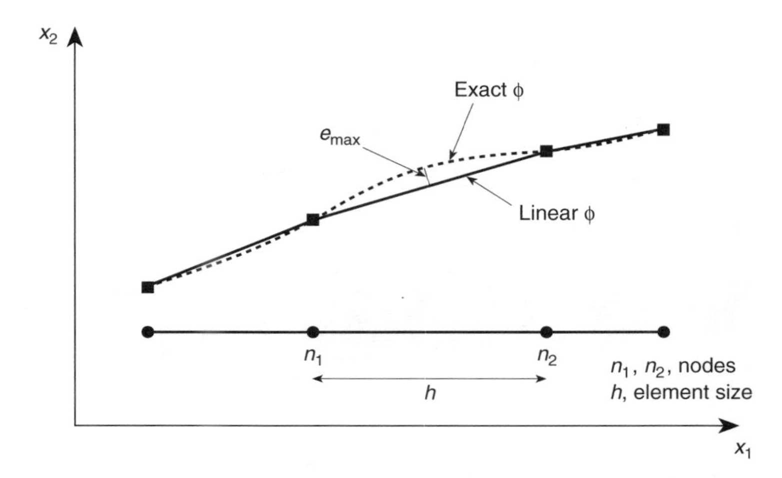

Fig. 4.12 Interpolation error in a one-dimensional problem with linear shape functions.

interpolation error as (see Fig. 4.12)

$$e_{\text{max}} = \frac{1}{8}h^2 \frac{\mathrm{d}^2\phi^h}{\mathrm{d}x^2} \tag{4.22}$$

or an RMS departure error as

$$e_{\text{RMS}} = \frac{1}{\sqrt{120}}h^2 \frac{\mathrm{d}^2\phi^h}{\mathrm{d}x^2} \tag{4.23}$$

In deducing the expressions (4.22) and (4.23), we have assumed that the nodal values of the function ϕ are exact. As we have shown in Volume 1 this is true only for some types of interpolating functions and equations. However the nodal values are always more accurate than those noted elsewhere and it would be sensible even in one-dimensional problems to strive for equal distribution of such errors. This would mean that we would now seek an element subdivision in which

$$h^2 \frac{\mathrm{d}^2\phi^h}{\mathrm{d}x^2} = C \tag{4.24}$$

To appreciate the value of the arbitrary constant C occurring in expression (4.24) we can interpret this as giving a permissible value of the limiting interpolation error and simply insisting that

$$h^2 \frac{\mathrm{d}^2\phi^h}{\mathrm{d}x^2} \leqslant e_p \tag{4.25}$$

where $e_p = C$ is the user-specified error limit.

If we consider the shape functions of ϕ to be linear then of course second derivatives are difficult quantities to determine. These are clearly zero inside every element and infinity at the element nodes in the one-dimensional case or element interfaces in two or three dimensions. Some averaging process has therefore to be used to determine the curvatures from nodally computed values. Before discussing, however, such procedures used for this, we must note the situation which will occur in two or three dimensions.

The extension to two or three dimensions is of course necessary for practical engineering problems. In two and three dimensions the second derivatives (or curvatures) are tensor valued and given as

$$\frac{\partial^2 \phi}{\partial x_i \, \partial x_j} \tag{4.26}$$

and such definitions require the determination of the *principal values* and directions. These principal directions are necessary for element elongation which is explained in the following section.

The determination of the curvatures (or second derivatives) of ϕ^h needs of course some elaboration. With linear elements (e.g. simple triangles or tetrahedra) the curvatures of ϕ^h which are interpolated as

$$\phi^h = \mathbf{N}\tilde{\boldsymbol{\phi}} \tag{4.27}$$

are zero within the elements and become infinity at element boundaries. There are two convenient methods available for the determination of curvatures of the approximate solution which are accurate and effective. Both of these follow some of the matter discussed in Chapter 14 of Volume 1 and are concerned with recovery. We shall describe them separately.

Local patch interpolation. Superconvergent values

In the first method we simply assume that the values of the function such as pressure or velocity converge at a rate which is one order higher at nodes than that achieved at other points of the element. If indeed such values are more accurate it is natural that they should be used for interpreting the curvatures and the gradients. Here the simplest way is to assume that a second-order polynomial is used to interpolate the nodal values in an element patch which uses linear elements. Such a polynomial can be applied in a least square manner to fit the values at all nodal points occurring within a patch which assembles the approximation at a particular node. For triangles this rule requires at least five elements that are assembled in a patch but this is a matter easily achieved. The procedure of determining such least squares is given fully in Chapter 14 of Volume 1 and will not be discussed here. However once a polynomial distribution of say ϕ is available then immediately the second derivatives of that function can be calculated at any point, the most convenient one being of course the point referring to the node which we require.

On occasion, as we shall see in other processes of refinement, it is not the curvature which is required but the gradient of the function. Again the maximum value of the gradient, for instance of ϕ, can easily be determined in any point of the patch and in particular at the nodes.

Second method

In this method we assume that the second derivative is interpolated in exactly the same way as the main function and write the approximation as

$$\left(\frac{\partial^2 \phi}{\partial x_i \partial x_j}\right)^* = \mathbf{N}\overline{\left(\frac{\partial^2 \phi}{\partial x_i \partial x_j}\right)^*} \tag{4.28}$$

This approximation is made to be a least square approximation to the actual distribution of curvatures, i.e.

$$\int_\Omega \mathbf{N}^{\mathrm{T}}\left[\mathbf{N}\overline{\left(\frac{\partial^2\phi}{\partial x_i \partial x_j}\right)^*} - \frac{\partial^2\phi^h}{\partial x_i \partial x_j}\right] \mathrm{d}\Omega = 0 \tag{4.29}$$

and integrating by parts to give

$$\overline{\left(\frac{\partial^2\phi}{\partial x_i \partial x_j}\right)^*} = \mathbf{M}^{-1}\left(\int_\Omega \mathbf{N}^{\mathrm{T}}\frac{\partial^2\phi^h}{\partial x_i \partial x_j}\right)\mathrm{d}\Omega = -\mathbf{M}^{-1}\left(\int_\Omega \frac{\partial \mathbf{N}^{\mathrm{T}}}{\partial x_i}\frac{\partial \mathbf{N}}{\partial x_j}\,\mathrm{d}\Omega\right)\tilde{\phi} \tag{4.30}$$

where $\mathbf{M}$ is the mass matrix given by

$$\mathbf{M} = \int_\Omega \mathbf{N}^{\mathrm{T}}\mathbf{N}\,\mathrm{d}\Omega \tag{4.31}$$

which of course can be 'lumped'.

4.5.2 Element elongation

Elongated elements are frequently introduced to deal with 'one-dimensional' phenomena such as shocks, boundary layers, etc. The first paper dealing with such elongation was presented as early as 1987 by Peraire *et al.*[22] and later by many authors for fluid mechanics and other problems.[70–73] But the possible elongation was limited by practical considerations if a general mesh of triangles was to be used. An alternative to this is to introduce a locally structured mesh in shocks and boundary layers which connects to the completely unstructured triangles. This idea has been extensively used by Hassan *et al.*,[39,53,56] Zienkiewicz and Wu[50] and Marchant *et al.*[65] in the compressible flow context. In both procedures it is necessary to establish the desired elongation of elements. Obviously in completely parallel flow phenomena no limit on elongation exists but in a general field the elongation ratio defining the maximum to minimum size of the element can be derived by considering curvatures. Thus the local error is proportional to the curvature and making h^2 times the curvature equal to a constant, we immediately derive the ratio $h_{\max}/h_{\min}$.

In Fig. 4.13, X_1 and X_2 are the directions of the minimum and maximum principal values of the curvatures. Thus for an equal distribution of the interpolation error we can write for each node†

$$h_{\min}^2\left|\frac{\partial^2\phi}{\partial X_2^2}\right| = h_{\max}^2\left|\frac{\partial^2\phi}{\partial X_1^2}\right| = C \tag{4.32}$$

which gives us the stretching ratio s as

$$s = \frac{h_{\max}}{h_{\min}} = \sqrt{\frac{\left|\dfrac{\partial^2\phi}{\partial X_2^2}\right|}{\left|\dfrac{\partial^2\phi}{\partial X_1^2}\right|}} \tag{4.33}$$

† Principal curvatures and directions can be found in a manner analogous to that of the determination of principal stresses and their directions. Procedures are described in standard engineering texts.

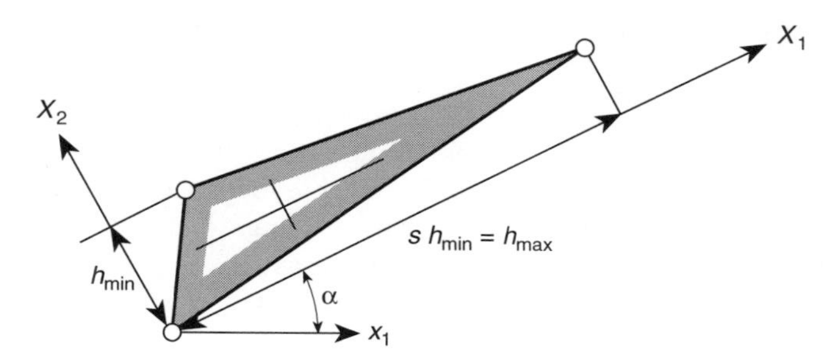

Fig. 4.13 Element elongation δ and minimum and maximum element sizes.

With the relations given above, we can formulate the following steps to adaptively refine a mesh:

1. Find the solution using an initial coarse mesh.
2. Select a suitable representative scalar variable and calculate the local maximum and minimum curvatures and directions of these at all nodes.
3. Calculate the new element sizes at all nodes from the maximum and minimum curvatures using the relation in Eq. (4.32).
4. Calculate the stretching ratio from the ratio of the calculated maximum to minimum element sizes (Eq. 4.33). If this is very high, limit it by a maximum allowable value.
5. Remesh the whole domain based on the new element size, stretching ratios and the direction of stretching.

To use the above procedure, an efficient unstructured mesh generator is essential. We normally use the advancing front technique operating on the background mesh principle[22] in most of the examples presented here.† The information from the previous solution in the form of local mesh sizes, stretching ratio and stretching direction are stored in the previous mesh and this mesh is used as a background mesh for the new mesh.

In the above steps of anisotropic mesh generation, to avoid very small and large elements (especially in compressible flows), the minimum and maximum allowable sizes of the elements are given as inputs. The maximum allowable stretching ratio is also supplied to the code to avoid bad elements in the vicinity of discontinuities. It is generally useful to know the minimum size of element used in a mesh as many flow solvers are conditionally stable. In such solvers the time step limitation depends very much on the element size.

The procedure just described for an elongated element can of course be applied for the generation of isotropic meshes simply by taking the maximum curvature at every point.

The matter to which we have not yet referred is that of suitably choosing the variable ϕ to which we will wish to assign the error. We shall come back to this matter later but it is clear that this has to be a well-representative quantity available from the choice of velocities, pressures, temperature, etc.

† Another successful unstructured mesh generator is based on Delaunay triangulation. The reader can obtain more information by consulting references 54, 59–62, 65–67, 74–85.

4.5.3 First derivative (gradient) based refinement

The nature of the fluid flow problems is elliptic in the vicinity of the boundaries often forming so-called viscous boundary layers though some distance from the boundaries the equations become almost hyperbolic. For such hyperbolic problems it is possible to express the propagation type error in terms of the gradient of the solution in the domain. In such cases the error can be considered as

$$h\frac{\partial\phi}{\partial n} = C \tag{4.34}$$

where n is the direction of maximum gradient and h is the element size (minimum size) in the same direction. The above expression can be used to determine the minimum element size at all nodes or other points of consideration in exactly the same manner as was done when using the curvature. However the question of stretching is less clear. At every point a maximum element size should be determined. One way of doing this is of course to return to the curvatures and find the curvature ratios. Another procedure to determine the maximum size of element is described by Zienkiewicz and Wu.[50] In this the curvature of the streamlines is considered and h_{max} is calculated as

$$h_{\text{max}} \leqslant \delta R \tag{4.35}$$

where R is the radius of curvature of the streamline and δ is a constant that varies between 0 and 1. Immediately the ratio between the maximum and minimum element size gives the stretching ratio.

4.5.4 Choice of variables

In both methods of mesh refinement, i.e. those following curvatures and those following gradients, a particular scalar variable needs to be chosen to define the mesh. The question of the suitable choice of the variable is an outstanding one and many authoritative procedures have been proposed. The simplest procedure is to consider only one of the many variables and here the one which is efficient is simply the absolute value of the velocity vector, i.e $|\mathbf{u}|$. Such a velocity is convenient both for problems of incompressible flow and, as we shall see later, for problems of compressible flow where local refinement is even more important than here. (Very often in compressible flows the Mach number, which in a sense measures the same quantity, has been used.)

Of course other variables can be chosen or any combination of variables such as velocities, pressures, temperatures, etc., can be used. Certainly in this chapter the absolute velocity is the most reasonable criterion. Some authors have considered using each of the problem variables to generate a new mesh.[59–62,69] However this is rather expensive and we believe velocity alone can give accurate results in most cases.

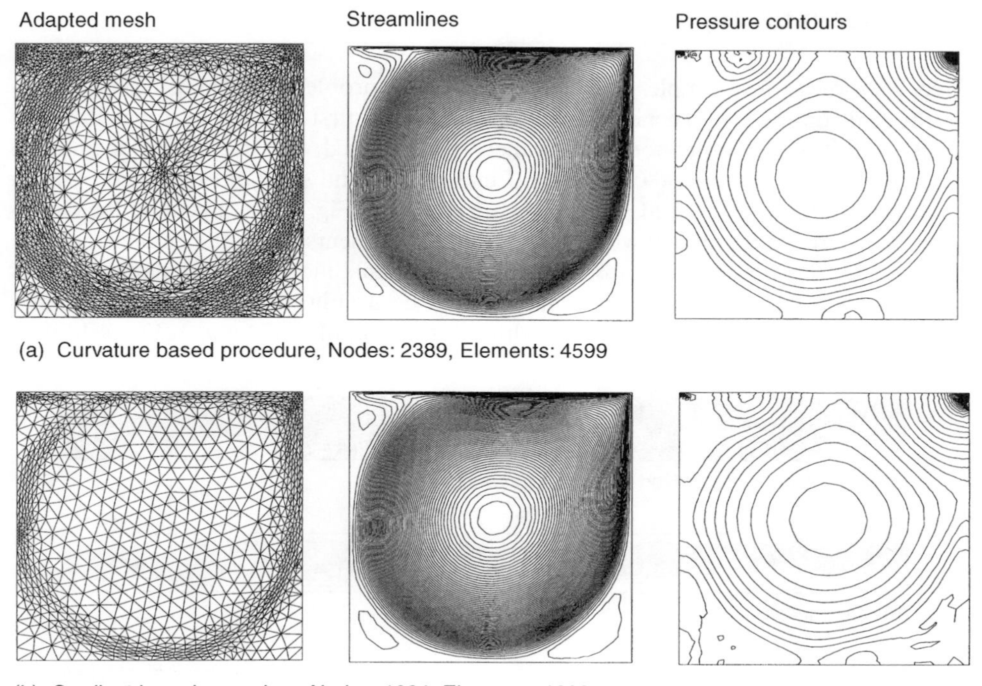

(a) Curvature based procedure, Nodes: 2389, Elements: 4599

(b) Gradient based procedure, Nodes: 1034, Elements: 1962

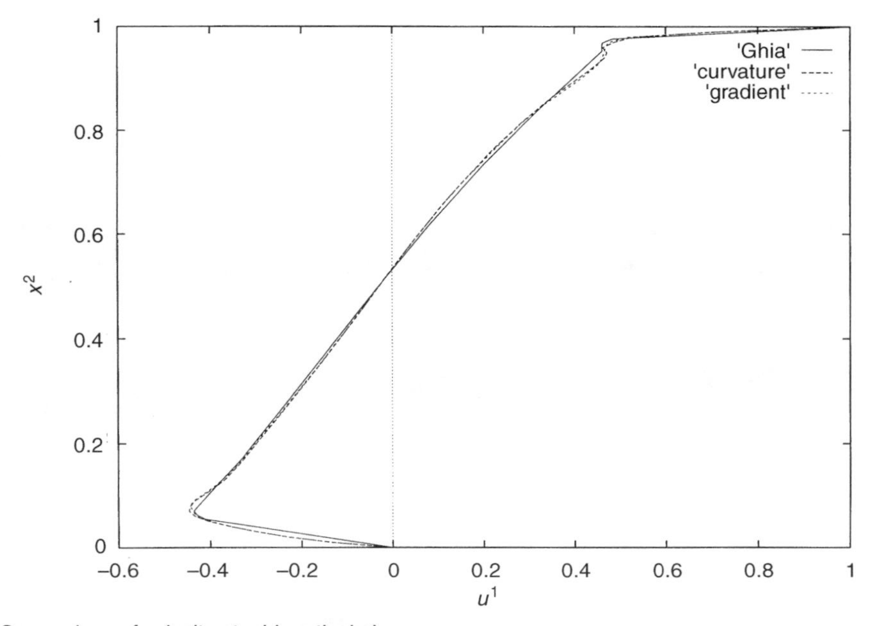

(c) Comparison of velocity at mid-vertical plane

Fig. 4.14 Lid-driven cavity, Re = 5000. Adapted meshes using curvature and gradient based refinements and solutions.

4.5.5 Some examples

Here we show some examples of incompressible flow problems solved using the above-mentioned adaptive mesh generation procedures. In the first problem of driven flow in a cavity which we have previously examined is again used. We use an initial uniform mesh with 481 nodes and 880 elements. Final meshes and solutions obtained by both curvature and gradient based procedures are shown in Fig. 4.14. In general the curvature based procedure gives a wide band of refined elements along the circulation path (Fig. 4.14a). However, the number of refined elements along the circulation path is smaller when the gradient based refinement is used (Fig. 4.14b). Both the meshes give excellent comparison with the benchmark solution of Ghia *et al.*[5] (Fig. 4.14c).

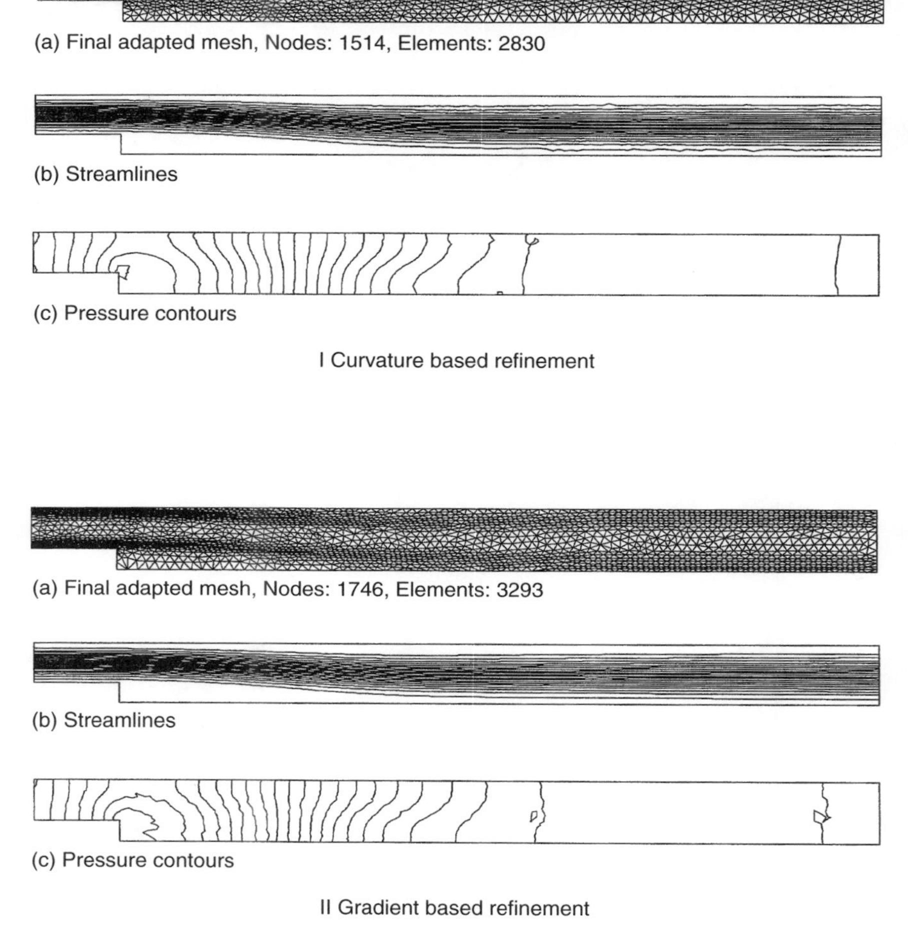

Fig. 4.15 Flow past a backward facing step, Re = 229. Adapted meshes using curvature and gradient based refinements and solutions.

A similar exercise has been carried out for the flow over a backward facing step which was also considered previously. Figure 4.15 shows the initial mesh and final meshes obtained from curvature and gradient based procedures. As we can see, a more meaningful mesh is obtained using the gradient based procedure in this case.

In the adaptive solutions shown here we have not used any absolute value of the desired error norm as the definition of a suitable norm presents certain difficulties, though of course the use of energy norm in the manner suggested in Volume 1 could be adopted. We shall use such an error requirement in some later problems.

4.6 Adaptive mesh generation for transient problems

In the preceding sections we have indicated various adaptive methods using complete mesh regeneration with error indicators of the interpolation kind. Obviously other methods of mesh refinement can be used (mesh enrichment or r refinement) and other procedures of error estimating can be employed if the problem is nearly elliptic. One such study in which the energy norm is quite effectively used is reported by Wu *et al.*[40] In that study the full transient behaviour of the *Von Karman* vortex street behind a cylinder is considered and the results are presented in Fig. 4.16.

In this problem, the mesh is regenerated at fixed time intervals using the energy norm error and the methodologies largely described in Chapter 15 of Volume 1.

Similar procedures have been used by others and the reader can refer to these works.[41,42]

4.7 Importance of stabilizing convective terms

We present here the effects of stabilizing terms introduced by the CBS algorithm at low and high Reynolds number flows. These terms are essential in compressible flow computations to suppress the oscillations. However, their effects are not clear in incompressible flow problems. To demonstrate the influence of stabilizing terms, the driven flow in a cavity is considered again for two different Reynold's numbers, 100 and 5000, respectively. Figure 4.17 shows the results obtained for these Reynolds numbers. The reader will notice only slight effects of stabilization terms at $\mathrm{Re} = 100$. However, at $\mathrm{Re} = 5000$, some oscillations in pressure in the absence of stabilization terms are noticed. These oscillations vanish in the presence of stabilization terms [namely terms proportional to Δt^2 in the momentum equations (3.23) and (3.24)]. In many problems of higher Reynold's number or compressibility the importance of stabilizing convective terms is more dramatic.

4.8 Slow flows – mixed and penalty formulations

4.8.1 Analogy with incompressible elasticity

Slow, viscous incompressible flow represents the extreme situation at the other end of the scale from the inviscid problem of Sec. 4.2. Here all dynamic (acceleration) forces

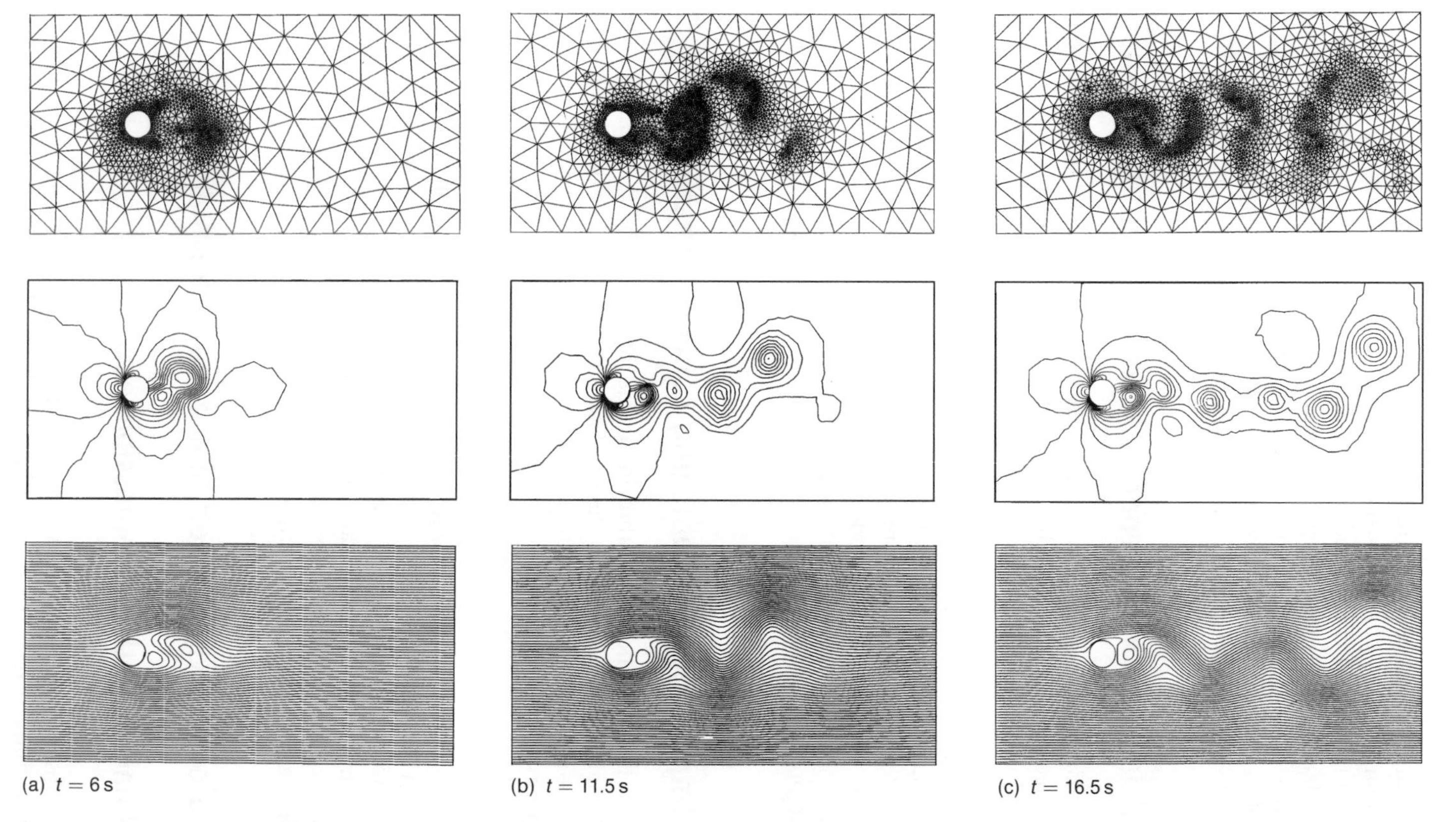

Fig. 4.16 Transient incompressible flow around a cylinder at Re = 250. Adaptively refined mesh. Pressure contours and streamlines at various times after initiation of 'vortex shedding'.

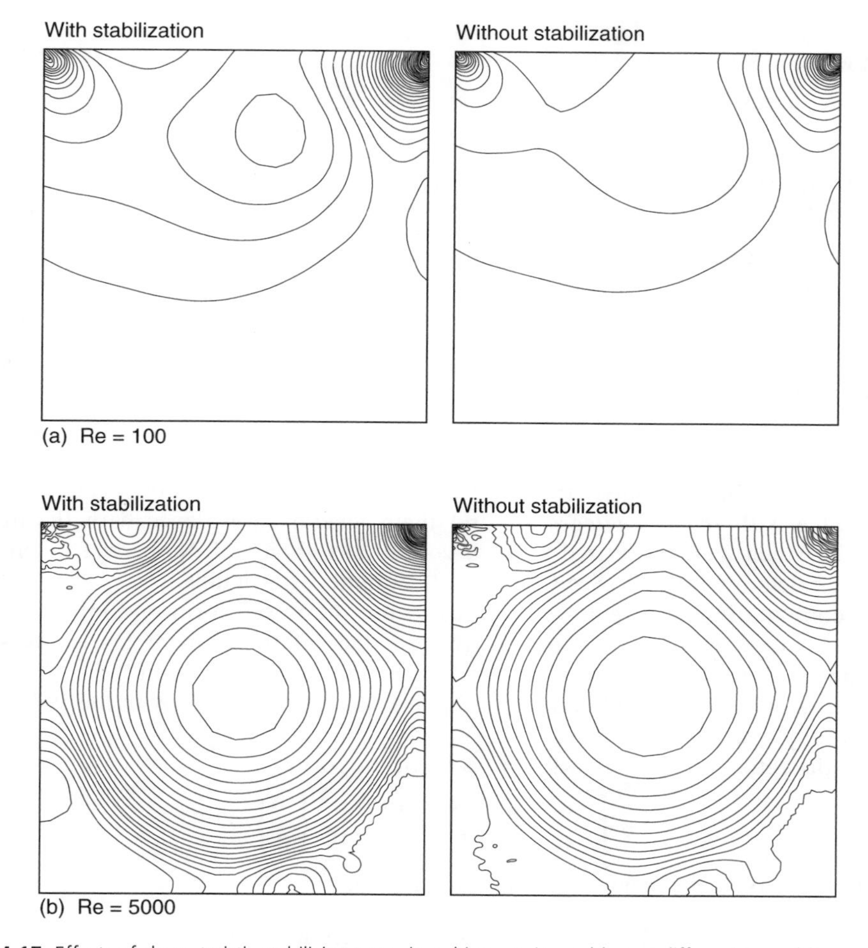

Fig. 4.17 Effects of characteristic stabilizing terms in a driven cavity problem at different Reynold's numbers.

are, *a priori*, neglected and Eqs (4.1) and (4.2) reduce, in tensorial form, to

$$\frac{\partial u_i}{\partial x_i} \equiv \dot{\varepsilon}_v = 0 \tag{4.36}$$

and

$$\frac{\partial \tau_{ij}}{\partial x_j} - \frac{\partial p}{\partial x_i} - \rho g_i = 0 \tag{4.37}$$

The above are completed of course by the constitutive relation

$$\tau_{ij} = \mu\left(\frac{\partial u_i}{\partial x_j} + \frac{\partial u_j}{\partial x_i} - \delta_{ij}\frac{2}{3}\frac{\partial u_k}{\partial x_k}\right) \tag{4.38}$$

which is identical to the problem of incompressible elasticity in which we replace:

(a) the displacements by velocities,

(b) the shear modulus G by the viscosity μ and
(c) the mean stress by negative pressure.

We have discussed such equations in Chapters 1 and 3.

4.8.2 Mixed and penalty discretization

The discretization can be started from the *mixed form* with independent approximations of $\mathbf{u}$ and p, i.e.

$$\mathbf{u} = \mathbf{N}_u \tilde{\mathbf{u}} \qquad p = \mathbf{N}_p \tilde{\mathbf{p}} \tag{4.39}$$

or by a penalty form in which Eq. (4.36) is augmented by p/γ where γ is a large penalty parameter

$$\mathbf{m}^{\mathrm{T}} \mathbf{S} \mathbf{u} + \frac{p}{\gamma} = 0 \tag{4.40}$$

allowing p to be eliminated from the computation. Such penalty forms are only applicable with *reduced integration* and their general equivalence with the mixed form in which p is discretized by a discontinuous choice of $\mathbf{N}_p$ between elements has been demonstrated.[86] (See Chapter 12, Volume 1 for details.)

As computationally it is advantageous to use the mixed form and introduce the penalty parameter only to eliminate the p values at the element levels, we shall presume such penalization to be done after the mixed discretization.

The use of penalty forms in fluid mechanics was introduced early in the 1970s[87–89] and is fully discussed elsewhere.[90–92]

The discretized equations will always be of the form

$$\begin{bmatrix} \mathbf{K} & -\mathbf{G} \\ -\mathbf{G}^{\mathrm{T}} & -h^2 \mathbf{I}/\gamma \end{bmatrix} \begin{Bmatrix} \tilde{\mathbf{u}} \\ \tilde{\mathbf{p}} \end{Bmatrix} = \begin{Bmatrix} \bar{\mathbf{f}} \\ \mathbf{0} \end{Bmatrix} \tag{4.41}$$

where h is a typical element size, $\mathbf{I}$ an identity matrix,

$$\begin{aligned} \mathbf{K} &= \int_\Omega \mathbf{B}^{\mathrm{T}} \mu \mathbf{I}_0 \mathbf{B} \, \mathrm{d}\Omega \qquad \text{where } \mathbf{B} \equiv \mathbf{S}\mathbf{N}_u \\ \mathbf{G} &= \int_\Omega (\nabla \mathbf{N}_u)^{\mathrm{T}} \mathbf{N}_p \, \mathrm{d}\Omega \\ \bar{\mathbf{f}} &= \int_\Omega \mathbf{N}_u^{\mathrm{T}} \rho \mathbf{g} \, \mathrm{d}\Omega + \int_{\Gamma_t} \mathbf{N}_u^{\mathrm{T}} \bar{\mathbf{t}} \, \mathrm{d}\Gamma \end{aligned} \tag{4.42}$$

and the penalty number, γ, is introduced purely as a numerical convenience. This is taken generally as[90,92]

$$\gamma = (10^7\text{–}10^8)\mu$$

There is little more to be said about the solution procedures for creeping incompressible flow with constant viscosity. The range of applicability is of course limited to low velocities of flow or high viscosity fluids such as oil, blood in biomechanics applications, etc. It is, however, important to recall here that the mixed form allows only certain combinations of $\mathbf{N}_u$ and $\mathbf{N}_p$ interpolations to be used without violating the convergence conditions. This is discussed in detail in Chapter 12 of Volume 1, but for completeness Fig. 4.18 lists some of the available elements together

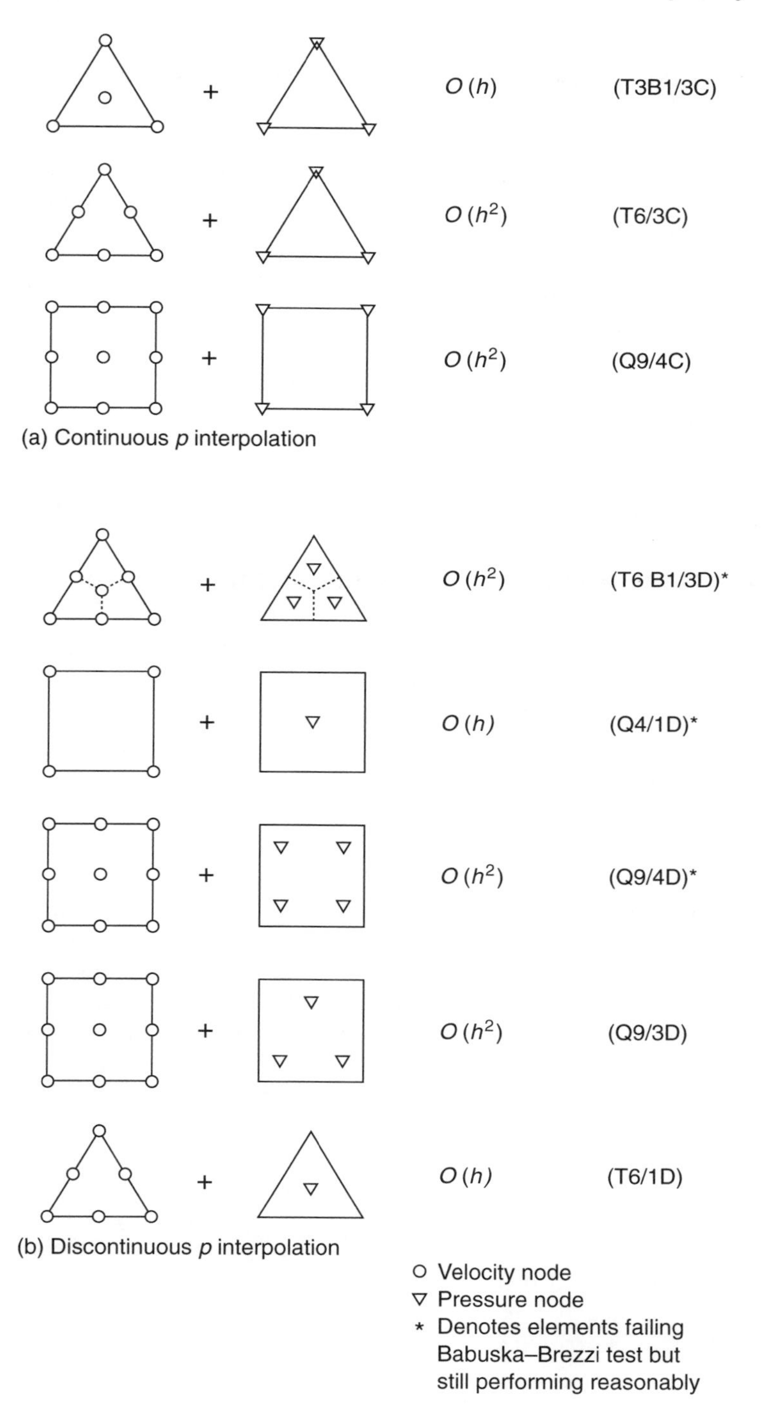

Fig. 4.18 Some useful velocity–pressure interpolations and their asymptotic, energy norm convergence rates.

with their asymptotic convergence rates.[93] Many other elements useful in fluid mechanics are documented elsewhere,[94–96] but those of proven performance are given in the table.

It is of general interest to note that frequently elements with C_0 continuous pressure interpolations are used in fluid mechanics and indeed that their performance is generally superior to those with discontinuous pressure interpolation on a given mesh, even though the cost of solution is marginally greater.

It is important to note that the recommendations concerning the element types for the Stokes problem carry over unchanged to situations in which dynamic terms are of importance.

The fairly obvious extension of the use of incompressible elastic codes to Stokes flow is undoubtedly the reason why the first finite element solutions of fluid mechanics were applied in this area.

4.9 Non-newtonian flows – metal and polymer forming

4.9.1 Non-newtonian flows including viscoplasticity and plasticity

In many fluids the viscosity, though isotropic, may be dependent on the rate of strain $\dot{\varepsilon}_{ij}$ as well as on the state variables such as temperature or total deformation. Typical here is, for instance, the behaviour of many polymers, hot metals, etc., where an exponential law of the type

$$\mu = \mu_0 \dot{\bar{\varepsilon}}^{(m-1)} \tag{4.43}$$

with

$$\mu_0 = \mu_0(T, \bar{\varepsilon})$$

governs the viscosity–strain rate dependence where m is a physical constant. In the above $\dot{\bar{\varepsilon}}$ is the second invariant of the deviatoric strain rate tensor defined from Eq. (3.34), T is the (absolute) temperature and $\bar{\varepsilon}$ is the total strain invariant.

This *secant* viscosity can of course be obtained by plotting the relation between the deviatoric stresses and deviatoric strains or their invariants, as Eq. (3.33) simply defines the viscosity by the appropriate ratio of the stress to strain rate. Such plots are shown in Fig. 4.19 where $\bar{\sigma}$ denotes the second deviatoric stress invariant. The above exponential relation of Eq. (4.43) is known as the Oswald de Wahle law and is illustrated in Fig. 4.19(b).

In a similar manner viscosity laws can be found for viscoplastic and indeed purely plastic behaviour of an incompressible kind. For instance, in Fig. 4.19(c) we show a viscoplastic Bingham fluid in which a threshold or yield value of the second stress invariant has to be exceeded before any strain rate is observed. Thus for the visco-plastic fluid illustrated it is evident that a highly non-linear viscosity relation is obtained. This can be written as

$$\mu = \frac{\bar{\sigma}_y + \gamma \dot{\bar{\varepsilon}}^m}{\dot{\bar{\varepsilon}}} \tag{4.44}$$

where $\bar{\sigma}_y$ is the value of the second stress invariant at yield.

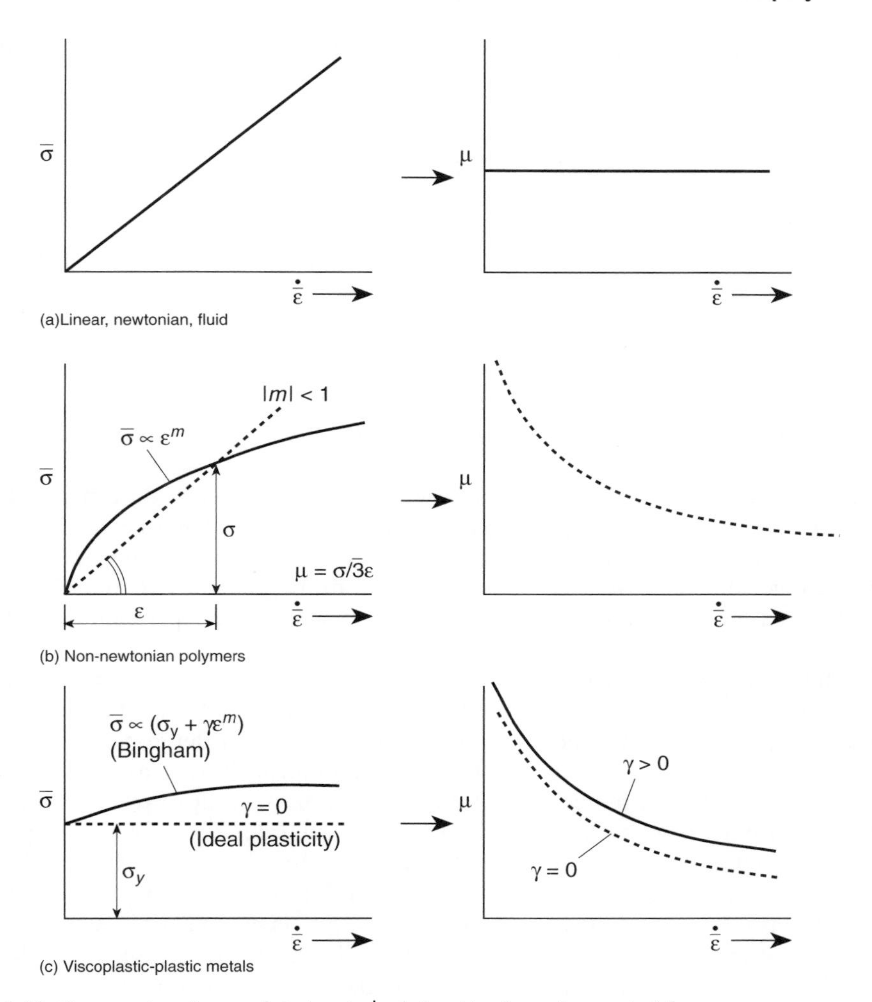

Fig. 4.19 Stress $\bar{\sigma}$, viscosity μ and strain rate $\dot{\bar{\varepsilon}}$ relationships for various materials.

The special case of pure plasticity follows of course as a limiting case with the fluidity parameter $\gamma = 0$, and now we have simply

$$\mu = \frac{\bar{\sigma}_y}{\dot{\bar{\varepsilon}}} \tag{4.45}$$

Of course, once again $\bar{\sigma}_y$ can be dependent on the *state* of the fluid, i.e.

$$\bar{\sigma}_y = \bar{\sigma}_y(T, \bar{\varepsilon}) \tag{4.46}$$

The solutions (at a given state of the fluid) can be obtained by various iterative procedures, noting that Eq. (4.41) continues to be valid but now with the matrix $\mathbf{K}$ being dependent on viscosity, i.e.

$$\mathbf{K} = \mathbf{K}(\mu) = \mathbf{K}(\dot{\bar{\varepsilon}}) = \mathbf{K}(\mathbf{u}) \tag{4.47}$$

thus being dependent on the solution.

The total iteration process can be used simply here (see Volume 2). Thus rewriting Eq. (4.41) as

$$\mathbf{A}\begin{Bmatrix} \tilde{\mathbf{u}} \\ \tilde{\mathbf{p}} \end{Bmatrix} = \begin{Bmatrix} \bar{\mathbf{f}} \\ \mathbf{0} \end{Bmatrix} \tag{4.48}$$

and noting that

$$\mathbf{A} = \mathbf{A}(\tilde{\mathbf{u}}, \tilde{\mathbf{p}})$$

we can write

$$\begin{Bmatrix} \tilde{\mathbf{u}} \\ \tilde{\mathbf{p}} \end{Bmatrix}^{i+1} = \mathbf{A}_i^{-1}\begin{Bmatrix} \bar{\mathbf{f}} \\ \mathbf{0} \end{Bmatrix} \qquad \mathbf{A}_i = \mathbf{A}(\tilde{\mathbf{u}}, \tilde{\mathbf{p}})^i \tag{4.49}$$

Starting with an arbitrary value of μ we repeat the solution until convergence is obtained.

Such an iterative process converges rapidly (even when, as in pure plasticity, μ can vary from zero to infinity), providing that the forcing $\bar{\mathbf{f}}$ is due to *prescribed boundary velocities* and thus immediately confines the variation of all velocities in a narrow range. In such cases, five to seven iterations are generally required to bring the difference of the ith and $(i+1)$th solutions to within the 1 per cent (euclidian) norm.

The first non-newtonian flow solutions were applied to polymers and to hot metals in the early 1970s.[97–99] Application of the same procedures to the forming of metals was introduced at the same time and has subsequently been widely developed.[89,100–127]

It is perhaps difficult to visualize steel or aluminium behaving as a fluid, being conditioned to use these materials as structural members. If, however, we note that during the forming process the elastic strains are of the order of 10^{-6} while the plastic strain can reach or exceed a value of unity, neglect of the former (which is implied in the viscosity definition) seems justifiable. This is indeed borne out by comparison of computations based on what we now call *flow formulation* with elastoplastic computation or experiment. The process has alternatively been introduced as a 'rigid-plastic' form,[105,106] though such modelling is more complex and less descriptive.

Today the methodology is widely accepted for the solution of metal and polymer forming processes, and only a limited selection of references of application can be cited. The reader would do well to consult references 115, 128, 129 for a complete survey of the field.

4.9.2 Steady-state problems of forming

Two categories of problems arise in forming situations. *Steady-state flow* is the first of these. In this, a real, continuing, flow is modelled, as shown in Fig. 4.20(a) and here velocity and other properties can be assumed to be fixed in a particular point of space. In Fig. 4.20(b) the more usual *transient* processes of forming are illustrated and we shall deal with these later. In a typical steady-state problem if the state parameters T and $\bar{\varepsilon}$ defining the temperature and viscosity are known in the whole field, the solution can be carried out in the manner previously described. We could, for instance, assume that the 'viscous' flow of the problem of Fig. 4.21 is that of an ideally plastic material under isothermal conditions modelling an extrusion process and obtain the solution shown in Table 4.1. For such a material exact extrusion forces can be calculated[130] and the table shows the errors obtained with the flow formulation using different triangular elements of Fig. 4.18 and two meshes.[93] The fine mesh here was arrived at using error estimates and a single adaptive remeshing.

In general the problem of steady-state flow is accompanied by the evolution of temperature (and other state parameters such as the total strain invariant $\bar{\varepsilon}$) and here it is necessary to couple the solution with the heat balance and possibly other evolution equations. The evolution of heat has already been discussed and the appropriate conservation equations such as Eq. (4.6). It is convenient now to rewrite this equation in a modified form.

Firstly, we note that the kinetic energy is generally negligible in the problems considered and that with a constant specific heat c per unit volume we can write

$$\rho E \approx \rho e = \hat{c}T \tag{4.50a}$$

where $\hat{c}$ is the specific heat. Secondly, we observe that the internal work dissipation

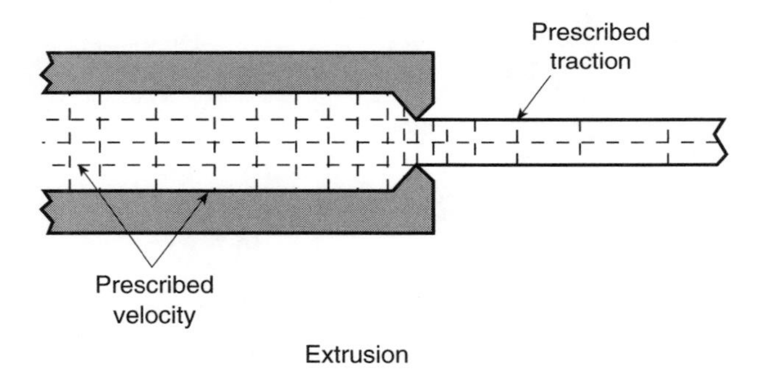

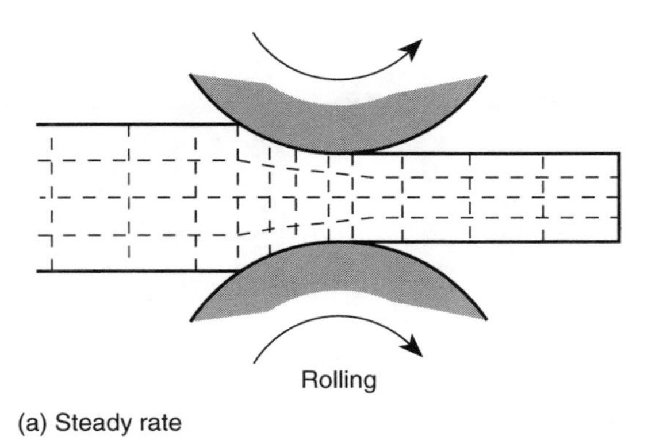

Fig. 4.20 Forming processes typically used in manufacture.

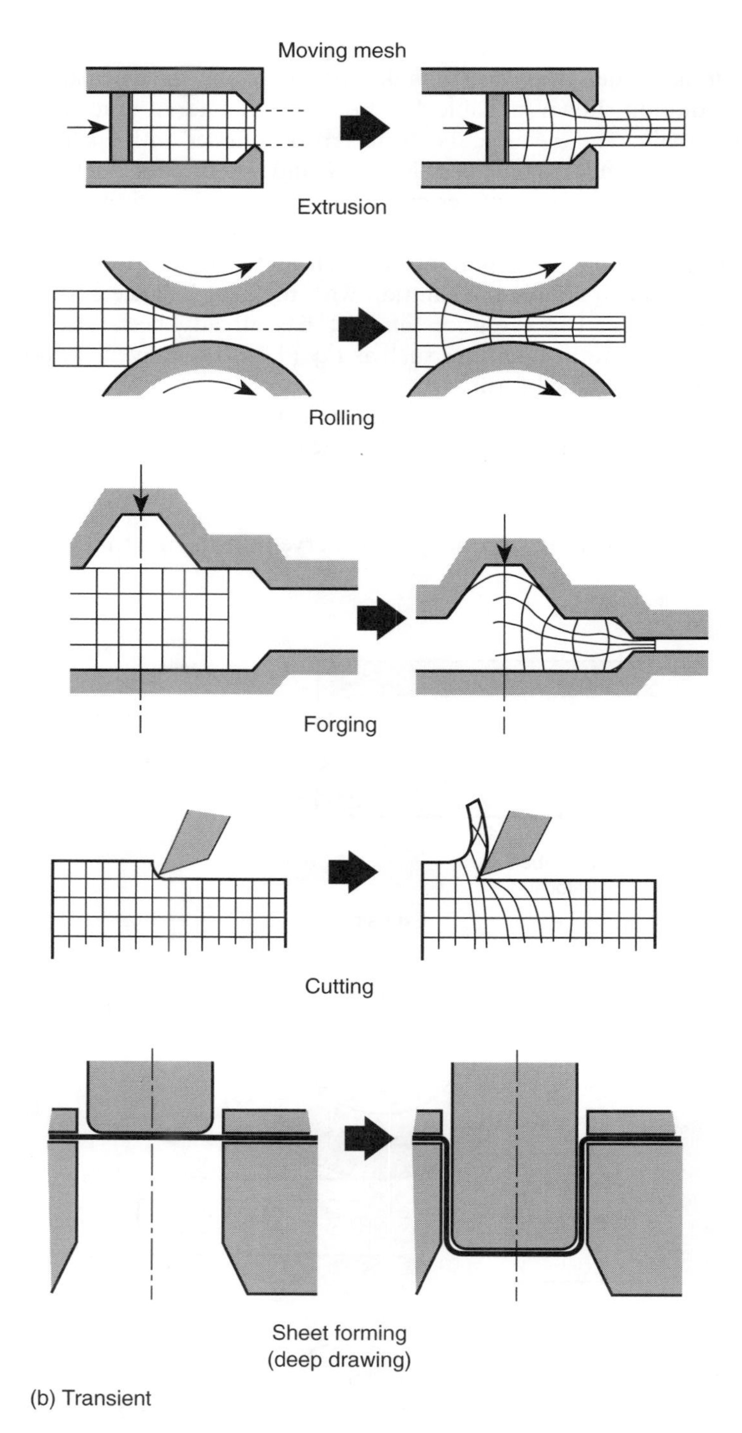

Fig. 4.20 Continued.

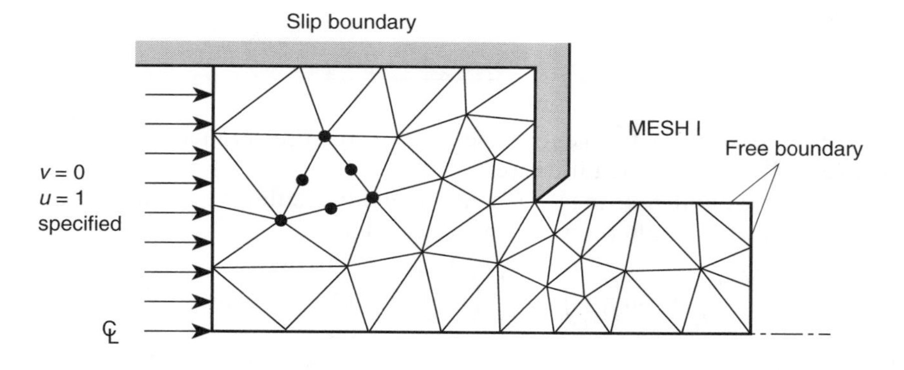

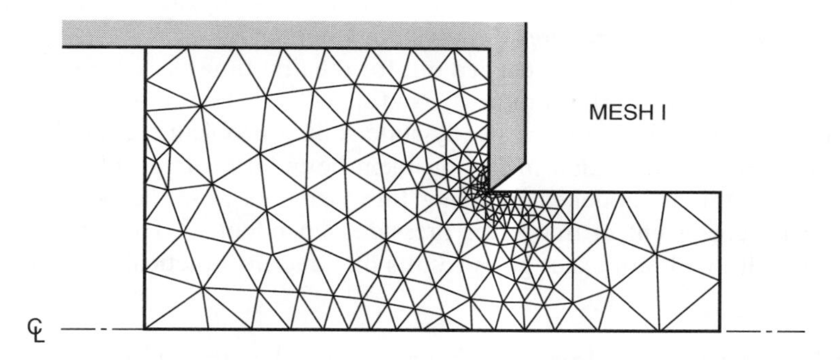

Fig. 4.21 Plane strain extrusion (extrusion ratio 2 : 1) with ideal plasticity assumed.

can be rewritten by the identity

$$\frac{\partial}{\partial x_i}(pu_i) - \frac{\partial}{\partial x_j}(\tau_{ji}u_i) \equiv -\sigma_{ji}\dot{\varepsilon}_{ji} \tag{4.50b}$$

where, by Eq. (1.9),

$$\sigma_{ji} = \tau_{ji} - \delta_{ji}p \tag{4.50c}$$

Table 4.1 Comparisons of performance of several triangular mixed elements of Fig. 4.21 in a plane extrusion problem (ideal plasticity assumed)[93]

	Mesh 1 (coarse)			Mesh 2 (fine)		
Element type	Ext. force	Force error %	CPU(s)	Ext. force	Force error %	CPU(s)
T6/1D	28 901.0	12.02	67.81	25 990.0	0.73	579.71
T6B1/3D	31 043.0	20.32	75.76	26 258.0	1.78	780.13
T6B1/3D*	29 031.0	12.52	73.08	26 229.0	1.66	613.92
T6/3C	27 902.5	8.15	87.62	25 975.0	0.67	855.38
Exact	25 800.0	0.00	—	25 800.0	0.00	—

and, by Eq. (1.2),

$$\dot{\varepsilon}_{ji} = \frac{\partial u_j/\partial x_i + \partial u_i/\partial x_j}{2} \tag{4.50d}$$

We note in passing that in general the effect of the pressure term in Eqs (4.50) is negligible and can be omitted if desired.

Using the above and inserting the incompressibility relation we can write the energy conservation as (for an alternative form see Eq. 4.6)

$$\left(\hat{c}\frac{\partial T}{\partial t} + \hat{c}u_i\frac{\partial T}{\partial x_i}\right) - \frac{\partial}{\partial x_i}\left(k\frac{\partial T}{\partial x_i}\right) - (\sigma_{ij}\dot{\varepsilon}_{ij} + \rho g_i u_i) = 0 \tag{4.51}$$

The solution of the coupled problem can be carried out iteratively. Here the term in the last bracket can be evaluated repeatedly from the known velocities and stresses from the flow solution. We note that the first bracketed term represents a total derivative of the convective kind which, even in the steady state, requires the use of the special weighting procedures discussed in Chapter 2.

Such coupled solutions were carried out for the first time as early as 1973 and later in 1978,[102,103] but are today practised routinely.[109,110] Figure 4.22 shows a typical thermally coupled solution for a steady-state rolling problem from reference 103.

It is of interest to note that in this problem boundary friction plays an important role and that this is modelled by using thin elements near the boundary, making the viscosity coefficient in that layer pressure dependent.[116] This procedure is very simple and although not exact gives results of sufficient practical accuracy.

4.9.3 Transient problems with changing boundaries

These represent the second, probably larger, category of forming problems. Typical examples here are those of forging, indentation, etc., and again thermal coupling can be included if necessary. Figures 4.23 and 4.24 illustrate typical applications.

The solution for velocities and internal stresses can be readily accomplished at a given configuration providing the temperatures and other state variables are known at that instant. This allows the new configuration to be obtained both for the boundaries and for the mesh by writing explicitly

$$\Delta x_i = u_i \Delta t \tag{4.52}$$

as the incremental relation.

If thermal coupling is important increments of temperature need also to be evaluated. However, we note that for convected coordinates Eq. (4.51) is simplified as the convected terms disappear. We can now write

$$\hat{c}\frac{\partial T}{\partial t} - \frac{\partial}{\partial x_i}\left(k\frac{\partial T}{\partial x_i}\right) - (\sigma_{ij}\dot{\varepsilon}_{ij} + \rho g_i u_i + g_h) = 0 \tag{4.53}$$

where the last term is the heat input known at the start of the interval and computation of temperature increments is made using either explicit or implicit procedures discussed in Chapter 3.

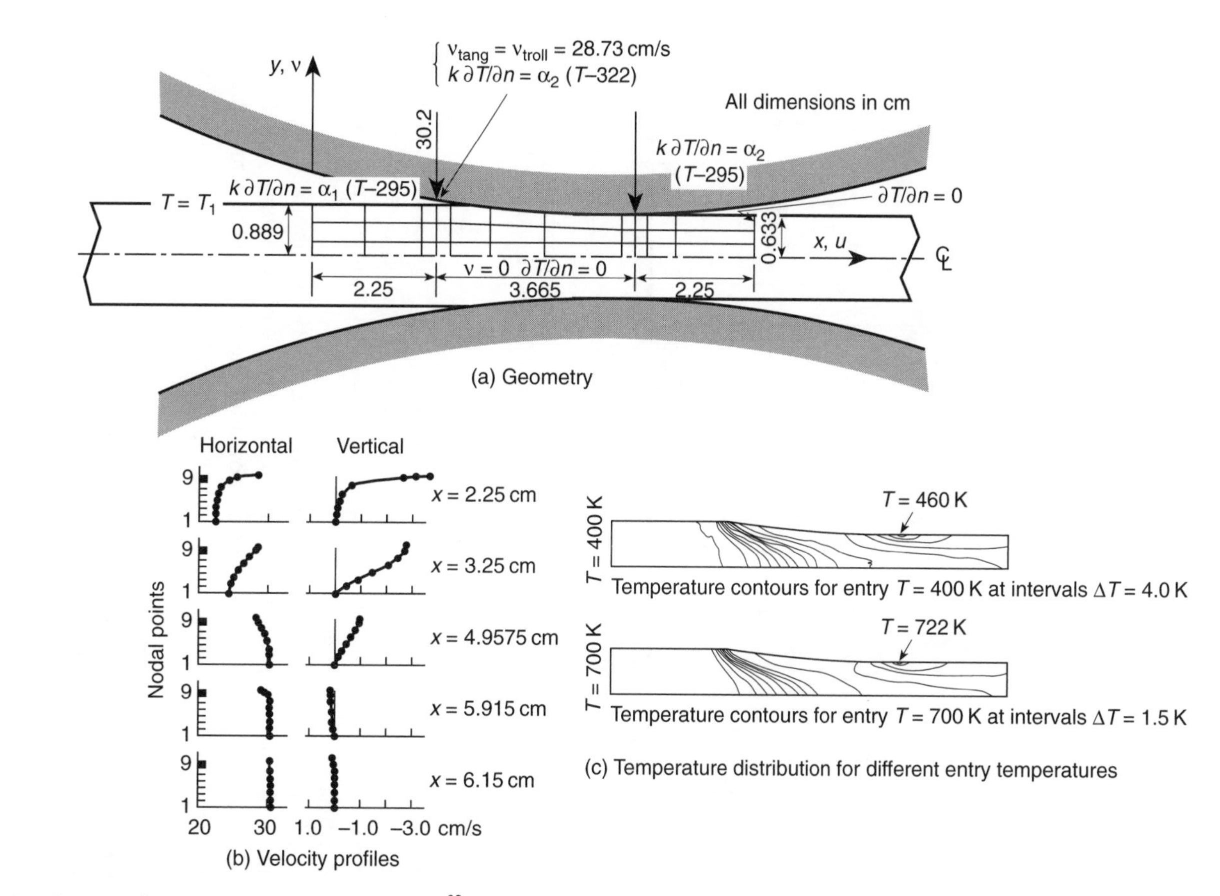

Fig. 4.22 Steady-state rolling process with thermal coupling.[93]

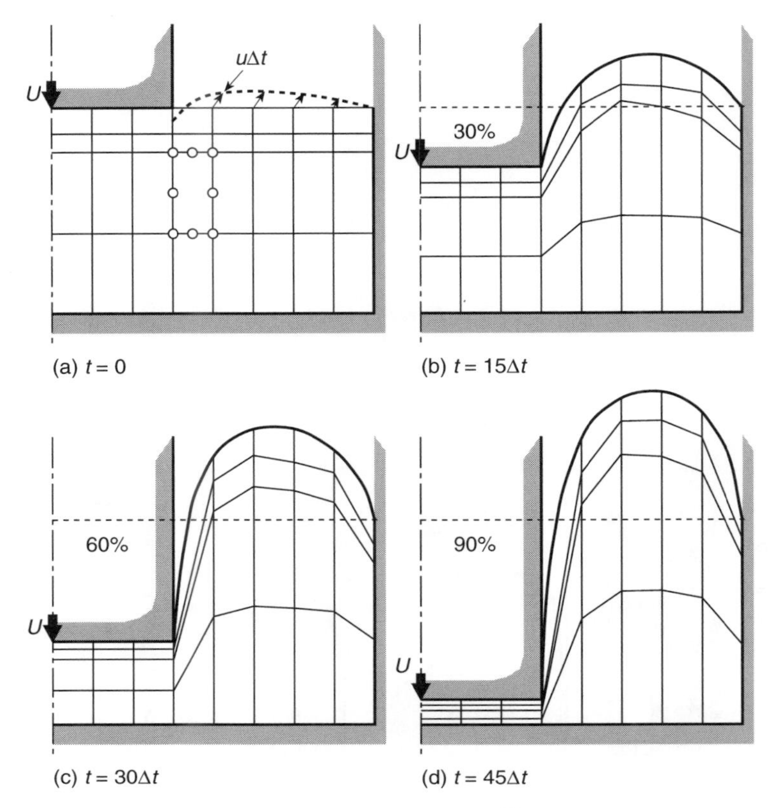

Fig. 4.23 Punch indentation problem (penalty function approach).[89] Updated mesh and surface profile with 24 isoparametric elements. Ideally plastic material; (a), (b), (c) and (d) show various depths of indentation (reduced integration is used here).

Indeed, both the coordinate and thermal updating can make use iteratively of the solution on the updated mesh to increase accuracy. However, it must be noted that any continuous mesh updating will soon lead to unacceptable meshes and some form of remeshing is necessary.

In the example of Fig. 4.23,[89] in which ideal plasticity was assumed together with isothermal behaviour, it is necessary only to keep track of boundary movements. As temperature and other state variables do not enter the problem the remeshing can be done simply – in the case shown by keeping the same vertical lines for the mesh position.

However, in the example of Fig. 4.24 showing a more realistic problem,[131,132] when a new mesh is created an interpolation of all the state parameters from the old to the new mesh positions is necessary. In such problems it is worthwhile to strive to obtain discretization errors within specified bounds and to remesh adaptively when these errors are too large.

We have discussed the problem of adaptive remeshing for linear problems in Chapter 15 of Volume 1. In the present examples similar methods have been adopted with success[133,134] and in Fig. 4.24 we show how remeshing proceeds during the

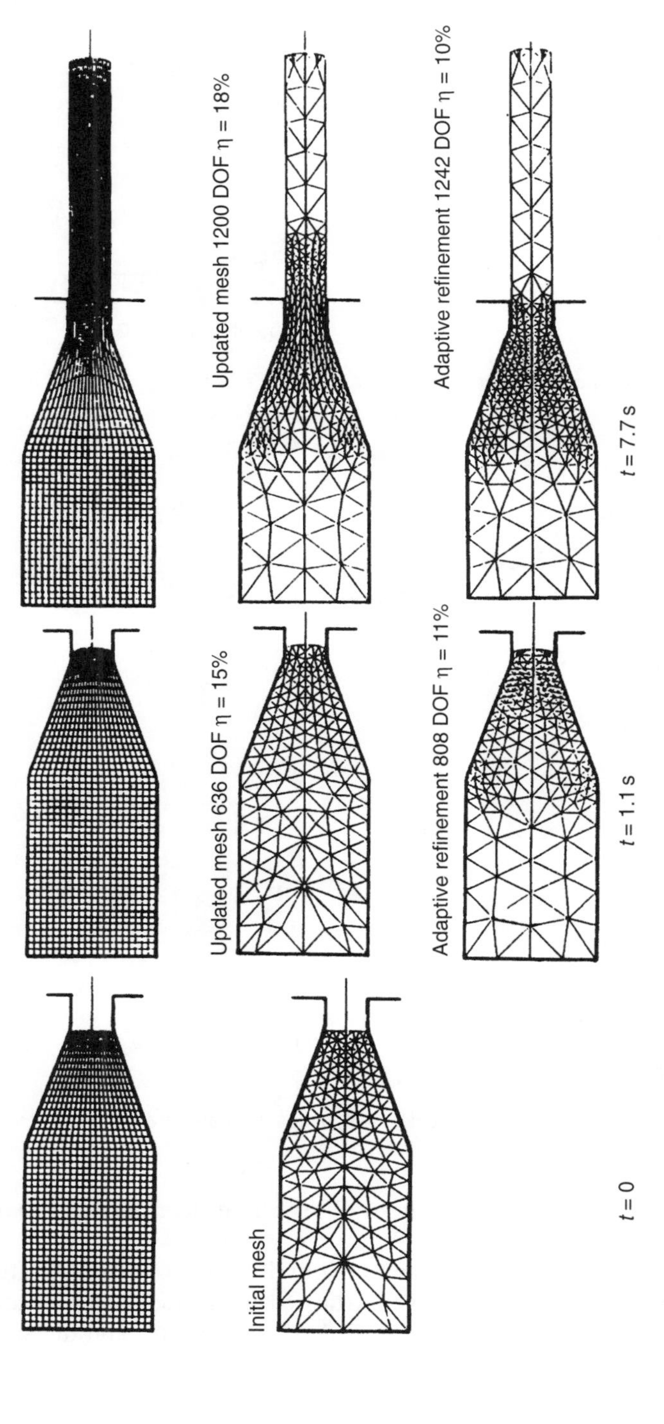

Fig. 4.24 (a) A material grid and updated and adapted meshes with material deformation (η percentage in energy norm).

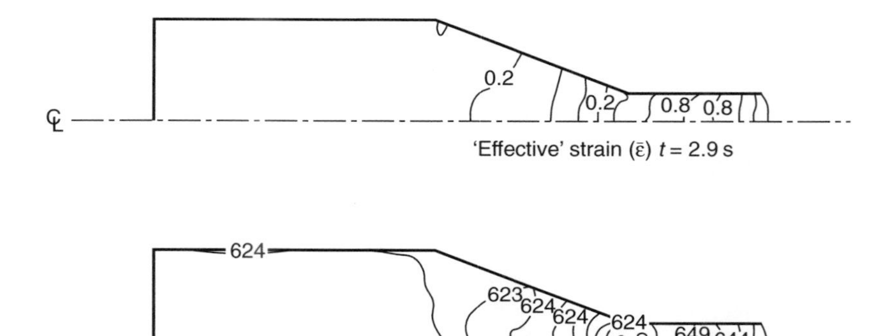

(b) Contours of state parameters at $t = 2.9$ s

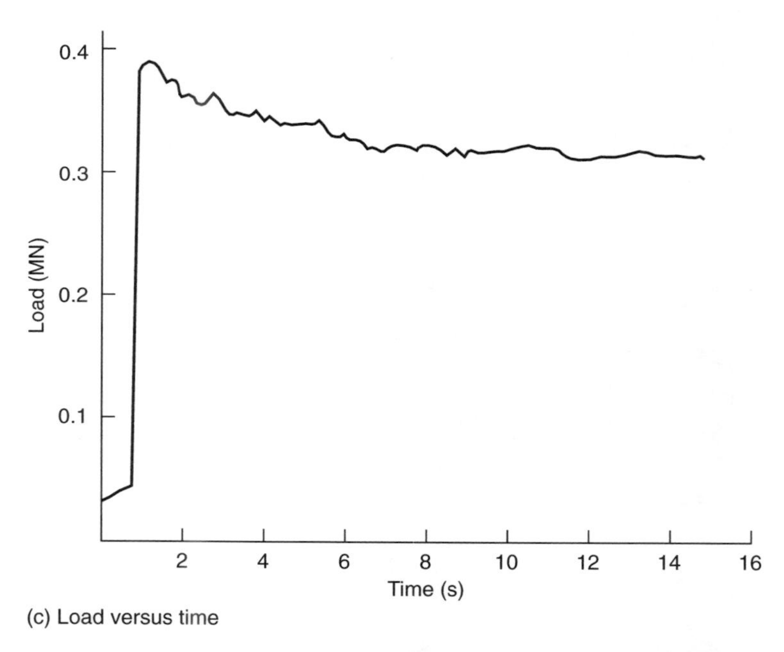

(c) Load versus time

Fig. 4.24 (continued) A transient extrusion problem with temperature and strain-dependent yield.[132] Adaptive mesh refinements uses T6/1D elements of Fig. 4.18.

forming process. It is of interest simply to observe that here the *energy norm* of the error is the measure used.

The details of various applications can be found in the extensive literature on the subject. This also deals with various sophisticated mesh updating procedures. One particularly successful method is the so-called ALE (arbitrary lagrangian–eulerian) method.[135–139] Here the original mesh is given some prescribed velocity $\bar{\mathbf{v}}$ in a manner fitting the moving boundaries, and the convective terms in the equations

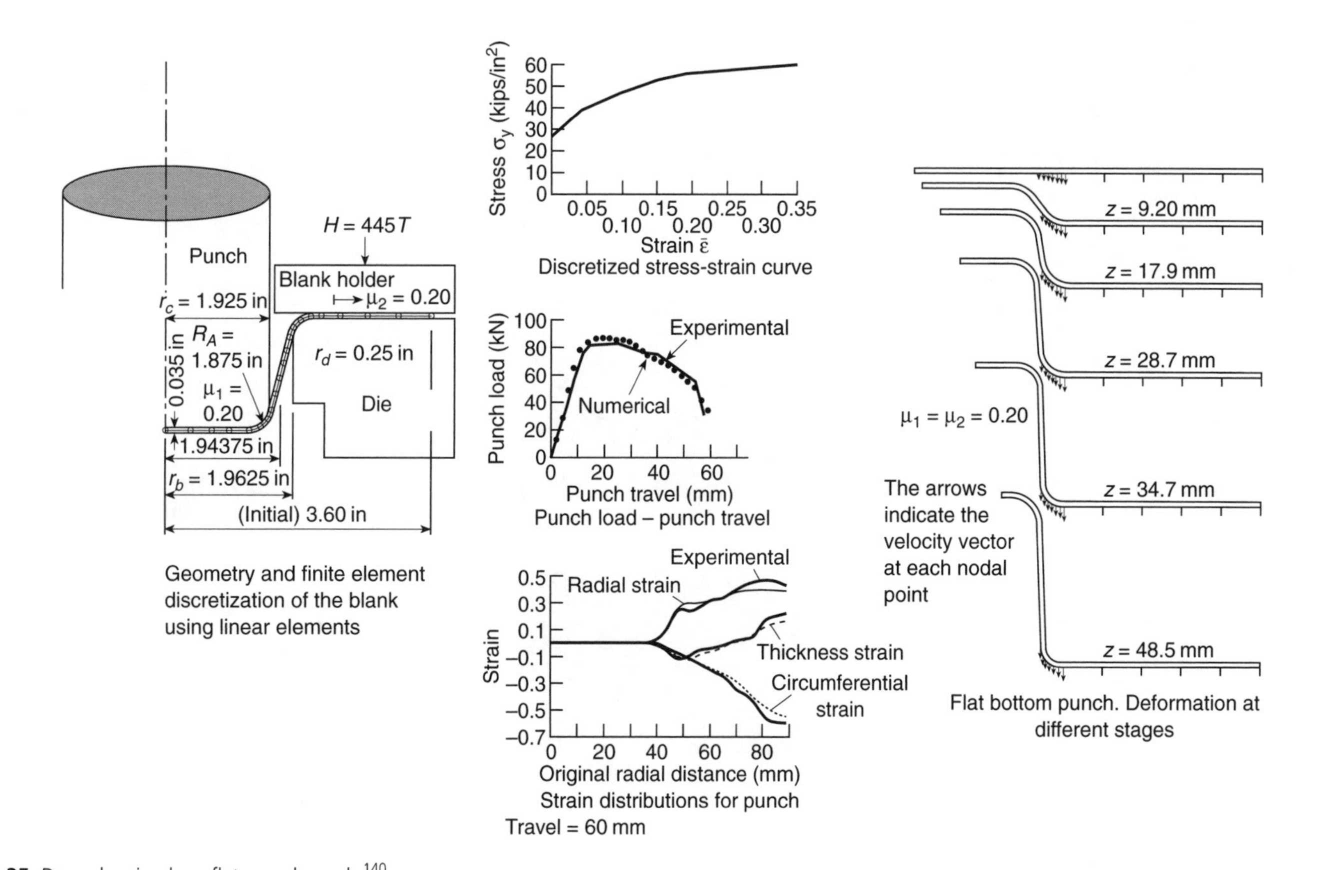

Fig. 4.25 Deep drawing by a flat nosed punch.[140]

are retained with reference to this velocity. In Eq. (4.52), for instance, in place of

$$\hat{c}u_i \frac{\partial T}{\partial x_i} \qquad \text{we write} \qquad \hat{c}(u_i - \bar{v}_i)\frac{\partial T}{\partial x_i}$$

etc., and the solution can proceed in a manner similar to that of steady state (with convection disappearing of course when $\bar{v}_i = u_i$; i.e. in the pure updating process).

It is of interest to observe that the flow methods can equally well be applied to the forming of thin sections resembling shells. Here of course all the assumptions of shell theory and corresponding finite element technology are applicable. Because of this, incompressibility constraints are no longer a problem but other complications arise. The literature of such applications is large, but much relevant information can be found in references 140–153. Practical applications ranging from the forming of beer cans to car bodies abound. Figures 4.25 and 4.26 illustrate some typical problems.

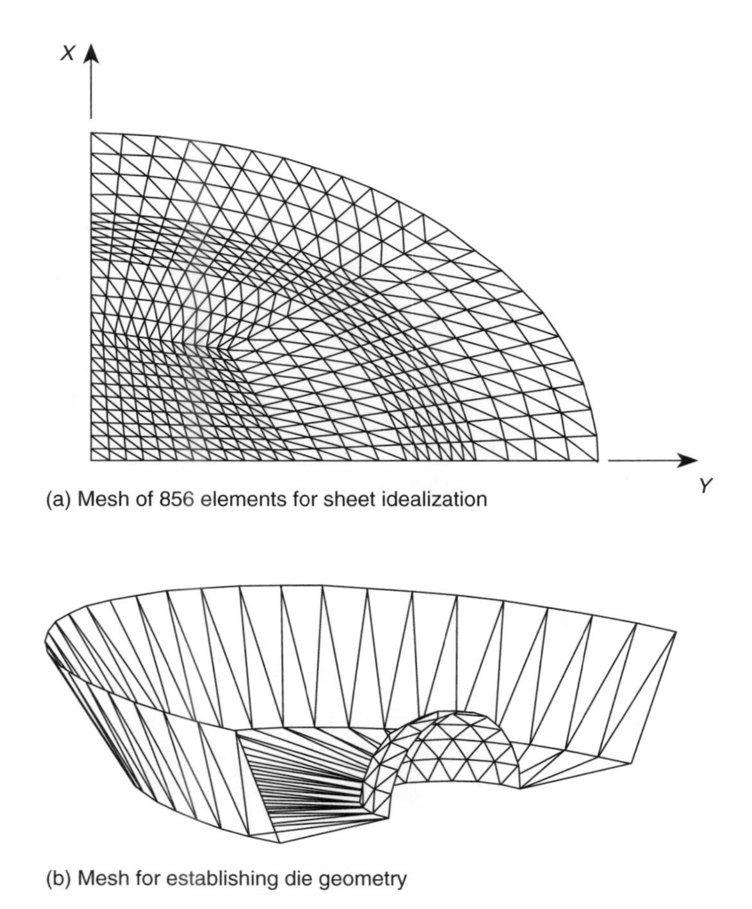

Fig. 4.26 Finite element simulation of the superplastic forming of a thin sheet component by air pressure application. This example considers the superplastic forming of a truncated ellipsoid with a spherical indent. The original flat blank was 150 × 100 mm. The truncated ellipsoid is 20 mm deep. The original thickness was 1 mm. Minimum final thickness was 0.53 mm; 69 time steps were used with a total of 285 Newton–Raphson iterations (complete equation solutions).[143]

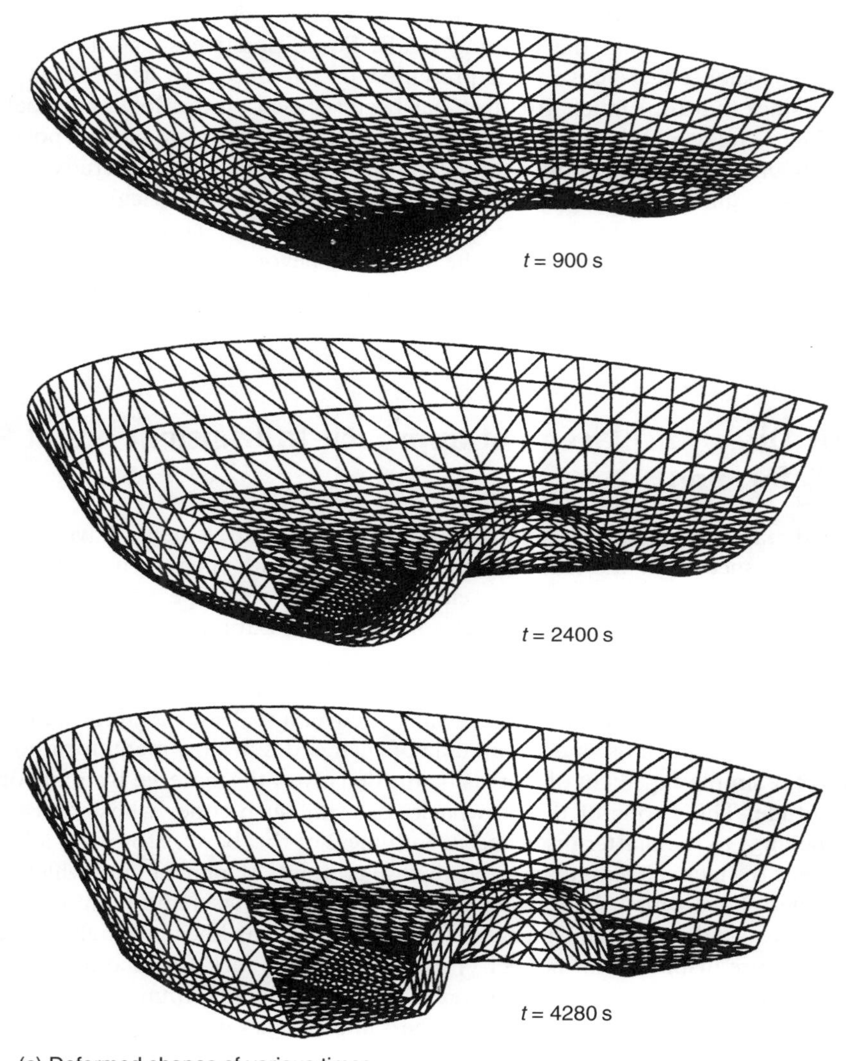

Fig. 4.26 Continued.

4.9.4 Elastic springback and viscoelastic fluids

In Sec. 4.9.1 we have argued that omission of elastic effects in problems of metal or plastic forming can be justified because of the small amount of elastic straining. This is undoubtedly true when we wish to consider the forces necessary to initiate large deformations and to follow these through. There are however a number of problems in which the inclusion of elasticity is important. One such problem is for instance that of 'spring-back' occurring particularly in metal forming of complex shapes. Here it is important to determine the amount of elastic recovery which may occur after removing the forming loads. Some possible suggestions for the treatment of such effects have been presented in reference 116 as early as 1984. However since that time much attention has been focused on the flow of viscoelastic fluids which is relevant to the above problem as well to the problem of transportation of fluids such as synthetic rubbers, etc. The procedures used in the study of such problems are quite complex and belong to the subject of numerical rheology. In this context the work of M. Crochet, K. Walters, P. Townsend and M.F. Webster[154–163] is notable. Obviously the subject is beyond the space limitations of the present book but an essential treatment can be found from the ideas discussed in this volume.

4.10 Direct displacement approach to transient metal forming

Explicit dynamic codes using quadrilateral or hexahedral elements have achieved considerable success in modelling short-duration impact phenomena with plastic deformation. The prototypes of finite element codes of this type are DYNA2d and DYNA3d developed at Lawrence Livermore National Laboratory.[164,165] For problems of relatively slow metal forming, such codes present some difficulties as in general the time step is governed by the elastic compressibility of the metal and a vast number of time steps would be necessary to cover a realistic metal forming problem. Nevertheless much use has been made of such codes in slow metal forming processes by the simple expedient of increasing the density of the material by many orders of magnitude. This is one of the drawbacks of using such codes whose description rightly belongs to the matter discussed in Volume 2 of this book. However, a further drawback is the lack of triangular or tetrahedral elements of a linear kind which could compete with linear quadrilaterals or hexahedra currently used and permit an easier introduction of adaptive refinement. It is well known that linear triangles or tetrahedra in a pure displacement (velocity) formulation will lock for incompressible or nearly incompressible materials. However we have already found that the CBS algorithm will avoid such locking when the same (linear) interpolation is used for both velocities and pressure.[166]

It is therefore possible to proceed in each step by solving a simple Stokes problem to evaluate the lagrangian velocity increment. We have described the use of such velocity formulation in the previous chapter. The update of the displacement allows new stresses to be evaluated by an appropriate plasticity law and the method can be used without difficulty as shown by Zienkiewicz *et al.*[166] In Fig. 4.27, we show a comparison between various methods of solving the impact of a circular bar made of an elastoplastic metal using an axisymmetric formulation. In this figure we show the results of a linear triangle displacement (Fig. 4.27b) form with a single integrating point for each element and a similar study again using displacement linear quadrilaterals (Fig. 4.27c) also with a single integration point. This figure also shows the same triangles and quadrilaterals solved using the CBS algorithm and very accurate final results (Fig. 4.27d and e).

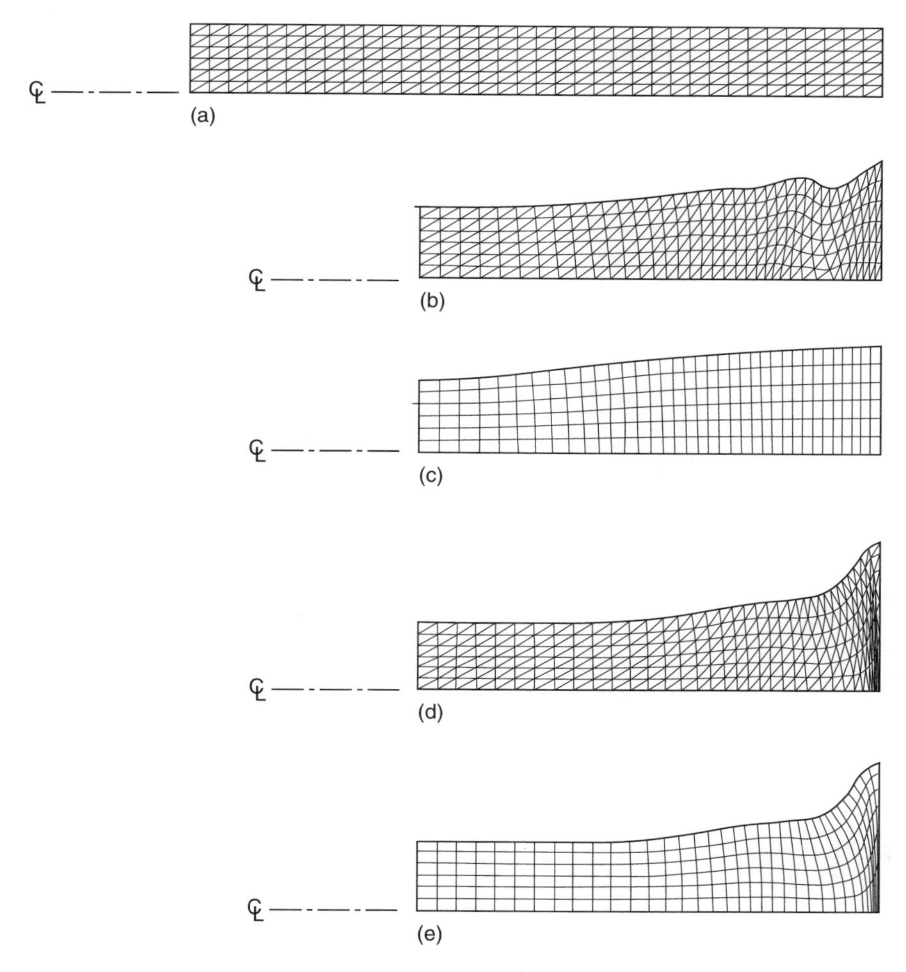

Fig. 4.27 Axisymmetric solutions to the bar impact problem: (a) initial shape; (b) linear triangles – displacement algorithm; (c) bilinear quadrilaterals – displacement algorithm; (d) linear triangles – CBS algorithm; (e) bilinear quadrilaterals – CBS algorithm.

In Fig. 4.28 we show similar results obtained with a full three-dimensional analysis. Similar methods for this problem have been presented by Bonet and Burton.[167]

4.11 Concluding remarks

The range of examples for which an incompressible formulation applies is very large as we have shown in this chapter. Indeed many other examples could have been included but for lack of space we proceed directly to Chapter 5 where the incompressible formulation is used for problems in which free surface or buoyancy occurs with gravity forces being the most important factor.

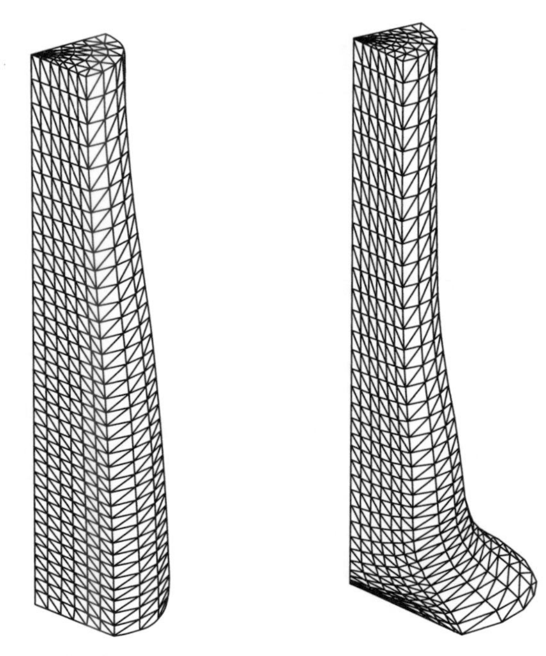

Fig. 4.28 Three-dimensional solution: (a) tetrahedral elements – standard displacement algorithm; (b) tetrahedral elements – CBS algorithm.

References

1. J.H. Argyris and G. Mareczek. Potential flow analysis by finite elements. *Ing. Archiv*, **41**, 1–25, 1972.
2. P. Bettess and J.A. Bettess. Analysis of free surface flows using isoparametric finite elements. *Int. J. Num. Meth. Eng.*, **19**, 1675–89, 1983.
3. T. Sarpkaya and G. Hiriart. Finite element analysis of jet impingement on axisymmetric curved deflectors, in *Finite Elements in Fluids* (eds J.T. Oden, O.C. Zienkiewicz, R.H. Gallagher and C. Taylor), Vol. 2, chap. 14, pp. 265–79, Wiley, Chichester, 1975.
4. M.J. O'Carroll. A variational principle for ideal flow over a spillway. *Int. J. Num. Meth. Eng.*, **15**, 767–89, 1980.
5. U. Ghia, K.N. Ghia and C.T. Shin. High-Resolution for incompressible flow using the Navier–Stokes equations and multigrid method. *J. Comp. Phys.*, **48**, 387–411, 1982.
6. R. Codina, M. Vázquez and O.C. Zienkiewicz. General algorithm for compressible and incompressible flows, Part III – A semi-implicit form. *Int. J. Num. Meth. Fluids*, **27**, 13–32, 1998.
7. O.C. Zienkiewicz, P. Nithiarasu, R. Codina, M. Vázquez and P. Ortiz. The characteristic based split (CBS) procedure: An efficient and accurate algorithm for fluid problems. *Int. J. Num. Meth. Fluids*, **31**, 359–92, 1999.
8. R. Codina, M. Vázquez and O.C. Zienkiewicz. A fractional step method for the solution of compressible Navier–Stokes equations. Eds. M. Hafez and K. Oshima. *Computational Fluid Dynamics Review 1998*, Vol 1, 331–47, World Scientific Publishing, 1998.

9. O.C. Zienkiewicz, B.V.K.S. Sai, K. Morgan, R. Codina and M. Vázquez. A general algorithm for compressible and incompressible flow – Part II. Tests on the explicit form. *Int. J. Num. Meth. Fluids*, **20**, 887–913, 1995.
10. P. Nithiarasu and O.C. Zienkiewicz. On stabilization of CBS algorithm. Internal and external time steps. *Int. J. Num. Meth. Eng.*, **48**, 875–80, 2000.
11. P.A. Gnoffo. A finite volume, adaptive grid algorithm applied to planetary entry flow fields. *AIAA J.*, **29**, 1249–54, 1983.
12. R. Löhner, K. Morgan and O.C. Zienkiewicz. Adaptive grid refinement for the Euler and compressible Navier–Stokes equations, in *Proc. Int. Conf. Accuracy Estimates and Adaptive Refinement in Finite Element Computations*, Lisbon, 1984.
13. M.J. Berger and J. Oliger. Adaptive mesh refinement for hyperbolic partial differential equations. *J. Comp. Phy.*, **53**, 484–512, 1984.
14. G.F. Carey and H. Dinh. Grading functions and mesh redistribution. *SIAM J. Num. Anal.*, **22**, 1028–40, 1985.
15. O.C. Zienkiewicz, K. Morgan, J. Peraire, M. Vahdati and R. Löhner. Finite elements for compressible gas flow and similar systems, in *7th Int. Conf. in Computational Methods in Applied Sciences and Engineering*, Versailles, Dec. 1985.
16. P. Palmeiro, V. Billey, A. Derviaux and J. Periaux. Self adaptive mesh refinement and finite element methods for solving Euler equations, in *Proc. ICFD Conf. on Numerical Methods for Fluid Dynamics* (Eds. K. Morton and M.J. Baines), Clarendon Press, Oxford, 1986.
17. J. T. Oden, P. Devloo and T. Strouboulis. Adaptive finite element methods for the analysis of inviscid compressible flow. *Comp. Meth. App. Mech. Eng.*, **59**, 327–62, 1986.
18. O.C. Zienkiewicz, K. Morgan, J. Peraire and J.Z. Zhu. Some expanding horizons for computational mechanics; error estimates, mesh generation and hyperbolic problems, in *Computational Mechanics – Advances and Trends* (Ed. A.K. Noor), AMD 75, ASME, 1986.
19. R. Löhner, K. Morgan and O.C. Zienkiewicz. Adaptive grid refinement for the Euler and compressible Navier–Stokes equations. *Accuracy Estimates and Adaptive Refinement in Finite Element Computations*, Chapter 15, 281–98 (Ed. I. Babuska, O.C. Zienkiewicz, J. Gago and E.R. de A. Oliveira), 1986, Wiley, New York.
20. J.F. Dannenhofer and J.R. Baron. Robust grid adaptation for coupled transonic flows. *AIAA paper 86-0495*, 1986.
21. B.N. Jiang and G.F. Carey. Adaptive refinement for least-square finite elements with element by element conjugate gradient solution. *Int. J. Num. Meth. Eng.*, **24**, 580–96, 1987.
22. J. Peraire, M. Vahdati, K. Morgan and O.C. Zienkiewicz. Adaptive remeshing for compressible flow computations. *J. Comp. Phy.*, **72**, 449–66, 1987.
23. J. T. Oden, T. Strouboulis and P. Devloo. Adaptive finite element methods for high speed compressible flows. *Int. J. Num. Meth. Eng.*, **7**, 1211–28, 1987.
24. J.H.S. Lee, J. Peraire and O.C. Zienkiewicz. The characteristic Galerkin method for advection dominated problems – an assessment. *Comp. Meth. App. Mech. Eng.*, **61**, 359–69, 1987.
25. J. Peraire, K. Morgan, J. Peiro and O.C. Zienkiewicz. An adaptive finite finite element method for high speed flows, in *AIAA 25th Aerospace Science Meeting*, Reno, Nevada, AIAA 87-0558, 1987.
26. O.C. Zienkiewicz, J.Z. Zhu, Y.C. Liu, K. Morgan and J. Peraire. Error estimates and adaptivity; from elasticity to high speed compressible flow, in *The Mathematics of Finite Elements and Application* (*MAFELAP 87*). Ed. J. R. Whiteman, pp. 483–512, Academic Press, London, 1988.
27. L. Formaggia, J. Peraire and K. Morgan. Simulation of state separation using the finite element method. *Appl. Math. Modelling*, **12**, 175–81, 1988.

28. O.C. Zienkiewicz, K. Morgan, J. Peraire, J. Peiro and L. Formaggia. Finite elements in fluid mechanics. Compressible flow, shallow water equations and transport. *ASME Conf. on Recent Developments in Fluid Dynamics*. AMD 95, ASME, Dec. 1988.
29. B. Palmerio 'A two-dimensional FEM adaptive moving-node method for steady Euler flow simulation. *Comp. Meth. App. Mech. Eng.*, **71**, 315–40, 1988.
30. J. Peraire, J. Peiro, L. Formaggia, K. Morgan and O.C. Zienkiewicz. Finite element Euler computations in 3-dimensions. *Int. J. Num. Meth. Eng.*, **26**, 2135–59, 1989.
31. R. Löhner, Adaptive remeshing for transient problems. *Comp. Meth. App. Mech. Eng.*, **75**, 195–214, 1989.
32. L. Demkowicz, J.T. Oden, W. Rachowicz and O. Hardy. Toward a universal *h-p* adaptive finite element strategy, Part 1. Constrained approximation and data structure. *Comp. Meth. App. Mech. Eng.*, **77**, 79–112, 1989.
33. J.T. Oden, L. Demkowicz, W. Rachowicz and T.A. Westermann. Toward a universal *h-p* adaptive finite element strategy, Part 2. A posteriori error estimation. *Comp. Meth. App. Mech. Eng.*, **77**, 113–80, 1989.
34. L. Demkowicz, J.T. Oden and W. Rachowicz. A new finite element method for solving compressible Navier–Stokes equations based on an operator splitting method and *h-p* adaptivity. *Comp. Meth. App. Mech. Eng.*, **84**, 275–326, 1990.
35. O.C. Zienkiewicz. Adaptivity-fluids-localization – the new challenge to computational mechanics. *Tran. Can. Soc. Mech. Eng.*, **15**, 137–45, 1991.
36. E.J. Probert, O. Hassan, K. Morgan and J. Peraire. An adaptive finite element method for transient compressible flows with moving boundaries. *Int. J. Num. Meth. Eng.*, **32**, 751–65, 1991.
37. E.J. Probert, O. Hassan, J. Peraire and K. Morgan. An adaptive finite element method for transient compressible flows. *Int. J. Num. Meth. Eng.*, **32**, 1145–59, 1991.
38. A. Evans, M.J. Marchant, J. Szmelter and N.P. Weatherill. Adaptivity for compressible flow computations using point embedding on 2-d structured multiblock meshes. *Int. J. Num. Meth. Eng.*, **32**, 895–919, 1991.
39. O. Hassan, K. Morgan, J. Peraire, E.J. Probert and R.R. Thareja. Adaptive unstructured mesh methods for steady viscous flow. *AIAA paper*, 1991.
40. J. Wu, J.Z. Zhu, J. Szmelter and O.C. Zienkiewicz. Error estimation and adaptivity in Navier–Stokes incompressible flows. *Comp. Mech.*, **61**, 30–9, 1991.
41. J.F. Hetu and D.H. Pelletier. Adaptive remeshing for viscous incompressible flows. *AIAA J.*, **30**, 1986–92, 1992.
42. J.F. Hetu and D.H. Pelletier. Fast, adaptive finite element scheme for viscous incompressible flows. *AIAA J.*, **30**, 2677–82, 1992.
43. J. Peraire, J. Peiro and K. Morgan. Adaptive remeshing for 3-dimensional compressible flow computations. *J. Comp. Phy.*, **103**, 269–85, 1992.
44. E.J. Probert, O. Hassan, K. Morgan and J. Peraire. Adaptive explicit and implicit finite element methods for transient thermal analysis. *Int. J. Num. Meth. Eng.*, **35**, 655–70, 1992.
45. J. Szmelter, M.J. Marchant, A. Evans and N.P. Weatherill. 2- dimensional Navier–Stokes equations with adaptivity on structured meshes. *Comp. Meth. Appl. Mech. Eng.*, **101**, 355–68, 1992.
46. R. Löhner and J.D. Baum. Adaptive *h*-refinement on 3D unstructured grid for transient problems. *Int. J. Num. Meth. Fluids*, **14**, 1407–19, 1992.
47. M.J. Marchant and N.P. Weatherill. Adaptivity techniques for compressible inviscid flows. *Comp. Meth. Appl. Mech. Eng.*, **106**, 83–106, 1993.
48. W. Rachowicz. An anisotropic *h*-type mesh refinement strategy. *Comp. Meth. Appl. Mech. Eng.*, **109**, 169–81, 1993.
49. P.A.B. de Sampaio, P.R.M. Lyra, K. Morgan and N.P. Weatherill. Petrov–Galerkin solutions of the incompressible Navier–Stokes equations in primitive variables with adaptive remeshing. *Comp. Meth. Appl. Mech. Eng.*, **106**, 143–78, 1993.

50. O.C. Zienkiewicz and J. Wu. Automatic directional refinement in adaptive analysis of compressible flows. *Int. J. Num. Meth. Eng.*, **37**, 2189–210, 1994.
51. J.T. Oden, W. Wu and M. Ainsworth. An *a posteriori* error estimate for finite element approximations of the Navier–Stokes equations. *Comp. Meth. App. Mech. Eng.*, **111**, 185–202, 1994.
52. J.T. Oden, W. Wu and V. Legat. An *hp* adaptive strategy for finite element approximations of the Navier–Stokes equations. *Int. J. Num. Meth. Fluids*, **20**, 831–51, 1995.
53. O. Hassan, E.J. Probert, K. Morgan and J. Peraire. Mesh generation and adaptivity for the solution of compressible viscous high speed flows. *Int. J. Num. Meth. Eng.*, **38**, 1123–48, 1995.
54. M.J. Castro-Diaz, H. Borouchaki, P.L. George, F. Hecht and B. Mohammadi. Error interpolation minimization and anisotropic mesh generation. *Proc. Int. Conf. Finite Elements Fluids – New Trends and Applications*, 1139–48, Venezia, 15–21, Oct. 1995.
55. R. Löhner. Mesh adaptation in fluid mechanics. *Eng. Frac. Mech.*, **50**, 819–47, 1995.
56. O. Hassan, K. Morgan, E.J. Probert and J. Peraire. Unstructured tetrahedral mesh generation for three dimensional viscous flows. *Int. J. Num. Meth. Eng.*, **39**, 549–67, 1996.
57. D.D.D. Yahia, W.G. Habashi, A. Tam, M.G. Vallet and M. Fortin. A directionally adaptive methodology using an edge based error estimate on quadrilateral grids. *Int. J. Num. Meth. Fluids*, **23**, 673–90, 1996.
58. M. Fortin, M.G. Vallet, J. Dompierre, Y. Bourgault and W.G. Habashi. Anisotropic mesh adaption: theory, validation and applications. *Computational Fluid Dynamics '96*, 174–80, 1996.
59. M.J. Castro-Diaz, H. Borouchaki, P.L. George, F. Hecht and B. Mohammadi. Anisotropic adaptive mesh generation in two dimensions for CFD. *Computational Fluid Dynamics '96*, 181–86, 1996.
60. K.S.V. Kumar, A.V.R. Babau, K.N. Seetharamu, T. Sundararajan and P.A.A. Narayana. A generalized Delaunay triangulation algorithm with adaptive grid size control. *Comm. Num. Meth. Eng.*, **13**, 941–48, 1997.
61. H. Borouchaki, P.L. George, F. Hecht, P. Laug and E. Saltel. Delaunay mesh generation governed by metric specifications. Part I. Algorithms. *Finite Elem. Anal. Des.*, **25**, 61–83, 1997.
62. H. Borouchaki, P.L. George and B. Mohammadi. Delaunay mesh generation governed by metric specifications, Part II. Applications. *Finite Elem. Anal. Des.*, **25**, 85–109, 1997.
63. W. Rachowicz. An anisotropic *h*-adaptive finite element method for compressible Navier–Stokes equations. *Comp. Meth. Appl. Mech. Eng.*, **146**, 231–52, 1997.
64. J. Peraire and K. Morgan. Unstructured mesh generation including directional refinement for aerodynamic flow simulation. *Finite Elem. Anal. Des.*, **25**, 343–56, 1997.
65. M.J. Marchant, N.P. Weatherill and O. Hassan. The adaptation of unstructured grids for transonic viscous flow simulation. *Finite Elem. Anal. Des.*, **25**, 199–218, 1997.
66. H. Borouchaki, F. Hecht and P.J. Frey. Mesh gradation control. *Int. J. Num. Meth. Fluids*, **43**, 1143–65, 1998.
67. H. Borouchaki and P.J. Frey. Adaptive triangular-quadrilateral mesh generation. *Int. J. Num. Meth. Eng.*, **41**, 915–34, 1998.
68. H. Jin and S. Prudhomme. *A posteriori* error estimation of steady-state finite element solution of the Navier–Stokes equations by a subdomain residual method. *Comp. Meth. Appl. Mech. Eng.*, **159**, 19–48, 1998.
69. P. Nithiarasu and O.C. Zienkiewicz. Adaptive mesh generation procedure for fluid mechanics problems. *Int. J. Num. Meth. Eng.*, **47**, 629–62, 2000.
70. M. Pastor, J. Peraire and O.C.Zienkiewicz. Adaptive remeshing for shear band localization problems. *Arch. App. Mech.*, **61**, 30–9, 1991.

71. M. Pastor, C. Rubio, P. Mira, O.C. Zienkiewicz and J.P. Vilotte. Shearband computations: influence of mesh alignment. *Computer Methods and Advances in Geomechanics* (eds Sinwardane and Zaman), Balkema, pp. 2681–87, 1995.
72. O.C. Zienkiewicz, M. Pastor and M.S. Huang. Softening, localisation and adaptive remeshing. Capture of discontinuous solutions. *Comp. Mech.*, **17**, 98–106, 1995.
73. O.C. Zienkiewicz, M.S. Huang and M. Pastor. Localization problems in plasticity using finite elements with adaptive remeshing. *Int. J. Num. Anal. Meth. Geomech.*, **19**, 127–48, 1995.
74. P.J. Green and R. Sibson. Computing Dirichlet tessellations in the plane. *Comp. J.*, **21**, 168–73, 1978.
75. D.F. Watson. Computing the *N*-dimensional Delaunay tessellation with application to Voronoi polytopes. *Comp. J.*, **24**, 167–72, 1981.
76. J.C. Cavendish, D.A. Field and W.H. Frey. An approach to automatic three dimensional finite element mesh generation. *Int. J. Num. Meth. Eng.*, **21**, 329–47, 1985.
77. W.J. Schroeder and M.S. Shephard. A combined octree Delaunay method for fully automatic 3-D mesh generation. *Int. J. Num. Meth. Eng.*, **29**, 37–55, 1990.
78. N.P. Weatherill. Mesh generation for aerospace applications. *Sadhana – Acc. Proc. Eng. Sci.*, **16**, 1–45, 1991.
79. N.P. Weatherill, P.R. Eiseman, J. Hause and J.F. Thompson. *Numerical Grid Generation in Computational Fluid Dynamics and Related Fields*. Pineridge Press, Swansea, 1994.
80. N.P. Weatherill and O. Hassan. Efficient 3-dimensional Delaunay triangulation with automatic point generation and imposed boundary constraints. *Int. J. Num. Meth. Eng.*, **37**, 2005–39, 1994.
81. G. Subramanian, V.V.S. Raveendra and M.G. Kamath. Robust boundary triangulation and Delaunay triangulation of arbitrary planar domains. *Int. J. Num. Meth. Eng.*, **37**, 1779–89, 1994.
82. N.P.Weatherill, O. Hassan and D.L. Marcum. Compressible flow field solutions with unstructured grids generated by Delaunay triangulation. *AIAA J.*, **33**, 1196–204, 1995.
83. R.W. Lewis, Y. Zhang and A.S. Usmani. Aspects of adaptive mesh generation based on domain decomposition and Delaunay triangulation. *Finite Elem. Anal. Des.*, **20**, 47–70, 1995.
84. H. Borouchaki and P.L. George. Aspects of 2-d Delaunay mesh generation. *Int. J. Num. Meth. Eng.*, **40**, 1957–75, 1997.
85. J.F. Thompson, B.K. Soni and N.P. Weatherill (Ed.), *Handbook of Grid Generation*. CRC Press, 1999.
86. D.S. Malkus and T.J.R. Hughes. Mixed finite element method reduced and selective integration techniques: a unification of concepts. *Comp. Meth. Appl. Mech. Eng.*, **15**, 63–81, 1978.
87. O.C. Zienkiewicz and P.N. Godbole. Viscous, incompressible flow with special reference to non-Newtonian (plastic) fluids, in *Finite Elements in Fluids* (eds J.T. Oden *et al.*), Vol. 2, pp. 25–55, Wiley, Chichester, 1975.
88. T.J.R. Hughes, R.L. Taylor and J.F. Levy. A finite element method for incompressible flows, in *Proc. 2nd Int. Symp. on Finite Elements in Fluid Problems ICCAD*. Sta Margharita Ligure, Italy, pp. 1–6, 1976.
89. O.C. Zienkiewicz and P.N. Godbole. A penalty function approach to problems of plastic flow of metals with large surface deformation. *J. Strain Anal.*, **10**, 180–3, 1975.
90. O.C. Zienkiewicz and S. Nakazawa. The penalty function method and its applications to the numerical solution of boundary value problems. *ASME, AMD*, **51**, 157–79, 1982.
91. J.T. Oden. R.I.P. methods for Stokesian flow, in *Finite Elements in Fluids* (eds R.H. Gallagher, D.N. Norrie, J.T. Oden and O.C. Zienkiewicz), Vol. 4, chap. 15, pp. 305–18, Wiley, Chichester, 1982.
92. O.C. Zienkiewicz, J.P. Vilotte, S. Toyoshima and S. Nakazawa. Iterative method for constrained and mixed approximation. An inexpensive improvement of FEM performance. *Comp. Meth. Appl. Mech. Eng.*, **51**, 3–29, 1985.

93. O.C. Zienkiewicz, Y.C. Liu and G.C. Huang. Error estimates and convergence rates for various incompressible elements. *Int. J. Num. Meth. Eng.*, **28**, 2191–202, 1989.
94. M. Fortin. Old and new finite elements for incompressible flow. *Int. J. Num. Meth. Fluids*, **1**, 347–64, 1981.
95. M. Fortin and N. Fortin. Newer and newer elements for incompressible flow, in *Finite Elements in Fluids* (eds R.H. Gallagher, G.F. Carey, J.T. Oden and O.C. Zienkiewicz), Vol. 6, chap. 7, pp. 171–88, Wiley, Chichester, 1985.
96. M.S. Engleman, R.L. Sani, P.M. Gresho and H. Bercovier. Consistent v. reduced integration penalty methods for incompressible media using several old and new elements. *Int. J. Num. Meth. Fluids*, **4**, 25–42, 1982.
97. K. Palit and R.T. Fenner. Finite element analysis of two dimensional slow non-Newtonian flows. *AIChE J.*, **18**, 1163–9, 1972.
98. K. Palit and R.T. Fenner. Finite element analysis of slow non-Newtonian channel flow. *AIChE J.*, **18**, 628–33, 1972.
99. B. Atkinson, C.C.M. Card and B.M. Irons. Application of the finite element method to creeping flow problems. *Trans. Inst. Chem. Eng.*, **48**, 276–84, 1970.
100. G.C. Cornfield and R.H. Johnson. Theoretical prediction of plastic flow in hot rolling including the effect of various temperature distributions. *J. Iron and Steel Inst.*, **211**, 567–73, 1973.
101. C.H. Lee and S. Kobayashi. New solutions to rigid plastic deformation problems using a matrix method. *Trans. ASME J. Eng. for Ind.*, **95**, 865–73, 1973.
102. J.T. Oden, D.R. Bhandari, G. Yagewa and T.J. Chung. A new approach to the finite element formulation and solution of a class of problems in coupled thermoelastoviscoplasticity of solids. *Nucl. Eng. Des.*, **24**, 420, 1973.
103. O.C. Zienkiewicz, E. Oñate and J.C. Heinrich. *Plastic Flow in Metal Forming*, pp. 107–20, ASME, San Francisco, December 1978.
104. O.C. Zienkiewicz, E. Oñate and J.C. Heinrich. A general formulation for coupled thermal flow of metals using finite elements. *Int. J. Num. Meth. Eng.*, **17**, 1497–514, 1981.
105. N. Rebelo and S. Kobayashi. A coupled analysis of viscoplastic deformation and heat transfer: I. Thoretical consideration: II. Application. *Int. J. Mech. Sci.*, **22**, 699–705, 1980 and **22**, 707–18, 1980.
106. P.R. Dawson and E.G. Thompson. Finite element analysis of steady state elastoviscoplastic flow by the initial stress rate method. *Int. J. Num. Meth. Eng.*, **12**, 47–57, 382–3, 1978.
107. Y. Shimizaki and E.G. Thompson. Elasto-visco-plastic flow with special attention to boundary conditions. *Int. J. Num. Meth. Eng.*, **17**, 97–112, 1981.
108. O.C. Zienkiewicz, P.C. Jain and E. Oñate. Flow of solids during forming and extrusion: some aspects of numerical solutions. *Int. J. Solids Struct.*, **14**, 15–38, 1978.
109. S. Nakazawa, J.F.T. Pittman and O.C. Zienkiewicz. Numerical solution of flow and heat transfer in polymer melts, in *Finite Elements in Fluids* (ed. R.H. Gallagher *et al.*), Vol. 4, chap. 13, pp. 251–83, Wiley, Chichester, 1982.
110. S. Nakazawa, J.F.T. Pittman and O.C. Zienkiewicz. A penalty finite element method for thermally coupled non-Newtonian flow with particular reference to balancing dissipation and the treatment of history dependent flows, in *Int. Symp. of Refined Modelling of Flows*, 7–10 Sept. 1982, Paris.
111. R.E. Nickell, R.I. Tanner and B. Caswell. The solution of viscous incompressible jet and free surface flows using finite elements. *J. Fluid Mech.*, **65**, 189–206, 1974.
112. R.I. Tanner, R.E. Nickell and R.W. Bilger. Finite element method for the solution of some incompressible non-Newtonian fluids mechanics problems with free surface. *Comp. Meth. Appl. Mech. Eng.*, **6**, 155–74, 1975.
113. J.M. Alexander and J.W.H. Price. Finite element analysis of hot metal forming, in *18th MTDR Conf.*, pp. 267–74, 1977.

114. S.I. Oh, G.D. Lahoti and A.T. Altan. Application of finite element method to industrial metal forming processes, in *Proc. Conf. on Industrial Forming Processes*, pp. 146–53, Pineridge Press, Swansea, 1982.
115. J.F.T. Pittman, O.C. Zienkiewicz, R.D. Wood and J.M. Alexander (eds). *Numerical Analysis of Forming Processes*. Wiley, Chichester, 1984.
116. O.C. Zienkiewicz. Flow formulation for numerical solutions of forming problems, in *Numerical Analysis of Forming Processes* (eds J.F.T. Pittman *et al.*), chap. 1, pp. 1–44, Wiley, Chichester, 1984.
117. S. Kobayashi. Thermoviscoplastic analysis of metal forming problems by the finite element method, in *Numerical Analysis of Forming Processes* (eds J.F.T. Pittman *et al.*), chap. 2, pp. 45–70, Wiley, Chichester, 1984.
118. J.L. Chenot, F. Bay and L. Fourment. Finite element simulation of metal powder forming. *Int. J. Num. Meth. Eng.*, **30**, 1649–74, 1990.
119. P.A. Balaji, T. Sundararajan and G.K. Lal. Viscoplastic deformation analysis and extrusion die design by FEM. *J. Appl. Mech. – Trans. ASME*, **58**, 644–50, 1991.
120. K.H. Raj, J.L. Chenot and L. Fourment. Finite element modelling of hot metal forming. *Indian J. Eng. Mat. Sci.*, **3**, 234–38, 1996.
121. J.L. Chenot. Recent contributions to the finite element modelling of metal forming processes. *J. Mat. Proc. Tech.*, **34**, 9–18, 1992.
122. J. Bonet, P. Bhargava and R.D. Wood. The incremental flow formulation for the finite element analysis of 3-dimensional superplastic forming processes. *J. Mat. Proc. Tech.*, **45**, 243–48, 1994.
123. R.D. Wood and J. Bonet. A review of the numerical analysis of super-plastic forming. *J. Mat. Proc. Tech.*, **60**, 45–53, 1996.
124. J.L. Chenot and Y. Chastel. Mechanical, thermal and physical coupling methods in FE analysis of metal forming processes. *J. Mat. Proc. Tech.*, **60**, 11–18, 1996.
125. J. Rojek, E. Oñate and E. Postek. Application of explicit FE codes to simulation of sheet and bulk metal forming processes. *J. Mat. Proc. Tech.*, **80–1**, 620–7, 1998.
126. J. Bonet. Reent developments in the incremental flow formulation for the numerical simulation of metal forming processes. *Eng. Comp.*, **15**, 345–57, 1998.
127. J. Rojek, O.C. Zienkiewicz, E. Oñate and E. Postek. Advances in FE explicit formulation for simulation of metal forming processes. *Proc. International Conference on Advances in Materials and Processing Technologies, AMPT 1999 and 16th Annual Conference of the Irish Manufacturing Committee, IMC16*, **Vol. 1**, Ed. M.S.J. Hashmi and L. Looney, Aug. 1999, Dublin City University.
128. J.-L. Chenot, R.D. Wood and O.C. Zienkiewicz (Eds). *Numerical Methods in Industrial Forming Processes, NUMIFORM 92*. Valbonne, France, 14–19 Sept., 1992, A.A. Balkema, Rotterdam.
129. J. Huétink and F.P.T. Baaijens (Eds). *Simulation of Materials Processing: Theory, Methods and Applications, NUMIFORM 98*. Enschede, Netherlands, 22–25 June, 1998, A.A. Balkema, Rotterdam.
130. W. Johnson and P.B. Mellor. *Engineering Plasticity*. Van Nostrand-Reinhold, London, 1973.
131. G.C. Huang. Error estimates and adaptive remeshing in finite element anlysis of forming processes. Ph.D. thesis, University of Wales, Swansea, 1989.
132. G.C. Huang, Y.C. Liu and O.C. Zienkiewicz. Error control, mesh updating schemes and automatic adaptive remeshing for finite element analysis of unsteady extrusion processes, in *Modelling of Metal Forming Processes* (eds J.L. Chenot and E. Oñate), pp. 75–83, Kluwer Academic, Dordrecht, 1988.
133. O.C. Zienkiewicz, Y.C. Liu and G.C. Huang. An error estimate and adaptive refinement method for extrusion and other forming problems. *Int. J. Num. Meth. Eng.*, **25**, 23–42, 1988.

134. O.C. Zienkiewicz, Y.C. Liu, J.Z. Zhu and S. Toyoshima. Flow formulation for numerical solution of forming processes. II – Some new directions, in *Proc. 2nd Int. Conf. on Numerical Methods in Industrial Forming Processes, NUMIFORM 86* (eds K. Mattiasson, A. Samuelson, R.D. Wood and O.C. Zienkiewicz), A.A. Balkema, Rotterdam, 1986.
135. T. Belytchko, D.P. Flanagan and J.M. Kennedy. Finite element methods with user controlled mesh for fluid structure interaction. *Comp. Meth. Appl. Mech. Eng.*, **33**, 669–88, 1982.
136. J. Donea, S. Giuliani and J.I. Halleux. An arbitrary Lagrangian–Eulerian finite element for transient dynamic fluid-structure interaction. *Comp. Meth. Appl. Mech. Eng.*, **75**, 195–214, 1989.
137. P.J.G. Schreurs, F.E. Veldpaus and W.A.M. Brakalmans. An Eulerian and Lagrangian finite element model for the simulation of geometrical non-linear hyper elastic and elasto-plastic deformation processes, in *Proc. Conf. on Industrial Forming Processes*, pp. 491–500, Pineridge Press, Swansea, 1983.
138. J. Donea. Arbitrary Lagrangian–Eulerian finite element methods, in *Computation Methods for Transient Analysis* (eds T. Belytchko and T.J.R. Hughes), chap. 10, pp. 474–516, Elsevier, Amsterdam, 1983.
139. J. van der Lugt and J. Huetnik. Thermo-mechanically coupled finite element analysis in metal forming processes. *Comp. Meth. Appl. Mech. Eng.*, **54**, 145–60, 1986.
140. E. Oñate and O.C. Zienkiewicz. A viscous sheet formulation for the analysis of thin sheet metal forming. *Int. J. Mech. Sci.*, **25**, 305–35, 1983.
141. N.M. Wang and B. Budiansky. Analysis of sheet metal stamping by a finite element method. *Trans. ASME, J. Appl. Mech.*, **45**, 73, 1976.
142. A.S. Wifi. An incremented complete solution of the stretch forming and deep drawing of a circular blank using a hemispherical punch. *Int. J. Mech. Sci.*, **18**, 23–31, 1976.
143. J. Bonet, R.D. Wood and O.C. Zienkiewicz. Time stepping schemes for the numerical analysis of superplastic forming of thin sheets, in *Modelling of Metal Forming Processes* (eds J.L. Chenot and E. Oñate), pp. 179–86, Kluwer Academic, Dordrecht, 1988.
144. E. Massoni, M. Bellet and J.L. Chenot. Thin sheet forming numerical analysis with membrane approach, in *Modelling of Metal Forming Processes* (eds J.L. Chenot and E. Oñate), Kluwer Academic, Dordrecht, 1988.
145. R.D. Wood, J. Bonet and A.H.S. Wargedipura. Simulation of the superplastic forming of thin sheet components using the finite element method, in *NUMIFORM 89 Proc.*, pp. 85–94, Balkhema Press, 1989.
146. J. Bonet, R.D. Wood and O.C. Zienkiewicz. Finite element modelling of the superplastic forming of a thin sheet. *Proc. Conf. on Superquality and Superplastic Forming* (eds C.H. Hamilton and N.E. Paton), The Minerals, Metals and Materials Society, USA, 1988.
147. E. Oñate and C.A. Desaracibar. Finite element analysis of sheet metal forming problems using a selective bending membrane formulation. *Int. J. Num. Meth. Eng.*, **30**, 1577–93, 1990.
148. W. Sosnowski, E. Oñate and C.A. Desaracibar. Comparative study of sheet metal forming processes by numerical modelling and experiment. *J. Mat. Proc. Tech.*, **34**, 109–16, 1992.
149. J. Bonet and R.D. Wood. Incremental flow procedures for the finite element analysis of thin sheet superplastic forming processes. *J. Mat. Pro. Tech.*, **42**, 147–65, 1994.
150. W. Sosnowski and M. Kleiber. A study on the influence of friction evolution on thickness changes in sheet metal forming. *J. Mat. Proc. Tech.*, **60**, 469–74, 1996.
151. J. Bonet, P. Bhargava and R.W. Wood. Finite element analysis of the superplastic forming of thick sheet using the incremental flow formulation. *Int. J. Num. Meth. Eng.*, **40**, 3205–28, 1997.
152. M. Kawka, T. Kakita and A. Makinouchi. Simulation of multi-step sheet metal forming processes by a static explicit FEM code. *J. Mat. Proc. Tech.*, **80**, 54–59, 1998.

153. S.K. Esche, G.K. Kinzel and J.K. Lee. An axisymmetric membrane element with bending stiffness for static implicit sheet metal forming simulation. *J. Appl. Mech. – Trans. ASME*, **66**, 153–64, 1999.
154. M.J. Crochet and R. Keunings. Finite element analysis of die swell of highly elastic fluid. *J. Non-Newtonian Fluid Mech.*, **10**, 339–56, 1982.
155. M.J. Crochet, A.R. Davies and K. Walters. *Numerical Simulation of Non-Newtonian Flow*. Rheology series: Volume 1, Elsevier, Amsterdam, 1984.
156. R. Keunings and M.J. Crochet. Numerical simulation of the flow of a viscoelastic fluid through an abrupt contraction. *J. Non-Newtonian Fluid Mech.*, **14**, 279–99, 1984.
157. S. Dupont, J.M. Marchal and M.J. Crochet. Finite element simulation of viscoelastic fluids of the integral type. *J. Non-Newtonian Fluid Mech.*, **17**, 157–83, 1985.
158. J.M. Marchal and M.J. Crochet. A new mixed finite element for calculating visco elastic flow. *J. Non-Newtonian Fluid Mech.*, **26**, 77–114, 1987.
159. M.J. Crochet and V. Legat. The consistent streamline upwind Petrov–Galerkin method for viscoelastic flow revisited. *J. Non-Newtonian Fluid Mech.*, **42**, 283–99, 1992.
160. H.R. Tamaddonjahromi, D. Ding, M.F. Webster and P. Townsend. A Taylor–Galerkin finite element method for non-newtonian flows. *Int. J. Num. Meth. Eng.*, **34**, 741–57, 1992.
161. E.O.A. Carew, P. Townsend and M.F. Webster. A Taylor–Galerkin algorithm for visco-elastic flow. *J. Non-Newtonian Fluid Mech.*, **50**, 253–87, 1993.
162. D. Ding, P. Townsend and M.F. Webster. Finite element simulation of an injection moulding process. *Int. J. Num. Meth. Heat Fluid Flow*, **7**, 751–66, 1997.
163. H. Matallah, P. Townsend and M.F. Webster. Recovery and stress-splitting schemes for viscoelastic flows. *J. Non-Newtonian Fluid Mech.*, **75**, 139–66, 1998.
164. G.L. Goudreau and J.O. Hallquist. Recent developments in large scale Lagrangian hydro-codes. *Comp. Meth. Appl. Mech. Eng.*, **33**, 725–57, 1982.
165. J.O. Hallquist, G.L. Goudreau and D.J. Benson. Sliding interfaces with contact–impact in large scale lagrangian computations. *Comp. Meth. Appl. Mech. Eng.*, **51**, 107–37, 1985.
166. O.C. Zienkiewicz, J. Rojek, R.L. Taylor and M. Pastor. Triangles and tetrahedra in explicit dynamic codes for solids. *Int. J. Num. Meth. Eng.*, **43**, 565–83, 1998.
167. J. Bonet and A.J. Burton. A simple average nodal pressure tetrahedral element for incompressible and nearly incompressible dynamic explicit applications. *Comm. Num. Meth. Eng.*, **14**, 437–49, 1998.

5

Free surfaces, buoyancy and turbulent incompressible flows

5.1 Introduction

In the previous chapter we have introduced the reader to general methods of solving incompressible flow problems and have illustrated these with many examples of newtonian and non-newtonian flows. In the present chapter, we shall address three separate topics of incompressible flow which were not dealt with in the previous chapters. This chapter is thus divided into three parts. In the first two parts the common theme is that of the action of the body force due to gravity. We start with a section addressed to problems of free surfaces and continue with the second section which deals with buoyancy effects caused by temperature differences in various parts of the domain. The third part discusses the important topic of turbulence and we shall introduce the reader here to some general models currently used in such studies. This last section will inevitably be brief and we will simply illustrate the possibility of dealing with time averaged viscosities and Reynolds stresses. We shall have occasion later to use such concepts when dealing with compressible flows in Chapter 6. However the first two topics of incompressible flow are of considerable importance and here we shall discuss matters in some detail.

The first part of this chapter, Sec. 5.2, will deal with problems in which a free surface of flow occurs when gravity forces are acting throughout the domain. Typical examples here would be for instance given by the disturbance of the free surface of water and the creation of waves by moving ships or submarines. Of course other problems of similar kinds arise in practice. Indeed in Chapter 7, where we deal with shallow water flows, a free surface is an essential condition but other assumptions and simplifications have to be introduced. Here we deal with the full problem and include either complete viscous effects or simply deal with an inviscid fluid without further physical assumptions. There are other topics of free surfaces which occur in practice. One of these for instance is that of mould filling which is frequently encountered in manufacturing where a particular fluid or polymer is poured into a mould and solidifies. We shall briefly refer to such examples. Space does not permit us to deal with this important problem in detail but we give some references to the current literature.

In Sec. 5.3, we invoke problems of buoyancy and here we can deal with pure (natural) convection when the only force causing the flow is that of the difference between uniform density and density which has been perturbed by a given temperature field. In such examples it is a fairly simple matter to modify the equations so as to deal only with the perturbation forces but on occasion forced convection is coupled with such naturally occurring convection.

5.2 Free surface flows

5.2.1 General remarks and governing equations

In many problems of practical importance a free surface will occur in the fluid (liquid). In general the position of such a free surface is not known and the main problem is that of determining it. In Fig. 5.1, we show a set of typical problems of free surfaces; these range from flow over and under water control structures, flow

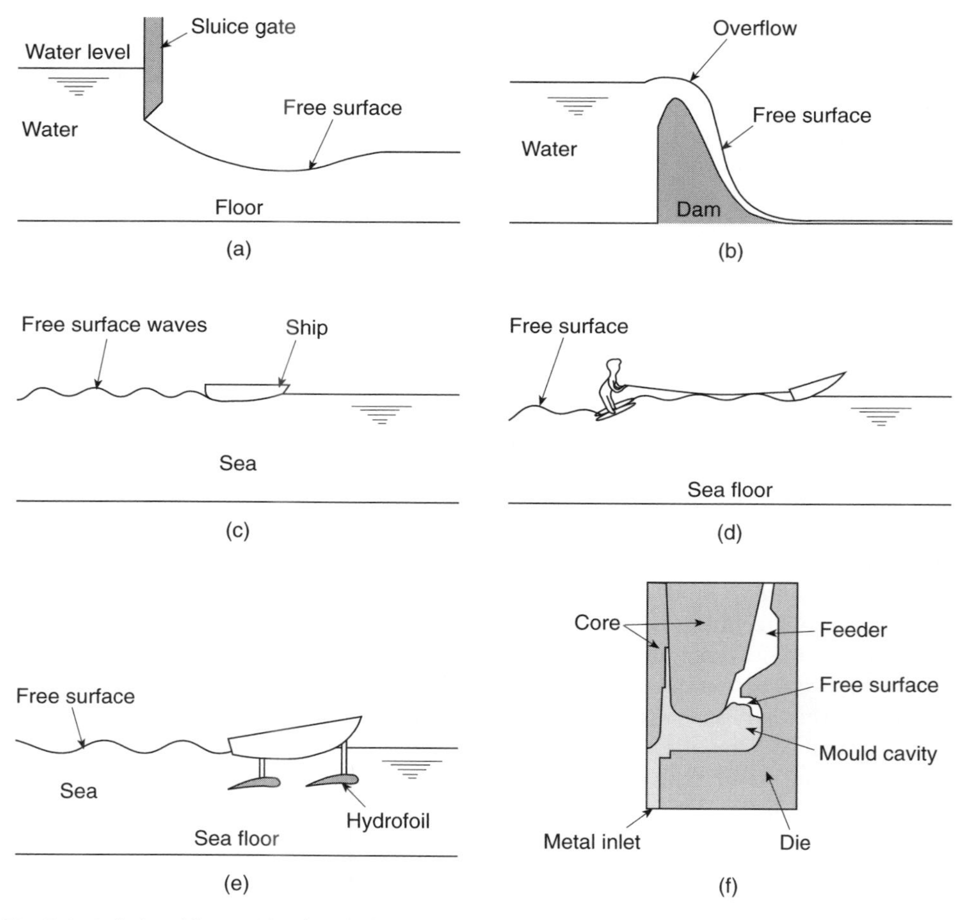

Fig. 5.1 Typical problems with a free surface.

around ships, to industrial processes such as filling of moulds. All these situations deal with a fluid which is incompressible and in which the viscous effects either can be important or on the other hand may be neglected. The only difference from solving the type of problem which we have discussed in the previous chapter is the fact that the position of the free surface is not known *a priori* and has to be determined during the computation.

On the free surface we have at all times to ensure that (1) the pressure (which approximates the normal traction) and tangential tractions are zero unless specified otherwise, and (2) that the material particles of the fluid belong to the free surface at all times.

Obviously very considerable non-linearities occur and the problem will have to be solved iteratively. We shall therefore concentrate in the following presentation on a typical situation in which such iteration can be used. The problem chosen for the more detailed discussion is that of *ship hydrodynamics* though the reader will obviously realize that for the other problems shown somewhat similar procedures of iteration will be applicable though details may well differ in each application.

5.2.2 Free surface wave problems in ship hydrodynamics

Figure 5.2 shows a typical problem of ship motion together with the boundaries limiting the domain of analysis. In the interior of the domain we can use either the

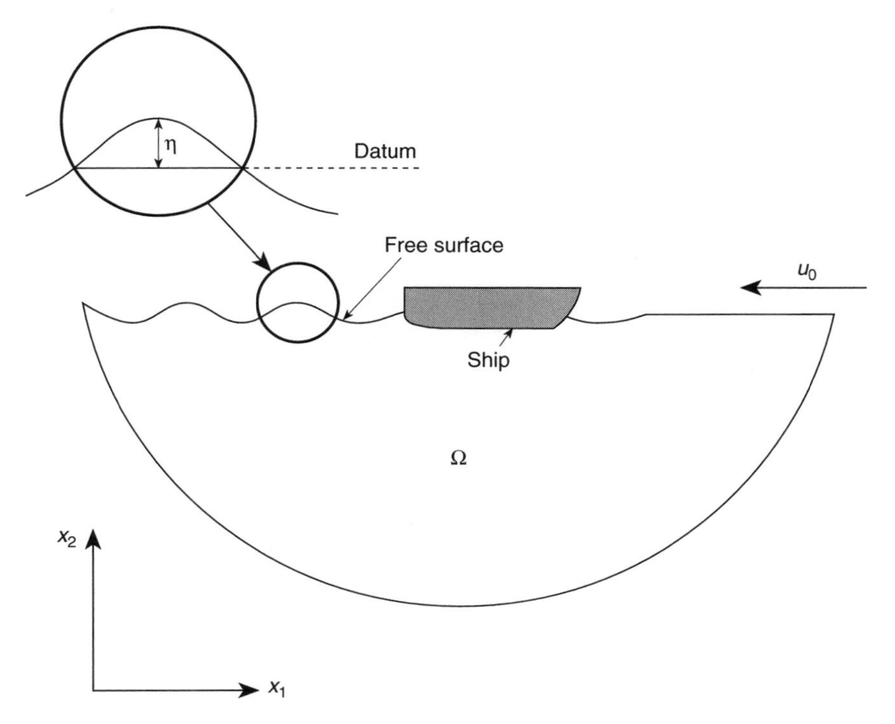

Fig. 5.2 A typical problem of ship motion.

full Navier–Stokes equations or, neglecting viscosity effects, a pure potential or Euler approximation. Both assumptions have been discussed in the previous chapter but it is interesting to remark here that the resistance caused by the waves may be four or five times greater than that due to viscous drag. Clearly surface effects are of great importance.

Historically many solutions that ignore viscosity totally have been used in the ship industry with good effect by involving so-called boundary solution procedures or panel methods.[1–11] Early finite element studies on the field of ship hydrodynamics have also used potential flow equations.[12] A full description of these is given in many papers. However complete solutions with viscous effects and full non-linearity are difficult to deal with. In the procedures that we present in this section, the door is opened to obtain a full solution without any extraneous assumptions and indeed such solutions could include turbulence effects, etc. We need not mention in any detail the question of the equations which are to be solved. These are simply those we have already discussed in Sec. 4.1 of the previous chapter and indeed the same CBS procedure will be used in the solution. However, considerable difficulties arise on the free surface, despite the fact that on such a surface both tractions are known (or zero). The difficulties are caused by the fact that at all times we need to ensure that this surface is a material one and contains the particles of the fluid.

Let us define the position of the surface by its elevation η relative to some previously known surface which we shall refer to as the reference surface (see Fig. 5.2). This surface may be horizontal and may indeed be the undisturbed water surface or may simply be a previously calculated surface. If η is measured in the direction of the vertical coordinate which we shall call x_3, we can write

$$\eta(t, x_1, x_2) = x_3 - x_{3\mathrm{ref}} \tag{5.1}$$

Noting that η is the position of the particle on the surface, we observe that

$$\frac{\mathrm{d}x_1}{\mathrm{d}t} = u_1, \qquad \frac{\mathrm{d}x_2}{\mathrm{d}t} = u_2, \qquad \frac{\mathrm{d}x_3}{\mathrm{d}t} = \frac{\mathrm{d}\eta}{\mathrm{d}t} = u_3 \tag{5.2}$$

and from Eq. (5.1,) we have finally

$$\frac{\mathrm{d}\eta}{\mathrm{d}t} = u_3 = \frac{\partial \eta}{\partial t} + u_1 \frac{\partial \eta}{\partial x_1} + u_2 \frac{\partial \eta}{\partial x_2} = \frac{\partial \eta}{\partial t} + \bar{\mathbf{u}}^{\mathrm{T}} \bar{\boldsymbol{\nabla}} \eta \tag{5.3}$$

where

$$\bar{\mathbf{u}} = [u_1, u_2]^{\mathrm{T}} \qquad \bar{\boldsymbol{\nabla}} = \left[\frac{\partial}{\partial x_1}, \frac{\partial}{\partial x_2}\right]^{\mathrm{T}} \tag{5.4}$$

We immediately observe that η obeys a pure convection equation (see Chapter 2) in terms of the variables t, u_1, u_2 and u_3 in which u_3 is a source term. At this stage it is worthwhile remarking that this surface equation has been known for a very long time and was dealt with previously by upwind differences, in particular those introduced on a regular grid by Dawson.[2] However in Chapter 2, we have already discussed other perfectly stable, finite element methods, any of which can be used for dealing with this equation. In particular the characteristic–Galerkin procedure can be applied most effectively.

It is important to observe that when the steady state is reached we simply have

$$u_3 = u_1 \frac{\partial \eta}{\partial x_1} + u_2 \frac{\partial \eta}{\partial x_2} \tag{5.5}$$

which ensures that the velocity vector is tangential to the free surface. The solution method for the whole problem can now be fully discussed.

5.2.3 Iterative solution procedures

An iterative procedure is now fairly clear and several alternatives are possible.

Mesh updating

The first of these solutions is that involving *mesh updatings*, where we proceed as follows. Assuming a known reference surface, say the original horizontal surface of the water, we specify that the pressure and tangential traction on this surface are zero and solve the resulting fluid mechanics problem by the methods of the previous chapter. Using the CBS algorithm we start with known values of the velocities and find the necessary increment obtaining $\mathbf{u}^{n+1}$ and p^{n+1} from initial values. At the same time we solve the increment of η using the newly calculated values of the velocities. We note here that this last equation is solved only in two dimensions on a mesh corresponding to the projected coordinates of x_1 and x_2.

At this stage the surface can be immediately updated to a new position which now becomes the new reference surface and the procedure can then be repeated.

Hydrostatic adjustment

Obviously the method of repeated mesh updating can be extremely costly and in general we follow the process described as *hydrostatic adjustment*. In this process we note that once the incremental η has been established, we can adjust the surface pressure at the reference surface by

$$\Delta p^n = \Delta \eta^n \rho g \tag{5.6}$$

Some authors say that this is a use of the Bernoulli equation but obviously it is a simple disregard of any acceleration forces that may exist near the surface and of any viscous stresses there. Of course this introduces an approximation but this approximation can be quite happily used for starting the following step.

If we proceed in this manner until the solution of the basic flow problem is well advanced and the steady state has nearly been reached we have a solution which is reasonably accurate for small waves but which can now be used as a starting point of the mesh adjustment if so desired.

In all practical calculations it is recommended that many steps of the *hydrostatic adjustment* be used before repeating the *mesh updating* which is quite expensive. In many ship problems it has been shown that with a single mesh quite good results can be obtained without the necessity of proceeding with mesh adjustment. We shall refer to such examples later.

The methodologies suggested here follow the work of Hino *et al.*, Idelshon *et al.*, Löhner *et al.* and Oñate *et al.*[13–18] The methods which we discussed in the context of ships here provide a basis on which other free surface problems could be started at all times and are obviously an improvement on a very primitive adjustment of surface by trial and error. However, some authors recommend alternatives such as pseudo-concentration methods,[19] which are more useful in the context of mould filling,[20–22] etc. We shall not go into that in detail further and interested readers can consult the necessary references.

5.2.4 Numerical examples

Example 1. A submerged hydrofoil We start with the two-dimensional problem shown in Fig. 5.3, where a NACA0012 aerofoil profile is used in submerged form as a hydrofoil which could in the imagination of the reader be attached to a ship. This is a model problem, as many two-dimensional situations are not realistic. Here the angle of attack of the flow is 5° and the Froude number is 0.5672. The Froude number is defined as

$$Fr = \frac{|\mathbf{U}_\infty|}{\sqrt{gL}} \tag{5.7}$$

In Fig. 5.4, we show the pressure distribution throughout the domain and the comparison of the computed wave profiles with the experimental[23] and other numerical solutions.[14] In Figs 5.3 and 5.4, the mesh is moved after a certain number of iterations using an advancing front technique.

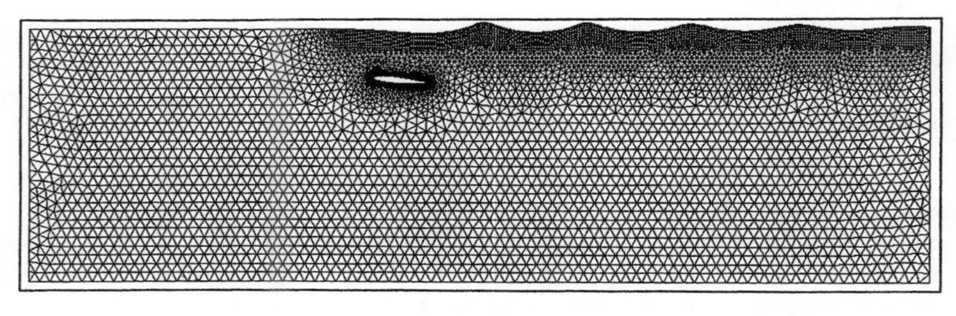

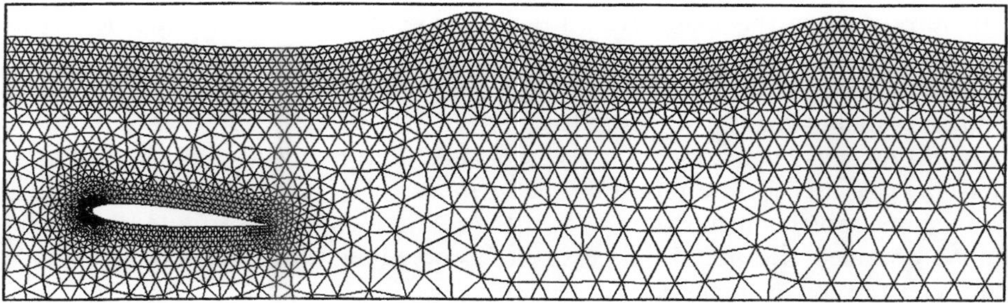

Fig. 5.3 A submerged hydrofoil. Mesh updating procedure. Euler flow. Mesh after 1900 iterations.

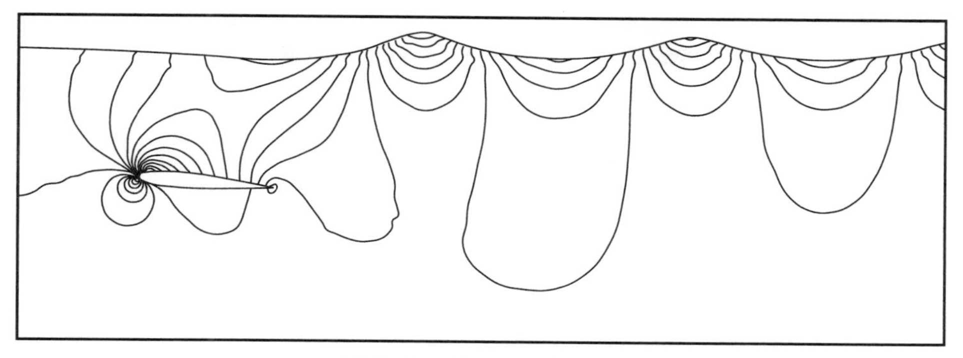

(a) Surface Pressure Contour

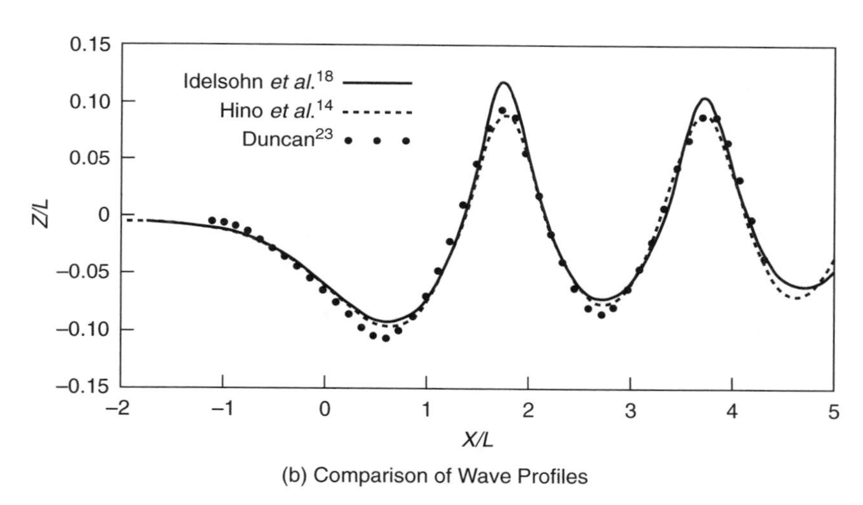

(b) Comparison of Wave Profiles

Fig. 5.4 A submerged hydrofoil. Mesh updating procedure. Euler flow. (a) Pressure distribution. (b) Comparison with experiment.

Figure 5.5 shows the same hydrofoil problem solved now using hydrostatic adjustment without moving the mesh. For the same conditions, the wave profile is somewhat under-predicted by the hydrostatic adjustment (Fig. 5.5(b)) while the mesh movement over-predicts the peaks (Fig. 5.4(b)).

In Fig. 5.6, the results for the same hydrofoil in the presence of viscosity are presented for different Reynolds numbers. As expected the wake is now strong as seen from the velocity magnitude contours (Fig. 5.6(a–d)). Also at higher Reynolds numbers (5000 and above), the solution is not stable behind the aerofoil and here an unstable vortex street is predicted as shown in Fig. 5.6(c) and 5.6(d). Figure 5.6(e) shows the comparison of wave profiles for different Reynolds numbers.

Example 2. Submarine In Fig. 5.7, we show the mesh and wave pattern contours for a submerged DARPA submarine model. Here the Froude number is 0.25. The converged solution is obtained by about 1500 time steps using a parallel computing environment. The mesh consisted of approximately 321 000 tetrahedral elements.

(a) Surface pressure contour

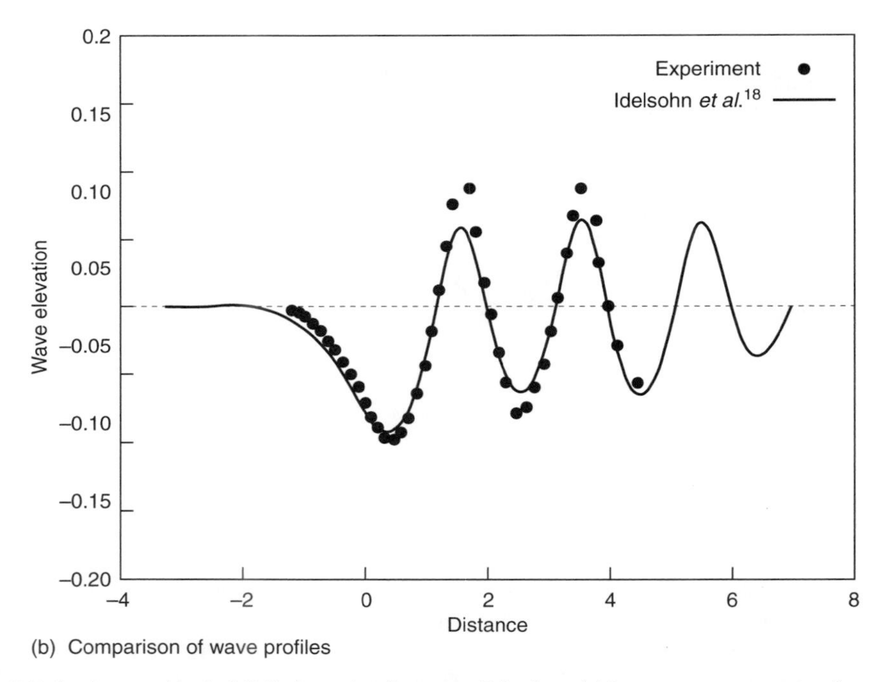

(b) Comparison of wave profiles

Fig. 5.5 A submerged hydrofoil. Hydrostatic adjustment. Euler flow. (a) Pressure contours and surface wave pattern. (b) Comparison with experiment.[23]

Example 3. Sailing boat The last example presented here is that of a sailing boat. In this case the boat has a 25° heel angle and a drift angle of 4°. Here it is essential to use either Euler or Navier–Stokes equations to satisfy the Kutta–Joukoski condition as the potential form has difficulty in satisfying these conditions on the trailing edge of the keel and rudder.

Here we used the Euler equations to solve this problem. Figure 5.8(a) shows a surface mesh of hull, keel, bulb and rudder. A total of 104 577 linear tetrahedral elements were used in the computation. Figure 5.8(b) shows the wave profile contours corresponding to a sailing speed of 10 knots.

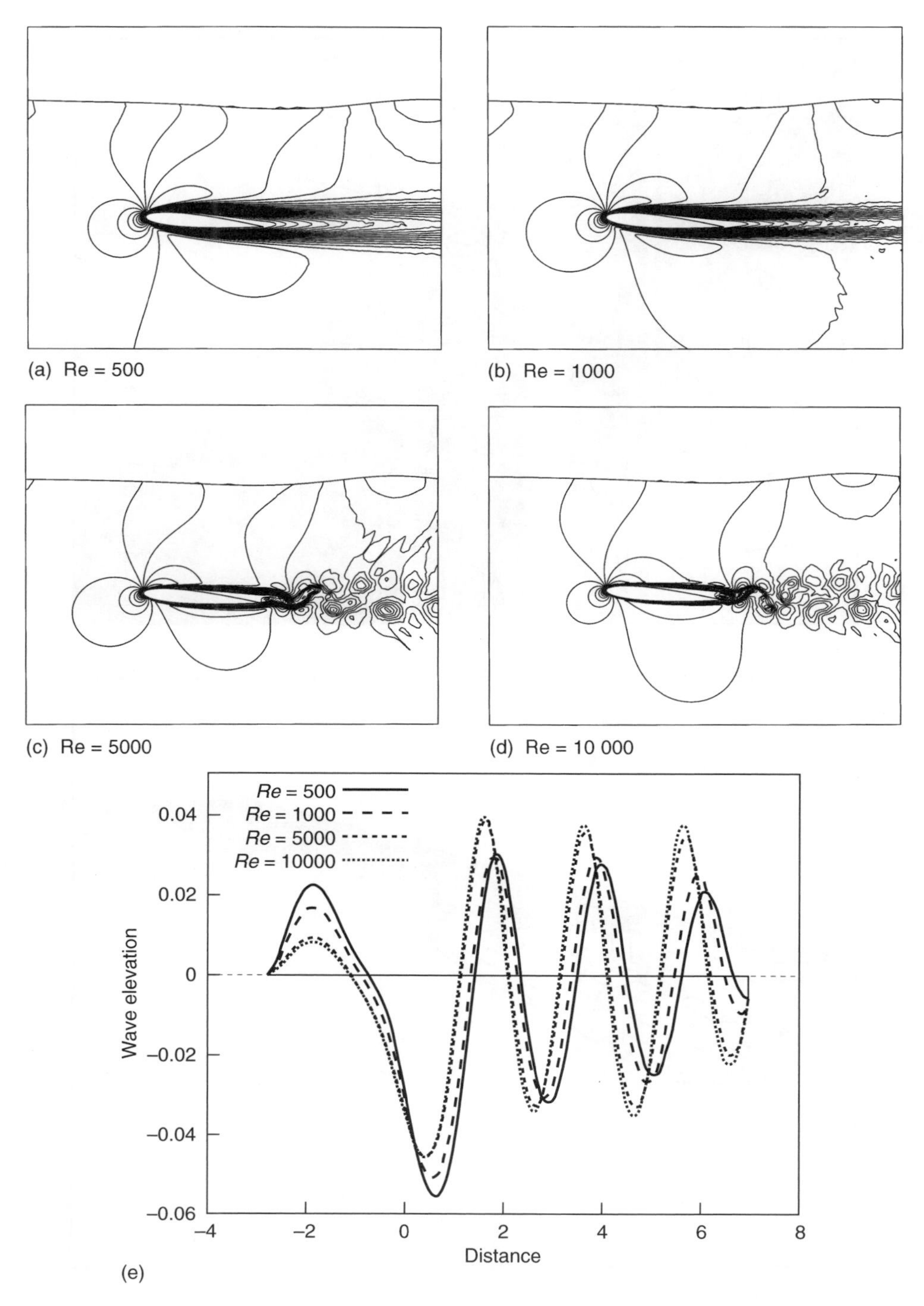

Fig. 5.6 A submerged hydrofoil. Hydrostatic adjustment. Navier–Stokes flow. (a)–(d) Magnitude of total velocity contours for different Reynolds numbers. (e) Wave profiles for different Reynolds numbers.

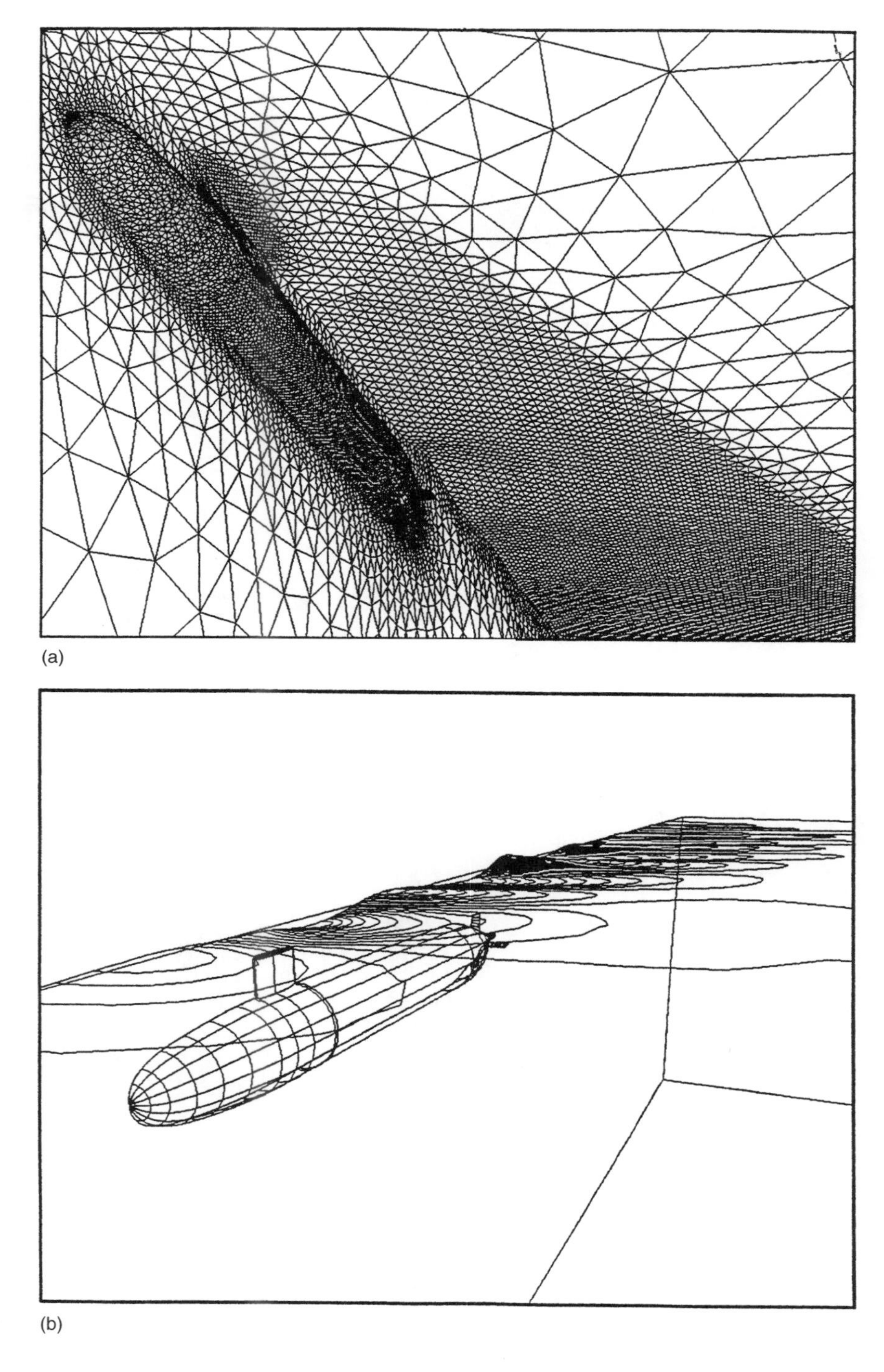

Fig. 5.7 Submerged DARPA submarine model. (a) Surface mesh. (b) Wave pattern.

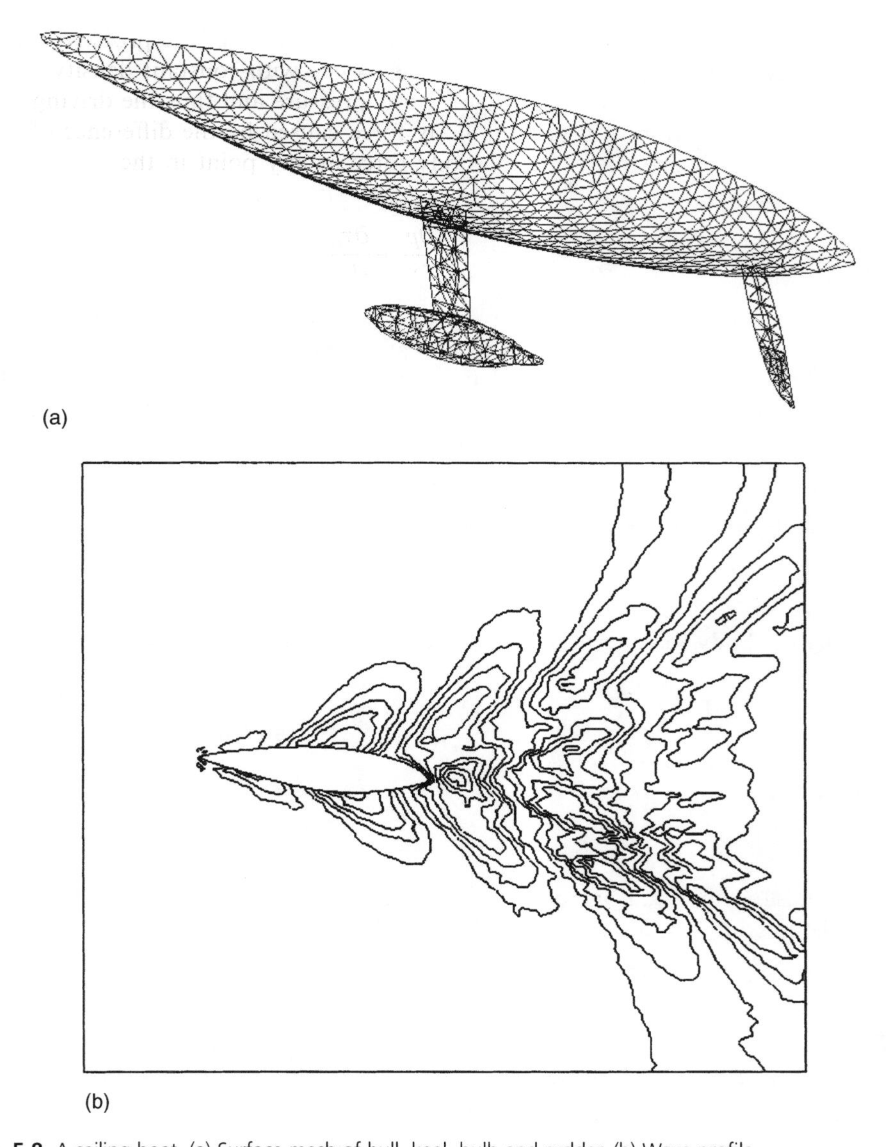

Fig. 5.8 A sailing boat. (a) Surface mesh of hull, keel, bulb and rudder. (b) Wave profile.

5.3 Buoyancy driven flows

5.3.1 General introduction and equations

In some problems of incompressible flow the heat transport equation and the equations of motion are weakly coupled. If the temperature distribution is known at any time, the density changes caused by this temperature variation can be evaluated. These may on occasion be the only driving force of the problem. In this situation it is convenient to note that the body force with constant density can be considered as balanced by an initial hydrostatic pressure and thus the driving force which causes the motion is in fact the body force caused by the difference of local density values. We can thus write the body force at any point in the equations of motion (4.2) as

$$\rho\left[\frac{\partial u_i}{\partial t}+\frac{\partial}{\partial x_j}(u_j u_i)\right]=-\frac{\partial p}{\partial x_i}+\frac{\partial \tau_{ij}}{\partial x_j}-g_i(\rho-\rho_\infty) \tag{5.8}$$

where ρ is the actual density applicable locally and ρ_∞ is the undisturbed constant density. The actual density entirely depends on the coefficient of thermal expansion of the fluid as compressibility is by definition excluded. Denoting the coefficient of thermal expansion as β_T, we can write

$$\beta_T=\frac{1}{\rho}\left(\frac{\partial \rho}{\partial T}\right) \tag{5.9}$$

where T is the absolute temperature. The above equation can be approximated to

$$\beta_T \approx \frac{1}{\rho}\frac{\rho-\rho_\infty}{T-T_\infty} \tag{5.10}$$

Replacing the body force term in the momentum equation by the above relation we can write

$$\rho\left[\frac{\partial u_i}{\partial t}+\frac{\partial}{\partial x_j}(u_j u_i)\right]=-\frac{\partial p}{\partial x_i}+\frac{\partial \tau_{ij}}{\partial x_j}+\beta_T g_i(T_\infty-T) \tag{5.11}$$

For perfect gases, we have

$$\rho=\frac{p}{RT} \tag{5.12}$$

and here R is the universal gas constant. Substitution of the above equation (assuming negligible pressure variation) into Eq. (5.9) leads to

$$\beta_T=\frac{1}{T} \tag{5.13}$$

The various governing non-dimensional numbers used in the buoyancy flow calculations are the Grashoff number (for a non-dimensionalization procedure see references 24, 25)

$$Gr=\frac{g\beta_T(T_\infty-T)L^3}{\nu\alpha} \tag{5.14}$$

and the Prandtl number

$$Pr=\frac{\nu}{\alpha} \tag{5.15}$$

where L is a reference dimension, and ν and α are the kinematic viscosity and thermal diffusivity respectively and are defined as

$$\nu=\frac{\mu}{\rho}, \qquad \alpha=\frac{k}{\rho c_p} \tag{5.16}$$

where μ is the dynamic viscosity, k is the thermal conductivity and c_p the specific heat at constant pressure. In many calculations of buoyancy driven flows, it is convenient to use another non-dimensional number called the Rayleigh number (Ra) which is the product of Gr and Pr.

In many practical situations, both buoyancy and forced flows are equally strong and such cases are often called mixed convective flows. Here in addition to the above-mentioned non-dimensional numbers, the Reynolds number also plays a role. The reader can refer to several available books and other publications to get further details.[26–39]

5.3.2 Natural convection in cavities

Fundamental buoyancy flow analysis in closed cavities can be classified into two categories. The first one is flow in closed cavities heated from the vertical sides and the second is bottom-heated cavities (Rayleigh–Benard convection). In the former, the CBS algorithm can be applied directly. However, the latter needs some perturbation to start the convective flow as they represent essentially an unstable problem.

Figure 5.9 shows the results obtained for a closed square cavity heated at a vertical side and cooled at the other.[24] Both the horizontal sides are assumed to be adiabatic. At all surfaces both of the velocity components are zero (no slip conditions). The nonuniform mesh used in this problem is the same as that in Fig. 4.3 of the previous chapter for all Rayleigh numbers considered.

As the reader can see, the essential features of a buoyancy driven flow are captured using the CBS algorithm. The quantitative results compare excellently with the available benchmark solutions as shown in Tables 5.1.[24]

Figure 5.10 shows the effect of directions of gravity at a Rayleigh number of 10^6.[31] The adapted meshes for two different Rayleigh numbers are shown in Fig. 5.11.

Another problem of buoyancy driven convection in closed cavities is shown in Fig. 5.12.[25] Here an 'L' shaped cavity is considered where part of the enclosure is heated from the side and another part from the bottom. As we can see, several vortices appear in the horizontal portion of the cavity while the vertical portion contains only one vortex.

Table 5.1 Natural convection in a square enclosure. Comparison with available numerical solutions.[24] References are shown in square brackets

Ra	ν			ψ_{max}			v_{max}		
	[40]	[41]	CBS	[40]	[41]	CBS	[40]	[41]	CBS
10^3	1.116	1.118	1.117	1.174	1.175	1.167	3.696	3.697	3.692
10^4	2.243	2.245	2.243	5.081	5.074	5.075	19.64	19.63	19.63
10^5	4.517	4.522	4.521	9.121	9.619	9.153	68.68	68.64	68.85
10^6	8.797	8.825	8.806	16.41	16.81	16.49	221.3	220.6	221.6
10^7	–	16.52	16.40	–	30.17	30.33	–	699.3	702.3
4×10^7	–	23.78	23.64	–	–	43.12	–	–	1417

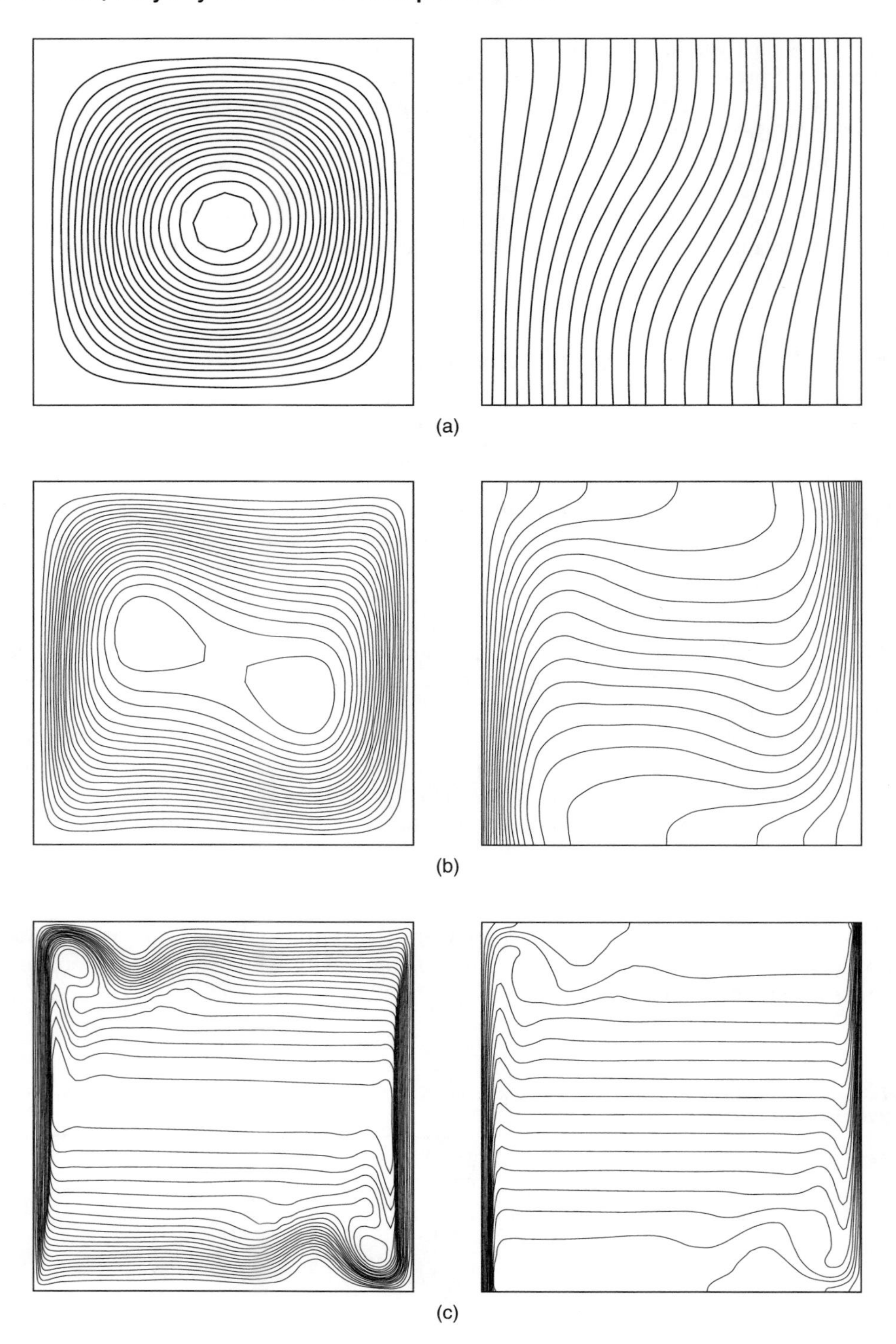

Fig. 5.9 Natural convection in a square enclosure. Streamlines and isotherms for different Rayleigh numbers.

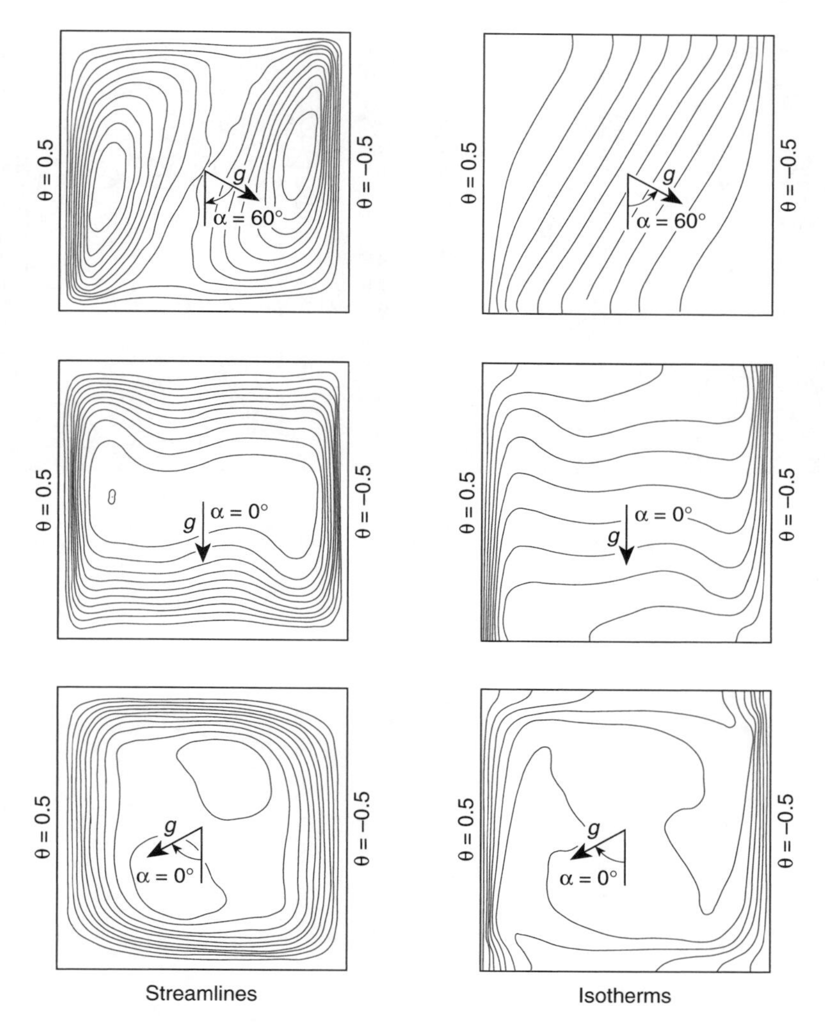

Fig. 5.10 Natural convection in a square enclosure. Streamlines and isotherms for different gravity directions, $Ra = 10^6$.

5.3.3 Buoyancy in porous media flows

Studies of convective motion and heat transfer in a porous medium are essential to understand many engineering problems including solidification of alloys, convection over heat exchanger tubes, thermal insulations, packed and fluidized beds, etc. We give a brief introduction to such flows in this section.

Porous medium flows are different from those of single-phase fluid flows due to the presence of the solid particles which for our purpose are considered as rigidly fixed in space. Many textbooks on porous medium flows are already available.[42,43] Similar porous media occur in problems of geomechanics in which generally the motion of

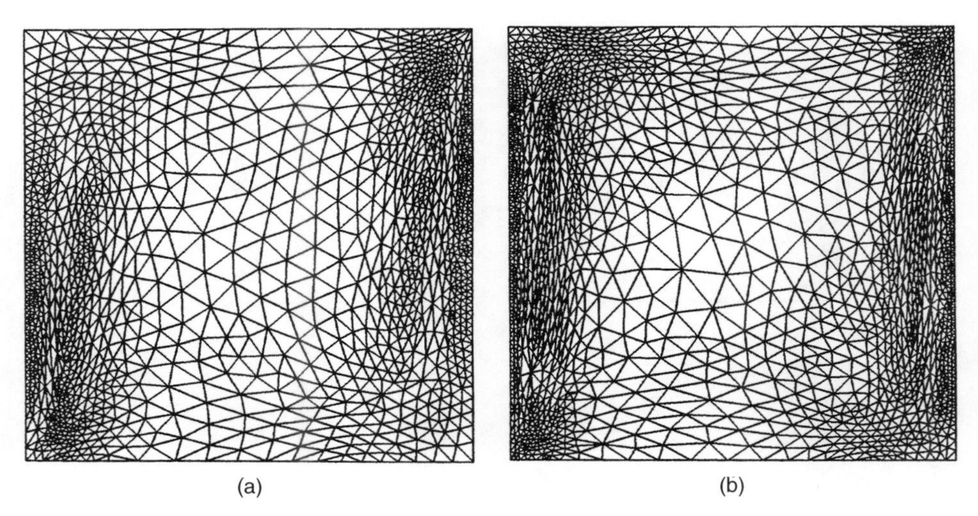

Fig. 5.11 Natural convection in a square enclosure. Adapted meshes for (a) $Ra = 10^5$ and (b) $Ra = 10^6$.

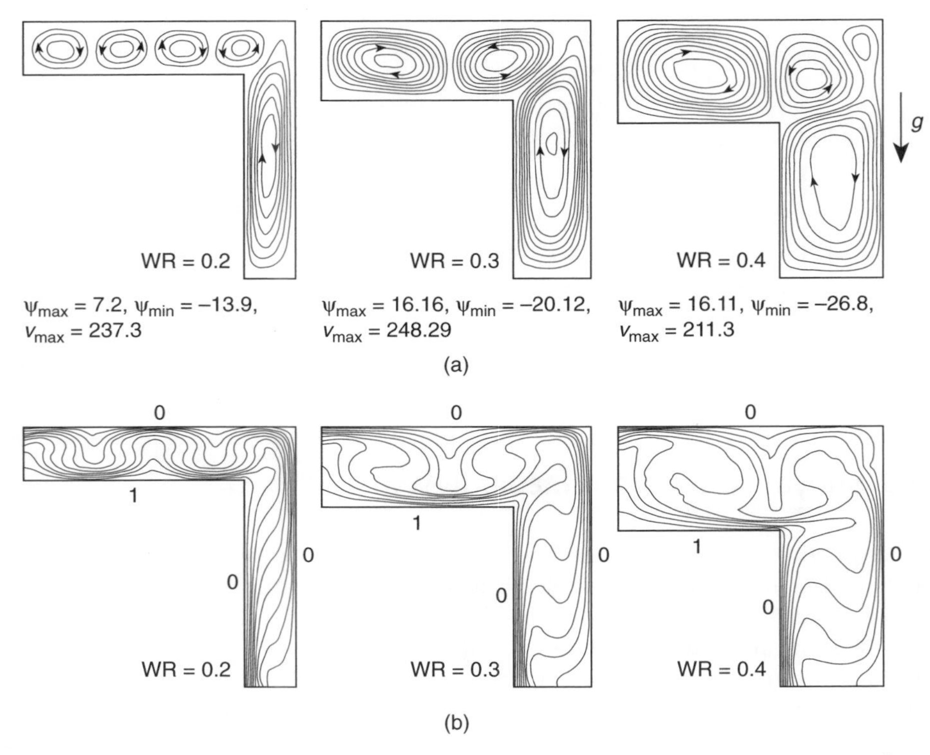

Fig. 5.12 Natural convection in an 'L' shaped enclosure. (a) Streamlines and (b) isotherms, $Ra = 10^6$.

the fluid and of the solid are coupled. For a survey of this problem the reader is referred to the recent book by Zienkiewicz *et al.*[44]

Here we use the averaged governing equations derived by many investigators to solve buoyancy driven convection in a porous medium.[45,46] These equations can be summarized for a variable porosity medium as[46]

continuity

$$\frac{\partial u_i}{\partial x_i} = 0 \tag{5.17}$$

momentum

$$\frac{\rho_f}{\varepsilon}\left[\frac{\partial u_i}{\partial t} + \frac{\partial}{\partial x_j}\left(\frac{u_j u_i}{\varepsilon}\right)\right] = -\frac{1}{\varepsilon}\frac{\partial}{\partial x_i}(p\varepsilon) + \frac{1}{\varepsilon}\frac{\partial}{\partial x_i}\left(\mu\frac{\partial u_i}{\partial x_i}\right) - \frac{\mu u_i}{\kappa} - C\frac{\rho_f}{\sqrt{\kappa}}\frac{\sqrt{u_k u_k}}{\varepsilon^{3/2}}u_i + g_i\beta(T_\infty - T) \tag{5.18}$$

energy

$$R_h\left[\frac{\partial T}{\partial t} + u_i\frac{\partial T}{\partial x_i}\right] = \frac{\partial}{\partial x_i}\left(k\frac{\partial T}{\partial x_i}\right) \tag{5.19}$$

where u_i are the averaged velocity components, ε is the porosity of the medium, κ is the medium permeability, C is a constant derived from experimental correlations and here we use Ergun's relations[47] in our calculations (some investigators vary the non-linear term using a non-dimensional parameter called the Forchheimer number; interested readers can consult reference 48), k is the thermal conductivity of the porous medium and R_h is the averaged heat capacity given as

$$R_h = \varepsilon(\rho c_p)_f + (1-\varepsilon)(\rho c_p)_s \tag{5.20}$$

In the above equations, subscripts f and s correspond to fluid and solid respectively. The following relation for permeability can be used if the porosity and average particle size are known

$$\kappa = \frac{\varepsilon^3 d_p^2}{150(1-\varepsilon^2)} \tag{5.21}$$

where d_p is the particle size. Some researchers use a value for μ different from the fluid viscosity. However, here we generally use the fluid viscosity. More details on the derivation of the above equations can be found in the cited articles.

As the structure of the above governing equations is similar to that of the single-phase flow equations, the application of the CBS algorithm is obvious.[49–55] However the fully explicit or semi-implicit forms cannot be used efficiently due to strong porous medium terms. Here, to overcome the time step limitations imposed by these terms (last two terms before the body force in the momentum equation) we need to solve them implicitly, though quasi-implicit schemes[47,49] can be used. Although the CBS algorithm is an obvious choice here, use of convection stabilizing terms can be neglected in low Rayleigh number (Reynolds number) porous media flows.

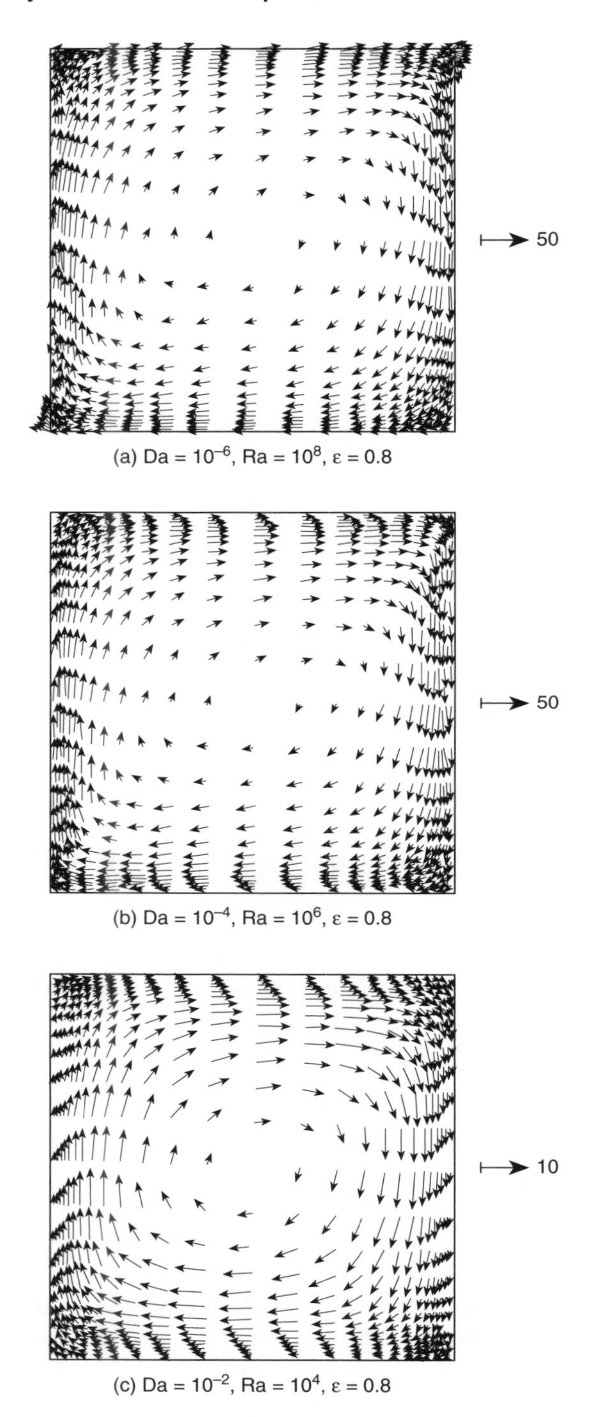

Fig. 5.13 Natural convection in a square enclosure filled with a fluid saturated porous medium, $Pr = 1$, velocity vector plots for different Rayleigh and Darcy numbers.

The Darcy number and thermal conductivity ratio are the two additional non-dimensional parameters used in porous media flows in addition to the Rayleigh and Prandtl numbers. The Darcy number and thermal conductivity ratio are defined respectively as

$$Da = \frac{\kappa}{L^2} \qquad k^* = \frac{k}{k_{\text{ref}}} \tag{5.22}$$

where k_{ref} is a reference thermal conductivity value (fluid value).

Figure 5.13 shows the velocity vector plots of buoyancy driven convection in a square cavity for different Darcy and Rayleigh numbers.[46] As we can see, at smaller Darcy numbers (10^{-6}) the velocity is higher near the walls and decreases towards the centre of the enclosure (Fig. 5.13(a)). However at higher Darcy numbers (10^{-2}), a pattern similar to single-phase flow is obtained with the velocity increasing from zero at the walls to a maximum value and then decreasing towards the centre of the cavity, indicating the viscous effects. Figure 5.13(b) shows a condition between Figs 5.13(a) and 5.13(b) and here the transition from the Darcy to non-Darcy flow regime occurs.

These governing equations approach a set of single-phase fluid equations when $\varepsilon \to 1$. Thus these equations are suitable for solving problems in which both a porous medium and single-phase domains are involved.

5.4 Turbulent flows

5.4.1 General remarks

We have observed that in many situations of viscous flow it is impossible to obtain steady-state results. The example of flow past a cylinder given in Fig. 4.16 illustrates the point well. As the speed increases the steady-state picture becomes oscillatory and the well-known von Karman street develops.

For higher speed the oscillations and eddies become smaller and distributed throughout the whole fluid domain. Whenever this happens the situation is that of turbulence and here unfortunately direct simulation is almost out of the question though many attempts at doing so are being made for realistic problems. It would be necessary to use many millions or hundreds of millions of elements to model reasonably the behaviour of the flow in real situations at high Reynolds numbers where turbulence is large, and for this reason attempts have been made to create approximate models which can be time averaged.[56–64] Here continuation of the direct numerical simulation (DNS)[65,66] is used in so-called large eddy simulation (LES)[67,68] but that is also very costly. For this reason simpler models involving n additional equations have been created and perform reasonably satisfactorily although they cannot always represent reality.

We do not have space in this book to implement and discuss all the above models in detail. The reader will observe that, in addition to solving the flow equations with viscosity which now varies from point to point, it is necessary to solve n additional effective transport equations each one corresponding to a specific defined parameter.

Such calculations can readily be carried out by the same algorithm as that used in CBS and indeed this was done for some problems.[64]

5.4.2 Averaged flow equations

The Reynolds averaged Navier–Stokes equations can be derived by considering the flow variables as

$$\phi = \bar{\phi} + \phi' \tag{5.23}$$

where $\bar{\phi}$ is the mean turbulent value and ϕ' is the fluctuating component. With such averaged quantities the governing equations can be rewritten as

continuity

$$\frac{\partial \overline{u_i}}{\partial x_i} = 0 \tag{5.24}$$

momentum

$$\frac{\partial \overline{u_i}}{\partial x_i} = \frac{\partial}{\partial x_j}(\overline{u_j}\,\overline{u_i}) - \frac{1}{\rho}\frac{\partial p}{\partial x_i} + \frac{1}{\rho}\frac{\partial \tau_{ij}}{\partial x_j} + \frac{\partial \tau_{ij}^R}{\partial x_j} + g_i \tag{5.25}$$

where τ_{ij} is the deviatoric stress tensor (Eq. 3.7) given as

$$\tau_{ij} = \mu\left(\frac{\partial u_i}{\partial x_j} + \frac{\partial u_j}{\partial x_i} - \frac{2}{3}\frac{\partial u_k}{\partial x_k}\delta_{ij}\right) \tag{5.26}$$

and τ_{ij}^R is the Reynolds stress tensor divided by the density. We use the first-order closure models and here the Boussinesq model is employed which relates the shear stresses and turbulent eddy viscosity ν_T. The turbulent viscosity can be calculated by different methods. We use the one and two equation models to demonstrate the application of the CBS algorithm. We can write the following relation from the Boussinesq model

$$\tau_{ij}^R = -\overline{u_i' u_j'} = \nu_T\left(\frac{\partial \overline{u_i}}{\partial x_j} + \frac{\partial \overline{u_j}}{\partial x_i} - \frac{2}{3}\frac{\partial \overline{u_k}}{\partial x_k}\delta_{ij}\right) - \frac{2}{3}\kappa\delta_{ij} \tag{5.27}$$

where ν_T is the turbulent eddy kinematic viscosity and κ is the turbulent kinetic energy. The reader will observe that the form of the original equations (governing laminar flow) is now reproduced in terms of the averaged quantities, thus confirming that the standard CBS algorithm can be used once again. Before proceeding further, it is necessary to define the turbulent eddy viscosity which we do below.

One-equation model

In the momentum equation the turbulent eddy viscosity is determined from the following relation

$$\nu_T = c_\mu^{1/4}\kappa^{1/2} l_m \tag{5.28}$$

where c_μ is a constant equal to 0.09, κ is the turbulent kinetic energy and l_m is the Prandtl mixing length (= $0.4y$, where y is the distance from the nearest wall). The Prandtl mixing length l_m is often related to the length scale of the turbulence L as

$$l_m = \left(\frac{c_\mu'^3}{C_D}\right)^{1/4} L \tag{5.29}$$

where C_D and c_μ' are constants.

The turbulent kinetic energy κ is calculated from the following transport equation

$$\frac{\partial \kappa}{\partial t} + \frac{\partial u_i \kappa}{\partial x_j} - \frac{\partial}{\partial x_i}\left(\nu + \frac{\nu_T}{\sigma_\kappa}\right)\frac{\partial \kappa}{\partial x_i} - \tau_{ij}^R \frac{\partial u_i}{\partial x_j} + \varepsilon = 0 \tag{5.30}$$

where σ_k is a constant generally equal to unity. Further,

$$\varepsilon = C_D \frac{\kappa^{3/2}}{L} \tag{5.31}$$

Two-equation models (κ-ε and κ-ω models)

Here in addition to the κ equation given above, another transport equation of the form

$$\frac{\partial \varepsilon}{\partial t} + \frac{\partial u_i \varepsilon}{\partial x_j} - \frac{\partial}{\partial x_i}\left(\nu + \frac{\nu_T}{\sigma_\varepsilon}\right)\frac{\partial \varepsilon}{\partial x_i} - C_{\varepsilon 1}\frac{\varepsilon}{\kappa}\tau_{ij}^R \frac{\partial u_i}{\partial x_j} + C_{\varepsilon 2}\frac{\varepsilon^2}{\kappa} = 0 \tag{5.32}$$

is solved and here $C_{\varepsilon 1}$ is a constant ranging between 1.45 and 1.55, $C_{\varepsilon 2}$ is a constant in the range 1.92–2.0 and σ_ε is also a constant equal to 1.3.

In the above two-equation model, ν_T is calculated as

$$\nu_T = c_\mu \frac{\kappa^2}{\varepsilon} \tag{5.33}$$

These models are not valid near walls. To model wall effects, either wall functions or low Reynolds number versions have to be employed. For further details on these models the reader can refer to the relevant works.[56,62] We give the following low Reynolds number versions for the sake of completeness.

Low Reynolds number models

For the one-equation model, the following form is suggested by Wolfstein[56]

$$\nu_t = c_\mu^{1/4} \kappa^{1/2} l_m f_\mu \tag{5.34}$$

$$\varepsilon = C_D \frac{\kappa^{3/2}}{L f_b} \tag{5.35}$$

and

$$f_\mu = 1 - \mathrm{e}^{-0.160 R_k}, \qquad f_b = 1 - \mathrm{e}^{-0.263 R_k}, \qquad R_k = \sqrt{\kappa}\frac{y}{\nu} \tag{5.36}$$

where y is the distance from the nearest wall.

For two-equation models, the coefficients c_μ, $C_{\varepsilon 1}$ and $C_{\varepsilon 2}$ appearing in the two-equation model discussed above are multiplied by damping functions f_μ, $f_{\varepsilon 1}$ and $f_{\varepsilon 2}$

respectively and these functions are given as[62]

$$f_\mu = (1 - e^{-0.0165 R_k})^2 \left(1 + \frac{20.5}{R_t}\right) \tag{5.37}$$

$$f_{\varepsilon 1} = 1 + \left(\frac{0.05}{f_\mu}\right)^3 \tag{5.38}$$

and

$$f_{\varepsilon 2} = 1 - e^{-R_t^2} \tag{5.39}$$

where $R_t = \kappa^2/\nu\varepsilon$. The wall boundary conditions are $\kappa = 0$ and $\partial\varepsilon/\partial y = 0$.

A model of somewhat similar form is known as the κ-ω model. This differs in the definition of the function ω which obeys a similar equation to that of ε now with a different parameter.[69]

The reader can now notice that the one and two equation models are again similar to the convection–diffusion equations discussed in Chapter 2 and thus the use of the CBS algorithm is obvious. A detailed study is described in reference 62. Here we give some results of flow past a backward facing step at a Reynolds number of 3025. In Fig. 5.14(a) the velocity profiles are compared with the experimental data of Denham *et al.*[70] As can be seen the agreement between the results is good. The streamlines and details of the recirculation are given in Fig. 5.14(b).

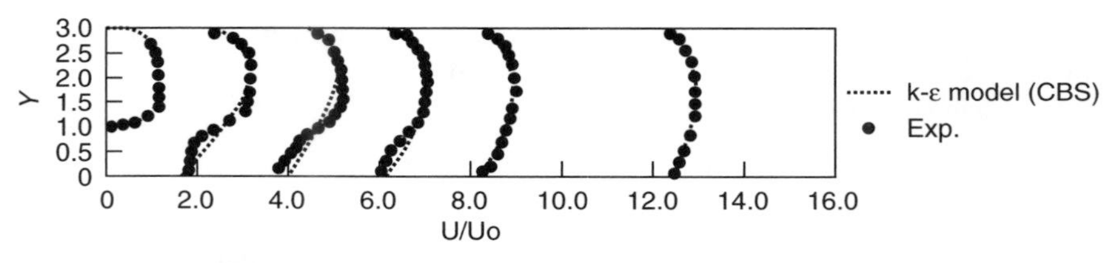

(a) Velocity profiles downward of the step

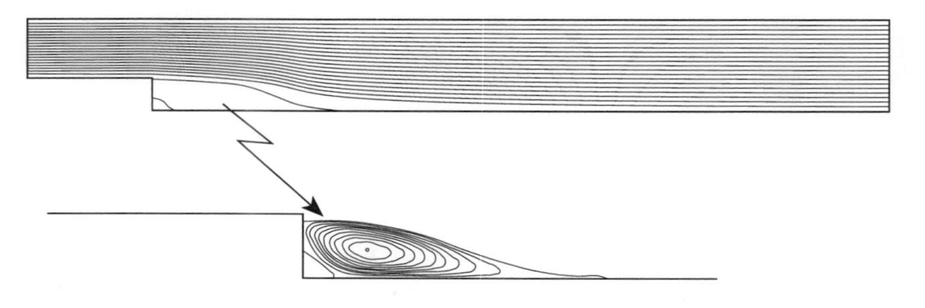

(b) Streamline pattern

Fig. 5.14 Turbulent flow past a backward facing step, velocity profiles and streamlines, $Re = 3025$.

References

1. J.V. Wehausen. The wave resistance of ships. *Adv. Appl. Mech.*, 1970.
2. C.W. Dawson. A practical computer method for solving ship wave problems. *Proc. of the 2nd Int. Conf. on Num. Ship Hydrodynamics*, USA, 1977.
3. K.J. Bai and J.H. McCarthy, Ed. *Proc. of the second DTNSRDC Workshop on Ship Wave Resistance Computations*, Bethesda, MD, USA, 1979.
4. F. Noblesse and J.H. McCarthy, Ed. *Proc. of the second DTNSRDC Workshop on Ship Wave Resistance Computations*, Bethesda, MD, USA, 1983.
5. G. Jenson and H. Soding. Ship wave resistance computation. *Finite Approximations in Fluid Mechanics II*, **25**, 1989.
6. Y.H. Kim and T. Lucas. Non-linear ship waves. *Proc. of the 18th Symp. on Naval Hydrodynamics*, MI, USA, 1990.
7. D.E. Nakos and Pd. Scavounos. On the steady and unsteady ship wave patterns. *J. Fluid Mech.*, **215**, 256–88, 1990.
8. H. Raven. A practical non-linear method for calculating ship wave making and wave resistance. *Proc. 19th Sym. on Ship Hydrodynamics*, Seoul, Korea, 1992.
9. R.F. Beck, Y. Cao and T.H. Lee. Fully non-linear water wave computations using the desingularized method. *Proc. of the 6th Sym. on Num. Ship Hydrodynamics*, Iowa City, Iowa, USA, 1993.
10. H. Soding. Advances in panel methods. *Proc. 21st Sym. on Naval Hydrodynamics*, Trondheim, Norway, 1996.
11. C. Janson and L. Larsson. A method for the optimization of ship hulls from a resistance point of view. *Proc. 21st Sym. on Naval Hydrodynamics*, Trondheim, Norway, 1996.
12. C.C. Mei and H.S. Chen. A hybrid element method for steady linearized free surface flows. *Int. J. Num. Meth. Eng.*, **10**, 1153–75, 1976.
13. T. Hino. Computation of free surface flow around an advancing ship by Navier–Stokes equations. *Proc. of the 5th Int. Conf. Num. Ship Hydrodynamics*, 103–17, Hiroshima, Japan, 1989.
14. T. Hino, L. Martinelli and A. Jameson. A finite volume method with unstructured grid for free surface flow. *Proc. of the 6th Int. Conf. on Num. Ship Hydrodynamics*, Iowa city, IA, 173–94, 1993.
15. T. Hino. An unstructured grid method for incompressible viscous flows with free surface. *AIAA-97-0862*, 1997.
16. R. Löhner, C. Yang, E. Oñate and I.R. Idelsohn. An unstructured grid based, parallel free surface solver. *AIAA 97-1830*, 1997.
17. R. Löhner, C. Yang and E. Oñate. Free surface hydrodynamics using unstructured grids. *4th ECCOMAS CFD Conf.*, Athens, Sept 7–11, 1998.
18. I.R. Idelsohn, E. Oñate and C. Sacco. Finite element solution of free surface ship wave problems. *Int. J. Num. Meth. Eng.*, **45**, 503–28, 1999.
19. C.W. Hirt and B.D. Nichols. Volume of fluid (VOF) method for the dynamics of free surface boundaries. *J. Comp. Phy.*, **39**, 210–25, 1981.
20. R. Codina, U. Schäfer and E. Oñate. Mould filling simulation using finite elements. *Int. J. Num. Meth. Heat Fluid Flow*, **4**, 291–310, 1994.
21. K. Ravindran and R.W. Lewis. Finite element modelling of solidification effects in mould filling. *Finite Elements in Analysis and Design*, **31**, 99–116, 1998.
22. R.W. Lewis and K. Ravindran. Finite element simulation of metal casting. *Int. J. Num. Meth. Eng.*, **47**, (to appear, 2000).
23. J.H. Duncan. The breaking and non-breaking wave resistance of a two dimensional hydrofoil. *J. Fluid Mech.*, **126**, 507–16, 1983.

24. N. Massarotti, P. Nithiarasu and O.C. Zienkiewicz. Characteristic-based-split (CBS) algorithm for incompressible flow problems with heat transfer. *Int. J. Num. Meth. Fluids*, **8**, 969–90, 1998.
25. P. Nithiarasu, T. Sundararajan and K.N. Seetharamu. Finite element analysis of transient natural convection in an odd-shaped enclosure. *Int. J. Num. Meth. Heat Fluid Flow*, **8**, 199–220, 1998.
26. F.P. Incropera and D.P. Dewitt. *Fundamentals of Heat and Mass Transfer*, 3rd Ed., Wiley, New York, 1990.
27. A. Bejan. *Heat Transfer*, Wiley, New York, 1993.
28. Y. Jaluria. *Natural Convection Heat Transfer*, Springer-Verlag, New York, 1980.
29. O.C. Zienkiewicz, R.H. Gallagher and P. Hood. Newtonian and non-Newtonian viscous incompressible flow. Temperature induced flows. Finite element solutions, in *Mathematics of Finite Elements and Applications*. Ed. J. Whiteman, Vol. II, Chap. 20, pp. 235–67, Academic Press, London, 1976.
30. J.C. Heinrich, R.S. Marshall and O.C. Zienkiewicz. Penalty function solution of coupled convective and conductive heat transfer, in *Numerical Methods in Laminar and Turbulent Flows*. Ed. C. Taylor, K. Morgan and C.A. Brebbia, pp. 435–47, Pentech Press, 1978.
31. M. Strada and J.C. Heinrich. Heat transfer rates in natural convection at high Rayleigh numbers in rectangular enclosures. *Numerical Heat Transfer*, **5**, 81–92, 1982.
32. J.C. Heinrich and C.C. Yu. Finite element simulation of buoyancy driven flow with emphasis on natural convection in horizontal circular cylinder. *Comp. Meth. Appl. Mech. Eng.*, **69**, 1–27, 1988.
33. B. Ramaswamy. Finite element solution for advection and natural convection flows. *Comp. Fluids*, **16**, 349–88, 1988.
34. B. Ramaswamy, T.C. Jue and J.E. Akin. Semi-implicit and explicit finite element schemes for coupled fluid thermal problems. *Int. J. Num. Meth. Eng.*, **34**, 675–96, 1992.
35. B.V.K. Sai, K.N. Seetharamu and P.A.A. Narayana. Finite element analysis of the effect of radius ratio on natural convection in an annular cavity. *Int. J. Num. Meth. Heat Fluid Flow*, **3**, 305–18, 1993.
36. C. Nonino and S. Delgiudice. Finite element analysis of laminar mixed convection in the entrance region of horizontal annular ducts. *Numerical Heat Transfer, Part A, Applications*, **29**, 313–30, 1996.
37. S.C. Lee and C.K. Chen. Finite element solutions of laminar and turbulent mixed convection in a driven cavity. *Int. J. Num. Meth. Fluids*, **23**, 47–64, 1996.
38. S.C. Lee, C.Y. Cheng and C.K. Chen. Finite element solutions of laminar and turbulent flows with forced and mixed convection in an air-cooled room. *Numerical Heat Transfer, Part A, Applications*, **31**, 529–50, 1997.
39. Y.T.K. Gowda, P.A.A. Narayana and K.N. Seetharamu. Finite element analysis of mixed convection over in-line tube bundles. *Int. J. Heat Mass Transfer*, **41**, 1613–19, 1998.
40. G. De Vahl Davis. Natural convection of air in a square cavity: A benchmark numerical solution. *Int. J. Num. Meth. Fluids*, **3**, 249–64, 1983.
41. P. Le Quere and T.A. De Roquefort. Computation of natural convection in two dimensional cavity with Chebyshev polynomials. *J. Comp. Phy.*, **57**, 210–28, 1985.
42. D.A. Nield and A. Bejan. *Convection in Porous Media*, Springer-Verlag, New York, 1992.
43. M. Kaviany. *Principles of Heat Transfer in Porous Media*, Springer-Verlag, New York, 1991.
44. O.C. Zienkiewicz, A.H.C. Chan, M. Pastor, B.A. Schrefler and T. Shiomi. *Computational Geomechanics with Special Reference to Earthquake Engineering*, Wiley, Chichester, 1999.
45. K. Vafai and C.L. Tien. Boundary and inertial effects of flow and heat transfer in porous media. *Int. J. Heat Mass Transfer*, **24**, 195–203, 1981.

46. P. Nithiarasu, K.N. Seetharamu and T. Sundararajan. Natural convective heat transfer in an enclosure filled with fluid saturated variable porosity medium. *Int. J. Heat Mass Transfer*, **40**, 3955–67, 1997.
47. S. Ergun. Fluid flow through packed columns. *Chem. Eng. Prog.*, **48**, 89–94, 1952.
48. G. Lauriat and V. Prasad. Non-Darcian effects on natural convection in a vertical porous enclosure. *Int. J. Heat Mass Transfer*, **32**, 2135–48, 1989.
49. P. Nithiarasu, K.N. Seetharamu and T. Sundararajan. Double-diffusive natural convection in an enclosure filled with fluid saturated porous medium – a generalised non-Darcy approach. *Numerical Heat Transfer, Part A, Applications*, **30**, 413–26, 1996.
50. P. Nithiarasu, K.N. Seetharamu and T. Sundararajan. Effects of porosity on natural convective heat transfer in a fluid saturated porous medium. *Int. J. Heat Fluid Flow*, **19**, 56–8, 1998.
51. P. Nithiarasu and K. Ravindran. A new semi-implicit time stepping procedure for buoyancy driven flow in a fluid saturated porous medium. *Comp. Meth. Appl. Mech. Eng.*, **165**, 147–54, 1998.
52. P. Nithiarasu, K.N. Seetharamu and T. Sundararajan. Numerical prediction of buoyancy driven flow in a fluid saturated non-Darcian porous medium. *Int. J. Heat Mass Transfer*, **42**, 1205–15, 1999.
53. P. Nithiarasu. Finite element modelling of migration of a third component leakage from a heat source buried into a fluid saturated porous medium. *Math. Comp. Modelling*, **29**, 27–39, 1999.
54. P. Nithiarasu, K.S. Sujatha, K.N. Seetharamu and T. Sundararajan. Buoyancy driven flow in a non-Darcian fluid saturated porous enclosure subjected to uniform heat flux – a numerical study. *Comm. Num. Meth. Eng.*, **15**, 765–76, 1999.
55. P. Nithiarasu, N. Massarotti and O.C. Zienkiewicz. Finite element computations of porous media flows, (to be published).
56. M. Wolfstein. Some solutions of plane turbulent impinging jets. *ASME J. Basic Eng.*, **92**, 915–22, 1970.
57. B.E. Launder and D.B. Spalding. *Mathematical Models of Turbulence*. Academic Press, N.Y., 1972.
58. P. Hood and C. Taylor. Navier–Stokes equations using mixed interpolation, J.T. Oden *et al.* (Ed.). *Finite Element Method in Flow Problems*, UAH Press, Huntsville, AL, pp. 121–32, 1974.
59. K. Morgan, T.G. Hughes and C. Taylor. An investigation of a mixing length and two-equation turbulence model utilising the finite element method. *App. Math. Modelling*, **1**, 395–9, 1977.
60. C. Taylor, T.G. Hughes and K. Morgan. Finite element solution of one equation models of turbulent flow. *J. Comp. Phys.*, **29**, 163–72, 1978.
61. A.J. Baker. Finite element analysis of turbulent flows. *Proc. Int. Conf. Num. Meth. in Laminar and Turbulent Flows*, Swansea, 1978.
62. C.K.G. Lam and K. Bermhorst. A modified form of κ-ε model for predicting wall turbulence. *J. Fluids Eng.*, **103**, 456–60, 1981.
63. M. Srinivas, M.S. Ravisanker, K.N. Seetharamu and P.A. Aswathanarayana. Finite element analysis of internal flows with heat transfer. *Sadhana – Academy Proc. Eng.*, **19**, 785–816, 1994.
64. O.C. Zienkiewicz, B.V.K.S. Sai, K. Morgan and R. Codina. Split characteristic based semi-implicit algorithm for laminar/turbulent incompressible flows. *Int. J. Num. Meth. Fluids*, **23**, 1–23, 1996.
65. J. Jiménez, A. Wray, P. Saffman and R. Rogallo. The structure of intense vorticity in isotropic turbulence. *J. Fluid Mech.*, **255**, 65–90, 1993.
66. P. Moin and K. Mahesh. Direct numerical simulation: A tool in turbulence research. *Ann. Rev. Fluid Mech.*, **30**, 539–78, 1998.

67. R. Rogallo and P. Moin. Numerical simulation of turbulent flows. *Ann. Rev. Fluid Mech.*, **16**, 99–137, 1984.
68. M. Lesieur and O. Métais. New trends in large eddy simulations of turbulence. *Ann. Rev. Fluid Mech.*, **28**, 45–82, 1996.
69. M.T. Manzari, O. Hassan, K. Morgan and N.P. Weatherill. Turbulent flow computations on 3D unstructured grids. *Finite Elements in Analysis and Design*, **30**, 353–63, 1998.
70. M.K. Denham, P. Briard and M.A. Patrick. A directionally sensitive laser anemometer for velocity measurements in highly turbulent flow. *J. Phys. E: Sci. Instrum.*, **8**, 681–83, 1975.

6

Compressible high-speed gas flow

6.1 Introduction

Problems posed by high-speed gas flow are of obvious practical importance. Applications range from the *exterior flows* associated with flight to *interior flows* typical of turbomachinery. As the cost of physical experiments is high, the possibilities of computations were explored early and the development concentrated on the use of finite difference and associated finite volume methods. It was only in the 1980s that the potential offered by the finite element forms were realized and the field is expanding rapidly.

One of the main advantages in the use of the finite element approximation here is its capability of fitting complex forms and permitting local refinement where required. However, the improved approximation is also of substantial importance as practical problems will often involve three-dimensional discretization with the number of degrees of freedom much larger than those encountered in typical structural problems (10^5–10^7 DOF are here quite typical).

For such large problems direct solution methods are obviously not practicable and iterative methods based generally on transient computation forms are invariably used. Here of course we follow and accept much that has been established by the finite difference applications but generally will lose some computational efficiency associated with *structured meshes* typically used here. However, the reduction of the problem size which, as we shall see, can be obtained by local refinement and adaptivity will more than compensate for this loss (though of course structured meshes are included in the finite element forms).

In Chapters 1 and 3 we have introduced the basic equations governing the flow of compressible gases as well as of incompressible fluids. Indeed in the latter, as in Chapter 4, we can introduce small amounts of compressibility into the procedures developed there specifically for incompressible flow. Here we shall deal with high-speed flows with Mach numbers generally in excess of 0.5. Such flows will usually involve the formation of shocks with characteristic discontinuities. For this reason we shall concentrate on the use of low-order elements and of explicit methods, such as those introduced in Chapters 2 and 3.

Here the pioneering work of the first author's colleagues Morgan, Löhner and Peraire must be acknowledged.[1–38] It was this work that opened the doors to practical finite element analysis in the field of aeronautics. We shall refer to their work frequently.

In the first practical applications the Taylor–Galerkin process outlined in Sec. 2.10 of Chapter 2 for vector-valued variables was used almost exclusively. Here we recommend however the CBS algorithm discussed in Chapter 3 as it presents a better approximation and has the advantage of dealing directly with incompressibility, which invariably occurs in small parts of the domain, even at high Mach numbers (e.g., in stagnation regions).

6.2 The governing equations

The Navier–Stokes governing equations for compressible flow were derived in Chapter 1. We shall repeat only the simplified form of Eqs (1.24) and (1.25) here again using indicial notation. We thus write, for $i = 1, 2, 3$,

$$\frac{\partial \mathbf{\Phi}}{\partial t} + \frac{\partial \mathbf{F}_i}{\partial x_i} + \frac{\partial \mathbf{G}_i}{\partial x_i} + \mathbf{Q} = 0 \tag{6.1}$$

with

$$\mathbf{\Phi}^{\mathrm{T}} = [\rho, \rho u_1, \rho u_2, \rho u_3, \rho E] \tag{6.2a}$$

$$\mathbf{F}_i^{\mathrm{T}} = [\rho u_i,\ \ \rho u_1 u_i + p\delta_{1i},\ \ \rho u_2 u_1 + p\delta_{2,i},\ \ \rho u_3 u_i + p\delta_{3i},\ \ \rho H u_i] \tag{6.2b}$$

$$\mathbf{G}_i^{\mathrm{T}} = \left[0, -\tau_{1i}, -\tau_{2i}, -\tau_{3i}, -\frac{\partial}{\partial x_i}(\tau_{ij} u_i) - k\left(\frac{\partial T}{\partial x_i}\right)\right] \tag{6.2c}$$

and

$$\mathbf{Q}^{\mathrm{T}} = [0, -\rho f_1, -\rho f_2, -\rho f_3, -\rho f_i u_i - q_H] \tag{6.2d}$$

In the above

$$\tau_{ij} = \mu\left[\left(\frac{\partial u_i}{\partial x_j} + \frac{\partial u_j}{\partial x_i}\right) - \delta_{ij}\frac{2}{3}\frac{\partial u_k}{\partial x_k}\right] \tag{6.2e}$$

The above equations need to be 'closed' by addition of the constitutive law relating the pressure, density and energy [see Eqs (1.16) and (1.17)]. For many flows the ideal gas law[39] suffices and this is

$$\rho = \frac{p}{RT} \tag{6.3}$$

where R is the universal gas constant.

In terms of specific heats

$$R = (c_p - c_v) = (\gamma - 1)c_v \tag{6.4}$$

where

$$\gamma = \frac{c_p}{c_v}$$

is the ratio of the constant pressure and constant volume specific heats.

The internal energy e is given as

$$e = c_v T = \left(\frac{1}{\gamma - 1}\right)\frac{p}{\rho} \tag{6.5}$$

and hence

$$\rho E = \left(\frac{1}{\gamma - 1}\right)p + \frac{u_i u_i}{2} \tag{6.6a}$$

$$\rho H = \rho E + p = \left(\frac{\gamma}{\gamma - 1}\right)p + \frac{u_i u_i}{2} \tag{6.6b}$$

The variables for which we shall solve are usually taken as the set of Eq. (6.2a), i.e.

$$\rho, \rho u_i \text{ and } \rho E$$

but of course other sets could be used, though then the conservative form of Eq. (6.1) could be lost.

In many of the problems discussed in this section inviscid behaviour will be assumed, with

$$\mathbf{G}_i = \mathbf{0}$$

and we shall then deal with the *Euler equations*.

In many problems the Euler solution will provide information about the main features of the flow and will suffice for many purposes, especially if augmented by separate boundary layer calculations (see Sec. 6.12). However, in principle it is possible to include the viscous effects without much apparent complication. Here in general steady-state conditions will never arise as the high speed of the flow will be associated with turbulence and this will usually be of a small scale capable of resolution with very small sized elements only. If a 'finite' size of element mesh is used then such turbulence will often be suppressed and steady-state answers will be obtained only in areas of no flow separation or oscillation. We shall in some examples include such full Navier–Stokes solutions using a viscosity dependent on the temperature according to Sutherland's law.[39] In the SI system of units for air this gives

$$\mu = \frac{1.45T^{3/2}}{T + 110} \times 10^{-6} \tag{6.7}$$

where T is in degrees Kelvin. Further turbulence modelling can be done by using the Reynolds' average viscosity and solving additional transport equations for some additional parameters in the manner discussed in Sec. 5.4, Chapter 5. We shall show some turbulent examples later.

6.3 Boundary conditions – subsonic and supersonic flow

The question of boundary conditions which can be prescribed for Euler and Navier–Stokes equations in compressible flow is by no means trivial and has been addressed in a general sense by Demkowicz *et al.*,[40] determining their influence on the existence and uniqueness of solutions. In the following we shall discuss the case of the inviscid Euler form and of the full Navier–Stokes problem separately.

We have already discussed the general question of boundary conditions in Chapter 3 dealing with numerical approximations. Some of these matters have to be repeated in view of the special behaviour of supersonic problems.

6.3.1 Euler equation

Here only first-order derivatives occur and the number of boundary conditions is less than that for the full Navier–Stokes problem.

For a *solid wall boundary*, Γ_u, only the normal component of velocity u_n needs to be specified (zero if the wall is stationary). Further, with lack of conductivity the energy flux across the boundary is zero and hence ρE (and ρ) remain unspecified.

In general the analysis domain will be limited by some arbitrarily chosen *external boundaries*, Γ_s, for exterior or internal flows, as shown in Fig. 6.1 (see also Sec. 3.6, Chapter 3).

Here, as discussed in Sec. 2.10.3, it will in general be necessary to perform a linearized Riemann analysis in the direction of the outward normal to the boundary $\mathbf{n}$ to determine the speeds of wave propagation of the equations. For this linearization of the Euler equations three values of propagation speeds will be found

$$\begin{aligned} \lambda_1 &= u_n \\ \lambda_2 &= u_n + c \\ \lambda_3 &= u_n - c \end{aligned} \tag{6.8}$$

where u_n is the normal velocity component and c is the compressible wave celerity (speed of sound) given by

$$c = \sqrt{\frac{\gamma p}{\rho}} \tag{6.9}$$

As of course no disturbances can propagate at velocities greater than those of Eqs (6.8) and in the case of supersonic flow, i.e. when the local Mach number is

$$M = \frac{|u_n|}{c} \geqslant 1 \tag{6.10}$$

we shall have to distinguish two possibilities:

(a) *supersonic inflow boundary* where

$$u_n < -c$$

and the analysis domain cannot influence the exterior, for such boundaries all components of the vector $\mathbf{\Phi}$ must be specified; and

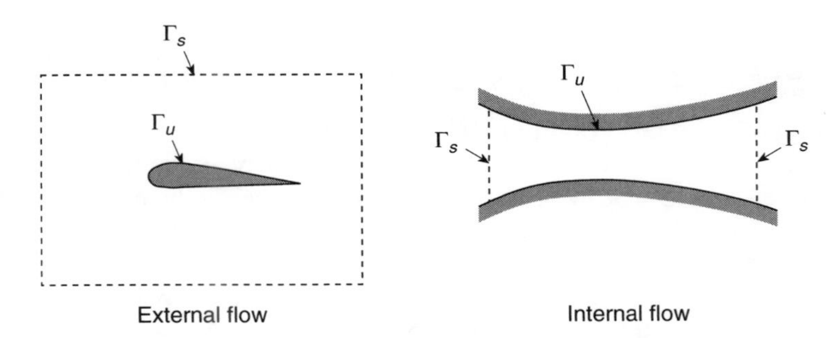

Fig. 6.1 Boundaries of a computation domain. Γ_u, wall boundary; Γ_s, fictitious boundary.

(b) *supersonic outflow boundaries* where

$$u_n > c$$

and here by the same reasoning no components of $\mathbf{\Phi}$ are prescribed.

For subsonic boundaries the situation is more complex and here the values of $\mathbf{\Phi}$ that can be specified are the components of the incoming Riemann variables. However, this may frequently present difficulties as the incoming wave may not be known and the usual compromises may be necessary as in the treatment of elliptic problems possessing infinite boundaries (see Chapter 3, Sec. 3.6).

6.3.2 Navier–Stokes equations

Here, due to the presence of second derivatives, additional boundary conditions are required.

For the *solid wall boundary*, Γ_u, all the velocity components are prescribed assuming, as in the previous chapter for incompressible flow, that the fluid is attached to the wall. Thus for a stationary boundary we put

$$u_i = 0$$

Further, if conductivity is not negligible, boundary temperatures or heat fluxes will generally be given in the usual manner.

For *exterior boundaries* Γ_s of the supersonic inflow kind, the treatment is identical to that used for Euler equations. However, for outflow boundaries a further approximation must be made, either specifying tractions as zero or making their gradient zero in the manner described in Sec. 3.6, Chapter 3.

6.4 Numerical approximations and the CBS algorithm

Various forms of finite element approximation and of solution have been used for compressible flow problems. The first successfully used algorithm here was, as we have already mentioned, the Taylor–Galerkin procedure either in its single-step or two-step form. We have outlined both of these algorithms in Chapter 2, Sec. 2.10. However the most generally applicable and advantageous form is that of the CBS algorithm which we have presented in detail in Chapter 3. We recommend that this be universally used as not only does it possess an efficient manner of dealing with the convective terms of the equations but it also deals successfully with the incompressible part of the problem. In all compressible flows in certain parts of the domain where the velocities are small, the flow is nearly incompressible and without additional damping the direct use of the Taylor–Galerkin method may result in oscillations there. We have indeed mentioned an example of such oscillations in Chapter 3 where they are pronounced near the leading edge of an aerofoil even at quite high Mach numbers (Fig. 3.4). With the use of the CBS algorithm such oscillations disappear and the solution is perfectly stable and accurate.

In the same example we have also discussed the single-step and two-step forms of the CBS algorithm. Both were found acceptable for use at lower Mach numbers.

However for higher Mach numbers we recommend the two-step procedure which is only slightly more expensive than the single-step version.

As we have already remarked if the algorithm is used for steady-state problems it is always convenient to use a localized time step rather than proceed with the same time step globally. The full description of the local time step procedure is given in Sec. 3.3.4 of Chapter 3 and this was invariably used in the examples of this chapter when only the steady state was considered.

We have mentioned in the same section, Sec. 3.3.4, the fact that when local time stepping is used nearly optimal results are obtained as Δt_{ext} and Δt_{int} are the same or nearly the same. However, even in transient problems it is often advantageous to make use of a different Δt in the interior to achieve nearly optimal damping there.

The only additional problem that we need to discuss further for compressible flows is that of the treatment of shocks which is the subject of the next section.

6.5 Shock capture

Clearly with the finite element approximation in which all the variables are interpolated using C_0 continuity the exact reproduction of shocks is not possible. In all finite element solutions we therefore represent the shocks simply as regions of very high gradient. The ideal situation will be if the rapid variations of variables are confined to a few elements surrounding the shock. Unfortunately it will generally be found that such an approximation of a discontinuity introduces local oscillations and these may persist throughout quite a large area of the domain. For this reason, we shall usually introduce into the finite element analysis additional viscosities which will help us in damping out any oscillations caused by shocks and, yet, deriving as sharp a solution as possible.

Such procedures using artificial viscosities are known as shock capture methods. It must be mentioned that some investigators have tried to allow the shock discontinuity to occur explicitly and thus allowed a discontinuous variation of an analytically defined kind. This presents very large computational difficulties and it can be said that to date such trials have only been limited to one-dimensional problems and have not really been used to any extent in two or three dimensions. For this reason we shall not discuss such *shock fitting* methods further.[41,42]

The concept of adding additional viscosity or diffusion to capture shocks was first suggested by von Neumann and Richtmyer[43] as early as 1950. They recommended that stabilization can be achieved by adding a suitable artificial dissipation term that mimics the action of viscosity in the neighbourhood of shocks. Significant developments in this area are those of Lapidus,[44] Steger,[45] MacCormack and Baldwin[46] and Jameson and Schmidt.[47] At Swansea, a modified form of the method based on the second derivative of pressure has been developed by Peraire *et al.*[16] and Morgan *et al.*[48] for finite element computations. This modified form of viscosity with a pressure switch calculated from the nodal pressure values is used subsequently in compressible flow calculations. Recently an anisotropic viscosity for shock capturing[49] has been introduced to add diffusion in a more rational way.

The implementation of artificial diffusion is very much simpler than shock filling and we proceed as follows. We first calculate the approximate quantities of the solution vector by using the direct explicit method. Now we modify each scalar component of these quantities by adding a correction which smoothes the result. Thus for instance if we consider a typical scalar component quantity ϕ and have determined the values of ϕ^{n+1}, we establish the new values as below.

$$\phi_s^{n+1} = \phi^{n+1} + \Delta t \mu_a \frac{\partial}{\partial x_i}\left(\frac{\partial \phi}{\partial x_i}\right) \tag{6.11}$$

where μ_a is an appropriate artificial diffusion coefficient. It is important that whatever the method used, the calculation of μ_a should be limited to the domain which is close to the shock as we do not wish to distort the results throughout the problem. For this reason many procedures add a *switch* usually activated by such quantities as gradients of pressure. In all of the procedures used we can write the quantity μ_a as a function of one or more of the independent variables calculated at time n. Below we only quote two of the possibilities.

Second derivative based methods

In these it is generally assumed that the coefficient μ_a must be the same for each of the equations dealt with and only one of the independent variables $\mathbf{\Phi}$ is important. It has usually been assumed that the most typical variable here is the pressure and that we should write[46]

$$\mu_a = C_e h^3 \frac{|\mathbf{u}| + c}{\bar{p}} \left| \frac{\partial^2 p}{\partial x_i \partial x_i} \right|_e \tag{6.12}$$

where C_e is a non-dimensional coefficient, $\mathbf{u}$ is the velocity vector, c the speed of sound, $\bar{p}$ is the average pressure and the subscript e indicates an element. In the above equation, the second derivative of pressure over an element can be established either by averaging the smoothed nodal pressure gradients or using any of the methods described in Chapter 4, Sec. 4.5.

A particular variant of the above method evaluates approximately the value of the second derivative of any scalar variable ϕ (e.g. p) as[48]

$$h^2 \overline{\frac{\partial^2 \phi}{\partial x^2}} \approx (\mathbf{M} - \mathbf{M}_L)\tilde{\mathbf{\phi}} \tag{6.13}$$

where $\mathbf{M}$ and $\mathbf{M}_L$ are consistent and lumped mass matrices respectively and the overline indicates a nodal value. Though the derivation of the above expression is not obvious, the reader can verify that in the one-dimensional finite difference approximation it gives the correct result. The heuristic extension to multidimensional problem therefore seems reasonable. Now μ_a for this approximate method can be rewritten in any space dimensions as (Eq. 6.12)

$$\tilde{\boldsymbol{\mu}}_a = C_e h \frac{|\mathbf{u}| + c}{\bar{p}} (\mathbf{M} - \mathbf{M}_L)\tilde{\mathbf{p}} \tag{6.14}$$

Note now that $\tilde{\boldsymbol{\mu}}_a$ is a nodal quantity. However a further approximation can give the following form of μ_a over elements:

$$\mu_{ae} = C_e h(|\mathbf{u}| + c) S_e \tag{6.15}$$

where S_e is the element pressure switch which is a mean of nodal switches S_i

calculated as[48]

$$S_i = \frac{|\Sigma_e(p_i - p_k)|}{\Sigma_e|p_i - p_k|} \tag{6.16}$$

It can be verified that $S_i = 1$ when the pressure has a local extremum at node i and $S_i = 0$ when the pressure at node i is the average of values for all nodes adjacent to node i (e.g. if p varies linearly). The user-specified coefficient C_e normally varies between 0.0 and 2.0.

The smoothed variables can now be rewritten with the Galerkin finite element approximations (from Eqs. 6.11, 6.13 and 6.15) as

$$\tilde{\boldsymbol{\phi}}_s^{n+1} = \tilde{\boldsymbol{\phi}}^{n+1} + \Delta t \mathbf{M}_L^{-1} \frac{C_e S_e}{\Delta t_e} (\mathbf{M} - \mathbf{M}_L)\tilde{\boldsymbol{\phi}}^n \tag{6.17}$$

Note that, in Eq. (6.15), $(|\mathbf{u}| + c)$ is replaced by $h/\Delta t_e$ to obtain the above equation. This method has been widely used and is very efficient. The cut-off localizing the effect of added diffusion is quite sharp. A direct use of second derivatives can however be employed without the above-mentioned modifications. In such a procedure, we have the following form of smoothing (from Eqs. 6.11 and 6.12)

$$\tilde{\boldsymbol{\phi}}_s^{n+1} = \tilde{\boldsymbol{\phi}}^{n+1} - \Delta t \mathbf{M}_L^{-1} C_e h^3 \frac{|\mathbf{u}| + c}{\bar{p}} \left| \frac{\partial^2 p}{\partial x_i^2} \right|_e \left(\int_\Omega \frac{\partial \mathbf{N}^{\mathrm{T}}}{\partial x_i} \frac{\partial \mathbf{N}}{\partial x_i} \, \mathrm{d}\Omega \right) \tilde{\boldsymbol{\phi}}^n \tag{6.18}$$

This method was successful in many viscous problems. Another alternative is to use residual based methods.

Residual based methods

In these methods $\mu_{a_i} = \mu(R_i)$, where R_i is the residual of the ith equation. Such methods were first introduced in 1986 by Hughes and Malett[50] and later used by many others.[51–54]

A variant of this was suggested by Codina.[49] We sometimes refer to this as anisotropic shock capturing. In this procedure the artificial viscosity coefficient is adjusted by subtracting the diffusion introduced by the characteristic–Galerkin method along the streamlines. We do not know whether there is any advantage gain in this but we have used the anisotropic shock capturing algorithm with considerable success. The full residual based coefficient is given by

$$\mu_{a_i} = C_e \frac{|R_i|}{|\nabla \phi_i|} \tag{6.19}$$

We shall not discuss here a direct comparison between the results obtained by different shock capturing diffusivities, and the reader is referred to various papers already published.[55,56]

6.6 Some preliminary examples for the Euler equation

The computation procedures outlined can be applied with success to many transient and steady-state problems. In this section we illustrate its performance on a few relatively simple examples.

6.6.1 Riemann shock tube – a transient problem in one dimension[1]

This is treated as a one-dimensional problem. Here an initial pressure difference between two sections of the tube is maintained by a diaphragm which is destroyed at $t = 0$. Figure 6.2 shows the pressure, velocity and energy contours at the seventieth time increment, and the effect of including consistent and lumped mass matrices is illustrated. The problem has an analytical, exact, solution presented by Sod[57] and the numerical solution is from reference 1.

6.6.2 Isothermal flow through a nozzle in one dimension

Here a variant of the Euler equation is used in which isothermal conditions are assumed and in which the density is replaced by ρa where a is the cross-sectional

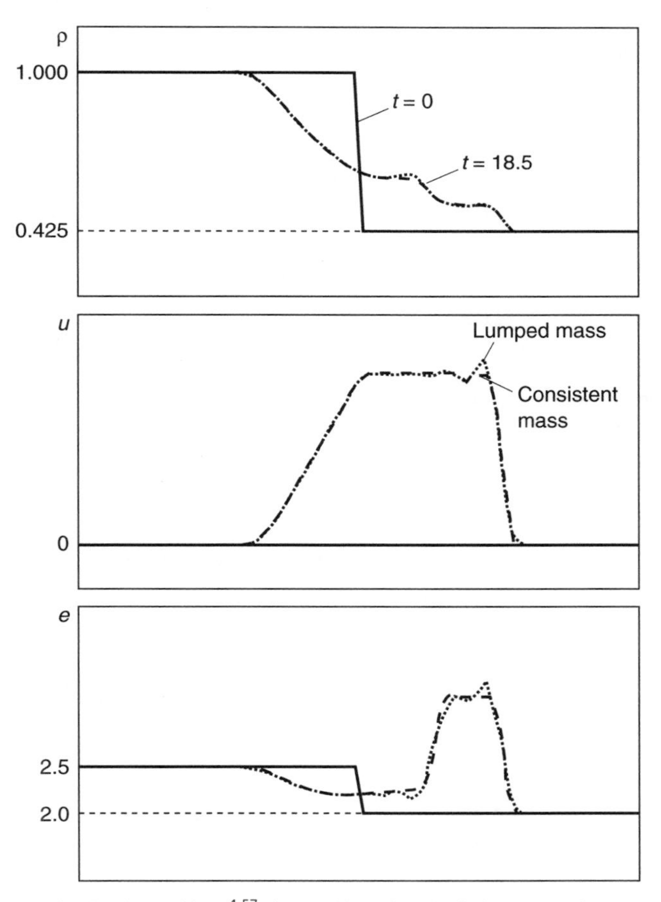

Fig. 6.2 The Riemann shock tube problem.[1,57] The total length is divided into 100 elements. Profile illustrated corresponds to 70 time steps ($\Delta t = 0.25$). Lapidus constant $C_{\mathrm{Lap}} = 1.0$.

area[1] assumed to vary as[58]

$$a = 1.0 + \frac{(x - 2.5)^2}{12.5} \qquad \text{for } 0 \leqslant x \leqslant 5 \tag{6.20}$$

The speed of sound is constant as the flow is isothermal and various conditions at inflow and outflow limits were imposed as shown in Fig. 6.3. In all problems

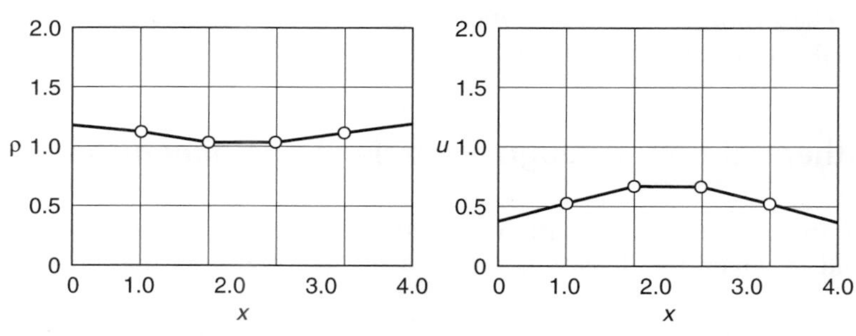

(a) Subsonic inflow and outflow

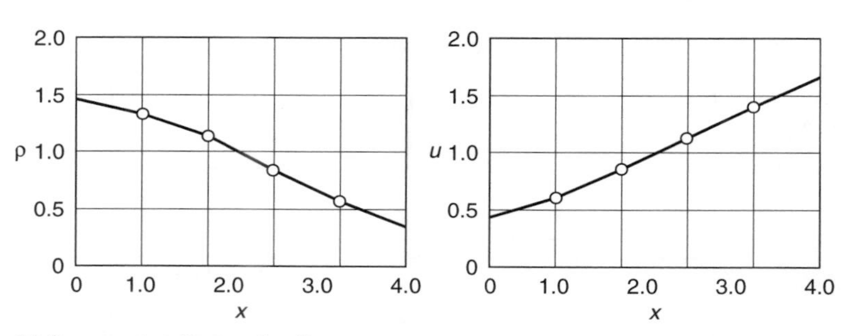

(b) Supersonic inflow and outflow

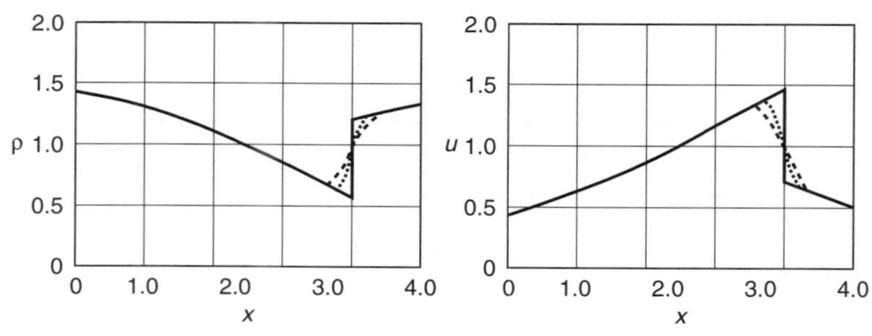

(c) Supersonic inflow–subbsonic outflow with shock

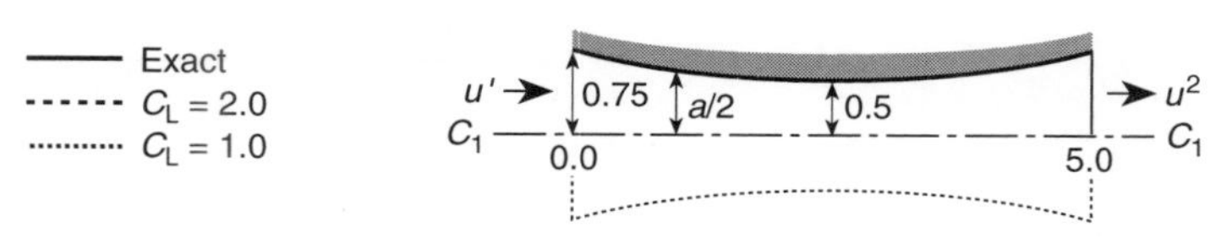

Fig. 6.3 Isothermal flow through a nozzle.[1] Forty elements of equal size used.

steady state was reached after some 500 time steps. For the case with supersonic inflow and subsonic outflow, a shock forms and Lapidus-type artificial diffusion was used to deal with it, showing in Fig. 6.3(c) the increasing amount of 'smearing' as the coefficient C_{Lap} is increased.

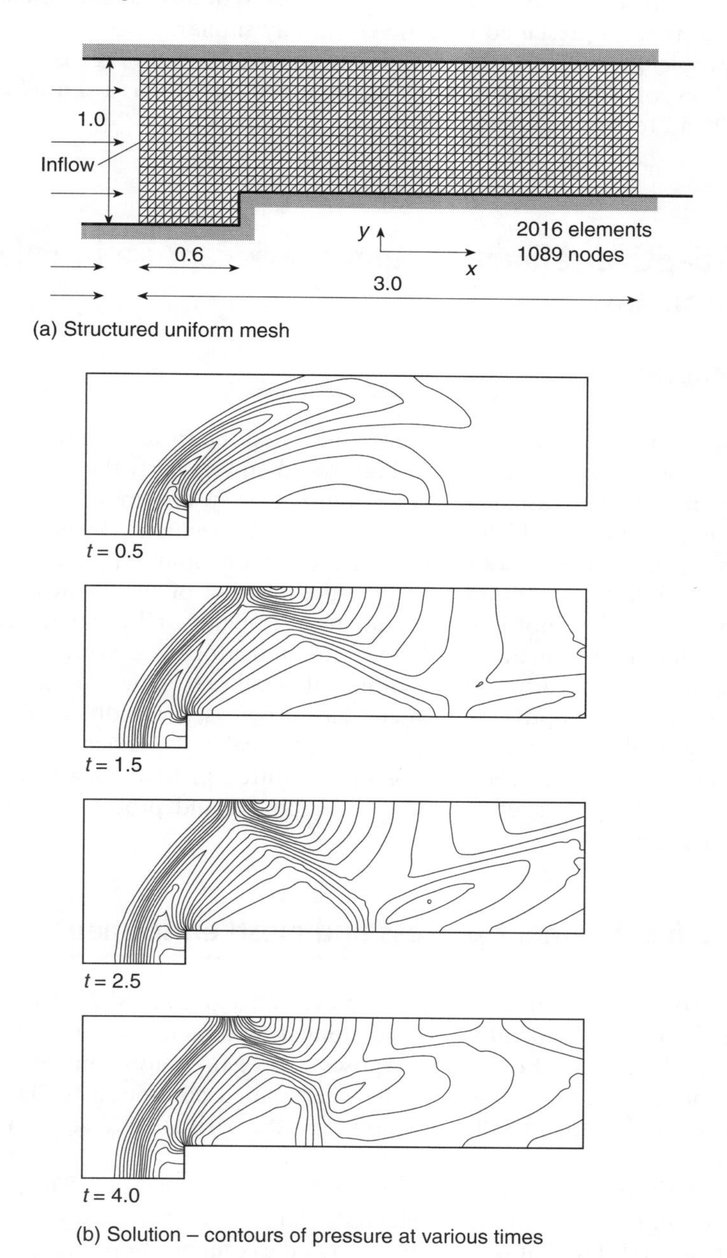

Fig. 6.4 Transient supersonic flow over a step in a wind tunnel[4] (problem of Woodward and Colella[59]). Inflow Mach 3 uniform flow.

6.6.3 Two-dimensional transient supersonic flow over a step

This final example concerns the transient initiation of supersonic flow in a wind tunnel containing a step. The problem was first studied by Woodward and Colella[59] and the results of reference 4 presented here are essentially similar.

In this problem a uniform mesh of linear triangles, shown in Fig. 6.4, was used and no difficulties of computation were encountered although a Lapidus constant $C_{\text{Lap}} = 2.0$ had to be used due to the presence of shocks.

6.7 Adaptive refinement and shock capture in Euler problems

6.7.1 General

The examples of the previous section have indicated the formation of shocks both in transient and steady-state problems of high-speed flow. Clearly the resolution of such discontinuities or near discontinuities requires a very fine mesh. Here the use of 'engineering judgement', which is often used in solid mechanics by designing *a priori* mesh refining near singularities posed by corners in the boundary, etc., can no longer be used. In problems of compressible flow the position of shocks, where the refinement is most needed, is not known in advance. For this and other reasons, the use of adaptive mesh refinement based on error indicators is essential for obtaining good accuracy and 'capturing' the location of shocks. It is therefore not surprising that the science of adaptive refinement has progressed rapidly in this area and indeed, as we shall see later, has been extended to deal with Navier–Stokes equations where a higher degree of refinement is also required in boundary layers. We have discussed the history of such adaptive development and procedures for its use in Sec. 4.5, Chapter 4.

6.7.2 The *h*-refinement process and mesh enrichment

Once an approximate solution has been achieved on a given mesh, the local errors can be evaluated and new element sizes (and elongation directions if used) can be determined for each element. For some purposes it is again convenient to transfer such values to the nodes so that they can be interpolated continuously. The procedure here is of course identical to that of smoothing the derivatives discussed in Sec. 4.5, Chapter 4.

To achieve the desired accuracy various procedures can be used. The most obvious is the process of *mesh enrichment* in which the existing mesh is locally subdivided into smaller elements still retaining the 'old' mesh in the configuration. Figure 6.5(a) shows how triangles can be readily subdivided in this way. With such enrichment an obvious connectivity difficulty appears. This concerns the manner in which the subdivided

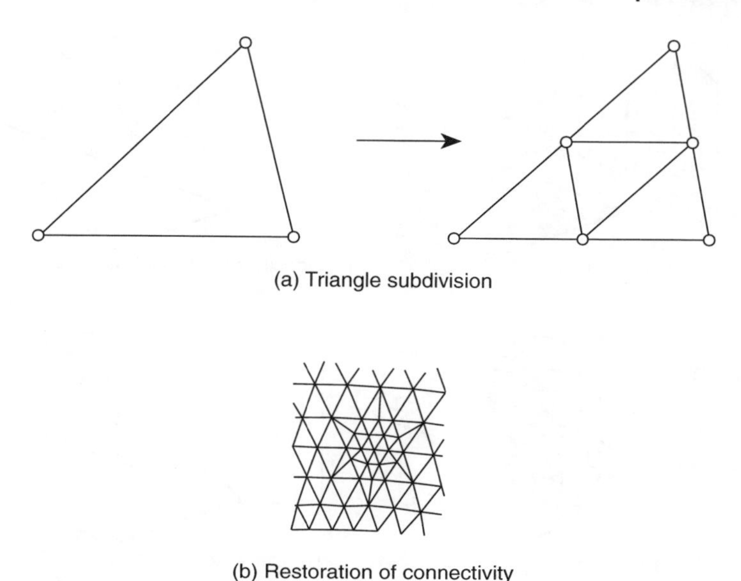

Fig. 6.5 Mesh enrichment. (a) Triangle subdivision. (b) Restoration of connectivity.

elements are connected to ones not so refined. A simple process is illustrated showing element halving in the manner of Fig. 6.5(b). Here of course it is fairly obvious that this process, first described in reference 9, can only be applied in a gradual manner to achieve the predicted subdivisions. However, element elongation is not possible with such mesh enrichment.

Despite such drawbacks the procedure is very effective in localizing (or capturing) shocks, as we illustrate in Fig. 6.6.

In Fig. 6.6, the theoretical solution is simply one of a line discontinuity shock in which a jump of all the components of $\mathbf{\Phi}$ occurs. The original analysis carried out on a fairly uniform mesh shows a very considerable 'blurring' of the shock. In Fig. 6.6 we also show the refinement being carried out at two stages and we see how the shock is progressively reduced in width.

In the above example, the mesh enrichment preserved the original, nearly equilateral, element form with no elongation possible.

Whenever a sharp discontinuity is present, local refinement will proceed indefinitely as curvatures increase without limit. Precisely the same difficulty indeed arises in mesh refinement near singularities for elliptic problems[60] if local refinement is the only guide. In such problems, however, the limits are generally set by the overall energy norm error consideration and the refinement ceases automatically. In the present case, the limit of refinement needs to be set and we generally achieve this limit by specifying the *smallest element size* in the mesh.

The h refinement of the type proposed can of course be applied in a similar manner to quadrilaterals. Here clever use of data storage allows the necessary refinement to be achieved in a few steps by ensuring proper transitions.[61]

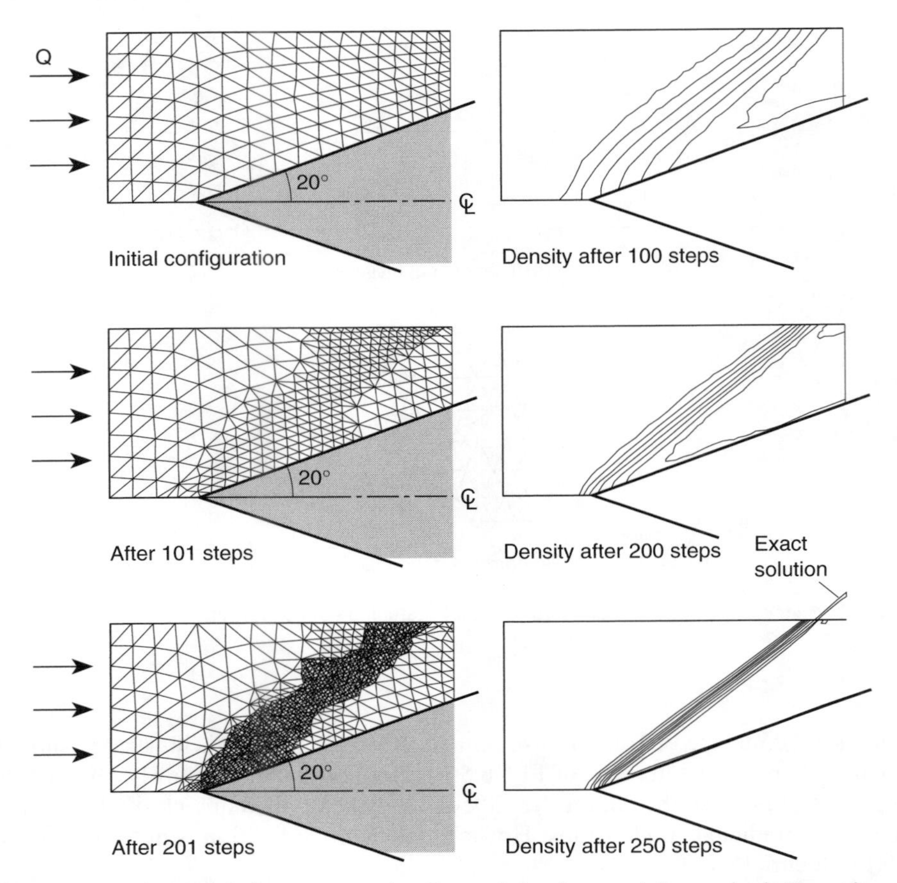

Fig. 6.6 Supersonic, Mach 3, flow past a wedge. Exact solution forms a stationary shock. Successive mesh enrichment and density contours.

6.7.3 *h*-refinement and remeshing in steady state two-dimensional problems

Many difficulties mentioned above can be resolved by *automatic generation of meshes of a specified density*. Such automatic generation has been the subject of much research in many applications of finite element analysis. We have discussed this subject in Sec. 4.5, Chapter 4. The closest achievement of a prescribed element size and directionality can be obtained for triangles and tetrahedra. Here the procedures developed by Peraire *et al.*[11,16] are most direct and efficient, allowing element stretching in prescribed directions (though of course the amount of such stretching is sometimes restricted by practical considerations).

We refer the reader for details of such mesh generation to the original publications. In the examples that follow we shall exclusively use this type of mesh adaptivity.

In Fig. 6.7 we show a simple example[11] of shock wave reflection from a solid wall. Here only a typical 'cut-out' is analysed with appropriate inlet and outlet conditions

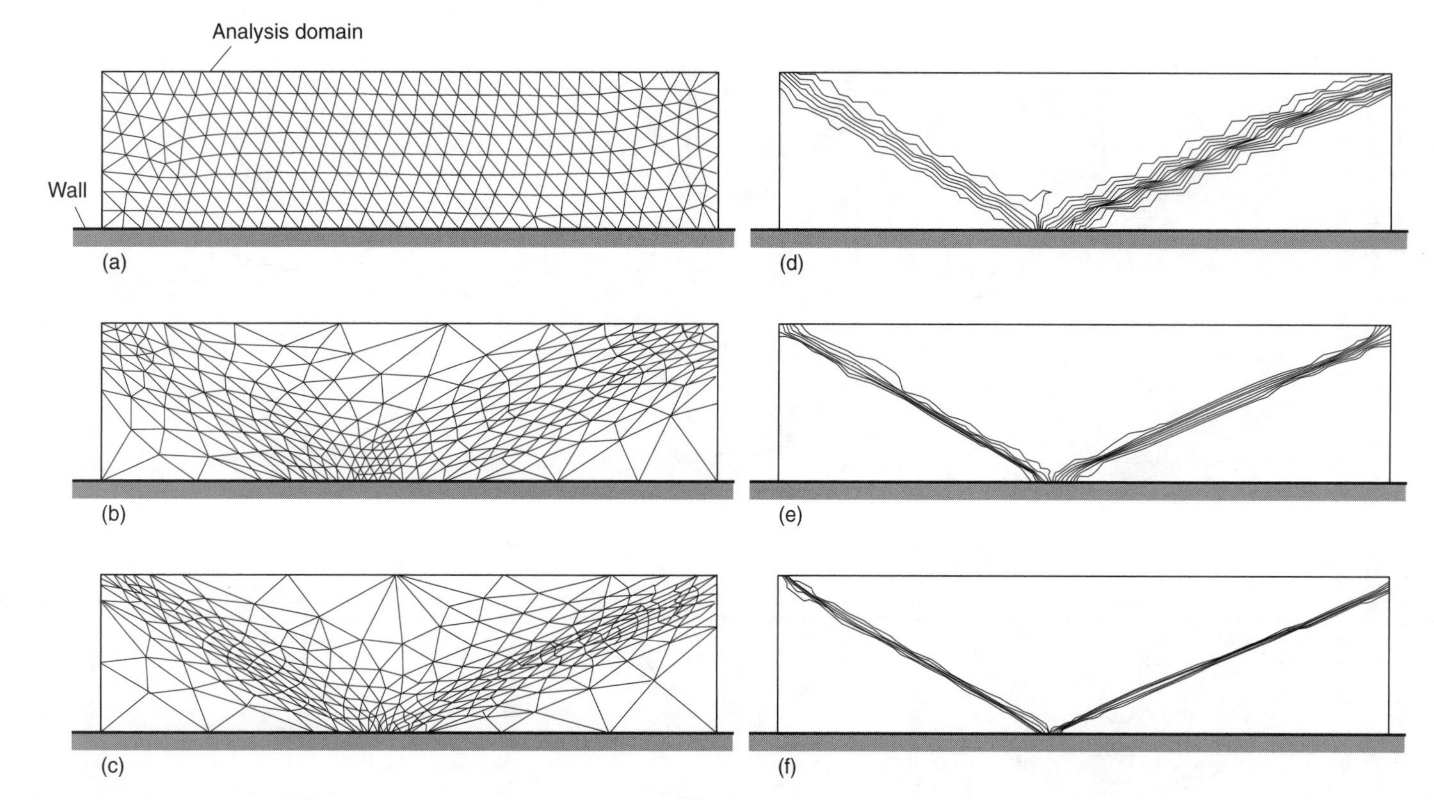

Fig. 6.7 Reflection of a shock wave at a wall[11]–Euler equations. A sequence of meshes, (a) nodes: 279, elements: 478, (b) nodes: 265, elements: 479, (c) nodes: 285, elements: 528 and corresponding pressure contours, (d) to (f).

imposed. The elongation of the mesh along the discontinuity is clearly shown. The solution was remeshed after the iterations nearly reached a steady state.

In Fig. 6.8 a somewhat more complex example of *hypersonic flow* around a blunt, two-dimensional obstacle is shown. Here it is of interest to note that:

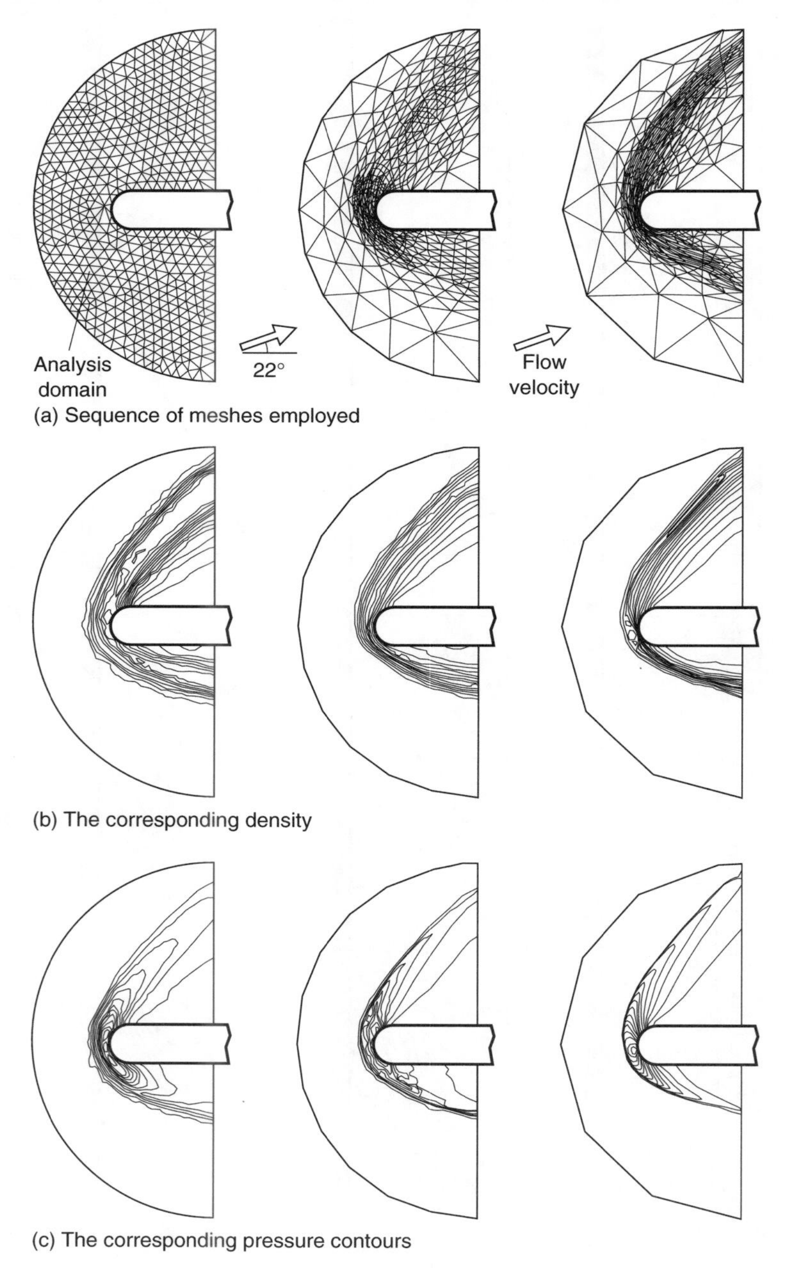

Fig. 6.8 Hypersonic flow past a blunt body[11] at Mach 25, 22° angle of attack. Initial mesh, nodes: 547, elements: 978; first mesh, nodes: 383, elements: 696; final mesh, nodes: 821, elements: 1574.

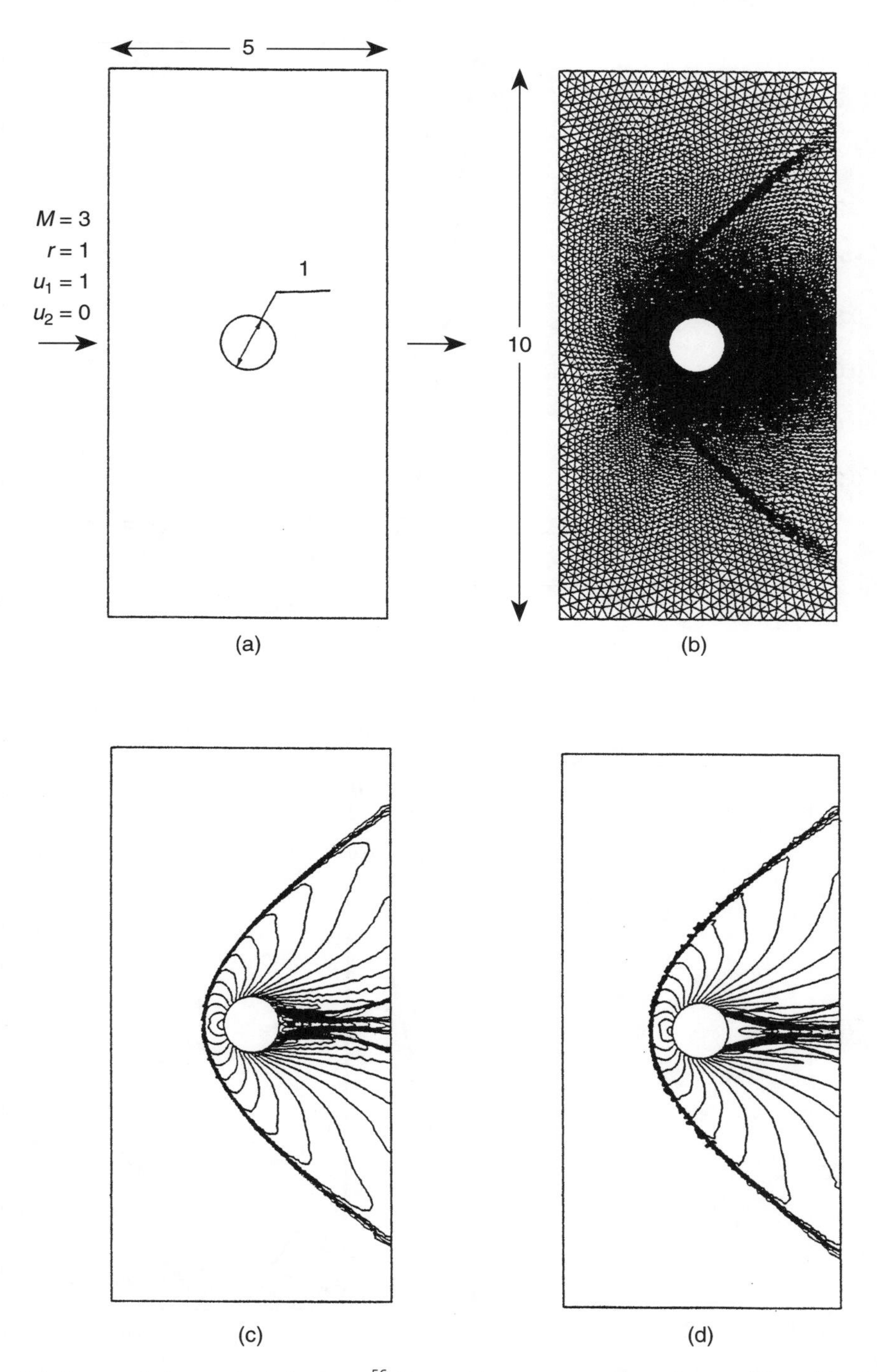

Fig. 6.9 Supersonic flow past a full cylinder.[56] $M = 3$, (a) geometry and boundary conditions, (b) adapted mesh, nodes: 12651, elements: 24 979, (c) Mach contours using second derivative shock capture and (d) Mach contours using anisotropic shock capture.

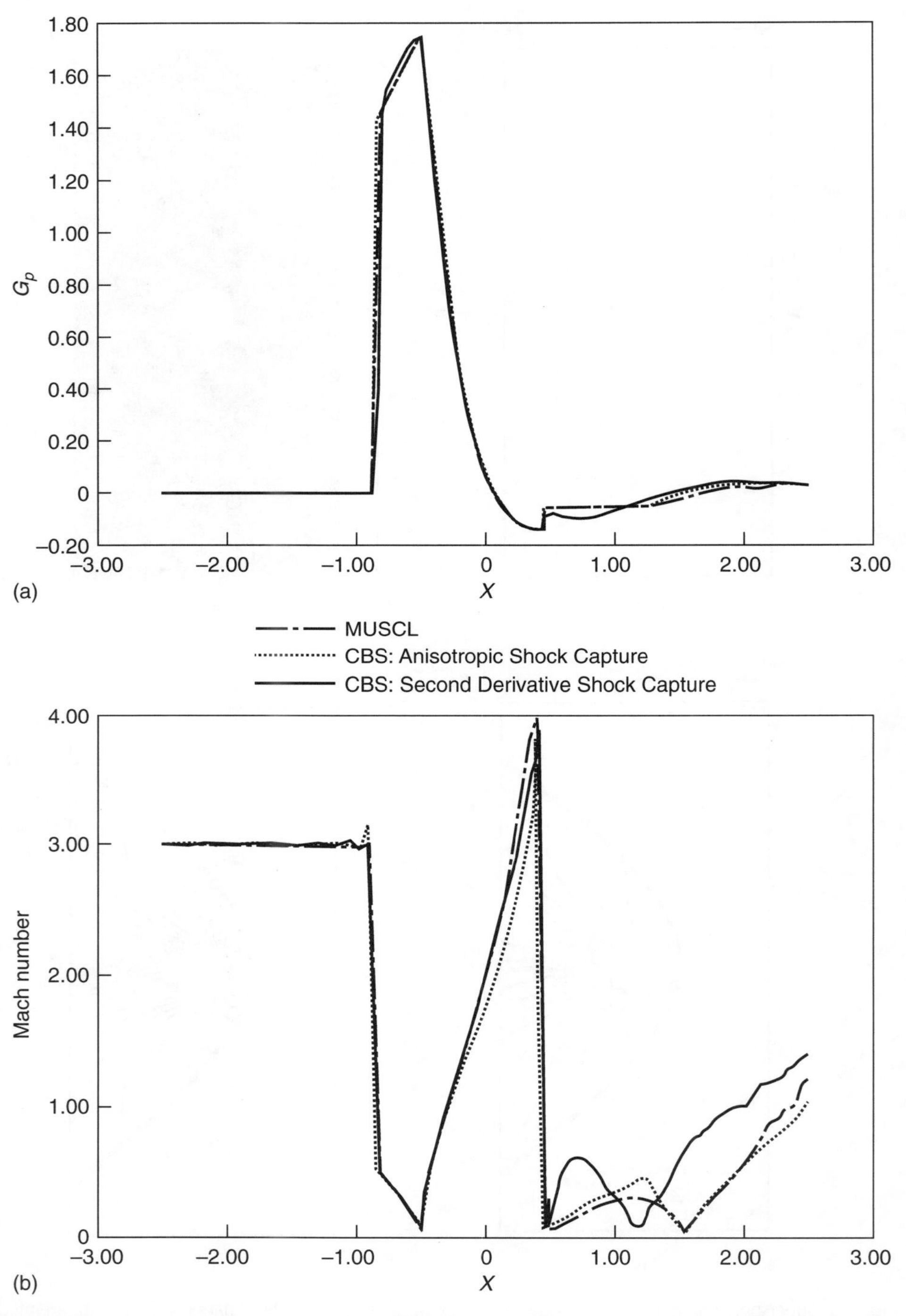

Fig. 6.10 Supersonic flow past a full cylinder.[56] $M = 3$, comparison of (a) coefficient of pressure, (b) Mach number distribution along the mid-height and cylinder surface.

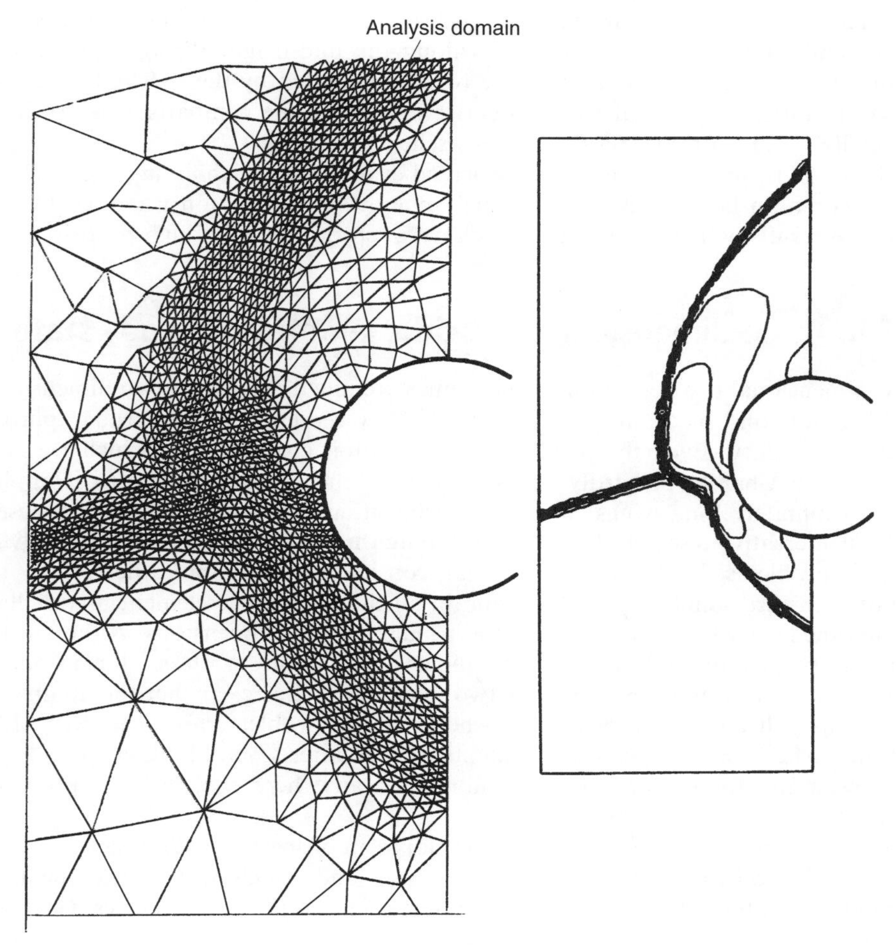

Fig. 6.11 Interaction of an impinging and bow shock wave.[17] Adapted mesh and pressure contours.

1. A detached shock forms in front of the body.
2. A very coarse mesh suffices in front of such a shock where simple free stream flow continues and the mesh is made 'finite' by a maximum element size prescription.
3. For the same minimum element size a reduction of degrees of freedom is achieved by refinement which shows much improved accuracy.

For such hypersonic problems, it is often claimed that special methodologies of solution need to be used. References 62–64 present quite sophisticated methods for dealing with such high-speed flows.

In Figs 6.9 and 6.10, we show the results of supersonic Mach 3 flow past a full cylinder.[56] The mesh (Fig. 6.9(b)) is adapted along the shock front to get a good resolution of the shock. The mesh behind the cylinder is very fine to capture the recirculatory motion. In Figs 6.9(c) and 6.9(d), the Mach contours are obtained using the CBS algorithm using the second derivative based shock capture and residual based shock capture respectively. In Fig. 6.10, the coefficient of pressure values and Mach number distribution along the mid-height through the surface of the cylinder are presented. Here the results generated by the MUSCL[62] scheme are also plotted for the sake of comparison. As seen the comparison is excellent, especially for anisotropic (residual-based) scheme.

Figure 6.11 shows a yet more sophisticated example in which an impinging shock interacts with a bow shock. An extremely fine mesh distribution was used here to compare results with experiment,[17] which were reproduced with high precision.

6.8 Three-dimensional inviscid examples in steady state

Two-dimensional problems in fluid mechanics are much rarer than two-dimensional problems in solid mechanics and invariably they represent a very crude approximation to reality. Even the problem of an aerofoil cross-section, which we have discussed in Chapter 3, hardly exists as a two-dimensional problem as it applies only to infinitely long wings. For this reason attention has largely been focused, and much creative research done, in developing three-dimensional codes for solving realistic problems. In this section we shall consider some examples derived by the use of such three-dimensional codes and in all these the basic element used will be the tetrahedron which now replaces the triangle of two dimensions. Although the solution procedure and indeed the whole formulation in three dimensions is almost identical to that described for two dimensions, it is clear that the number of unknowns will increase very rapidly when realistic problems are dealt with. It is common when using linear order elements to encounter several million variables as unknowns and for this reason here, more than anywhere else, iterative processes are necessary.

Indeed much effort has gone into the development of special procedures of solution which will accelerate the iterative convergence and which will reduce the total computational time. In this context we should mention three approaches which are of help.

The recasting of element formulation in an edge form

Here a considerable reduction of storage can be achieved by this procedure and some economies in computational time achieved. We have not discussed this matter in detail but refer the reader to reference 26 where the method is fully described and for completeness we summarize the essential features of edge formulation in Appendix C.

Multigrid approaches

In the standard iteration we proceed in a time frame by calculating point by point the changes in various quantities and we do this on the finest mesh. As we have seen this may become very fine if adaptivity is used locally. In the multigrid solution, as initially introduced into the finite element field, the solution starts on a coarse mesh, the results of which are used subsequently for generating the first approximation to the fine mesh. Several iterative steps are then carried out on the fine mesh. In general a return to the coarse mesh is then made to calculate the changes of residuals there and the process is repeated on several meshes done subsequently. This procedure can be used on several meshes and the iterative process is much accelerated. We discuss this process in Appendix D in a little more detail. However, we quote here several references[65–69] in which such multigrid procedures have been used and these are of considerable value.

Multigrid methods are obviously designed for meshes which are 'nested' i.e. in which coarser and finer mesh nodes coincide. This need not be the case generally. In many applications completely different meshes of varying density are used at various stages.

Parallel computation

The third procedure of reducing the solution time is to use parallization. We do not discuss it here in detail as the matter is potentially coupled with the computational aspects of the problem. Here the reader should consult the current literature on the subject.[36–38]

In what follows we shall illustrate three-dimensional applications on a few inviscid examples as this section deals with Euler problems. However in Sec. 6.10 we shall return to a fully three-dimensional formulation using viscous Navier–Stokes equations.

6.8.1 Solution of the flow pattern around a complete aircraft

In the early days of numerical analysis applied to computational fluid dynamics which used finite differences, no complete aircraft was analysed as in general only structured meshes were admissible. The analysis thus had to be carried out on isolated components of the aircraft. Later construction of distorted and partly structured meshes increased the possibility of analysis. Nevertheless the first complete aircraft analyses were done only in the mid-1980s. In all of these, finite elements using unstructured meshes were used (though we include here the finite volume formulation which was almost identical to finite elements and was used by Jameson *et al.*[70]). The very first aircraft was the one dealt with using potential theory in the Dassault establishment. The results were publisshed later by Periaux and coworkers.[71] Very shortly after that a complete supersonic aircraft was analysed by Peraire *et al.*[16] in Swansea in 1987.

Figure 6.12 shows the aircraft analysed in Swansea[16] which is a supersonic fighter of generic type at Mach 2. The analysis was made slightly adaptive though adaptivity was not carried very far (due to cost). Nevertheless the refinement localized the shocks which formed.

In the analysis some 125 000 elements were used with approximately 70 000 nodes and therefore some 350 000 variables. This of course is not a precise analysis and many more variables would be used currently to get a more accurate representation of flow and pressure variables.

A more sophisticated analysis is shown in the plate at the front of the book. Here a civil aircraft in subsonic flow is modelled and this illustrates the use of multigrid

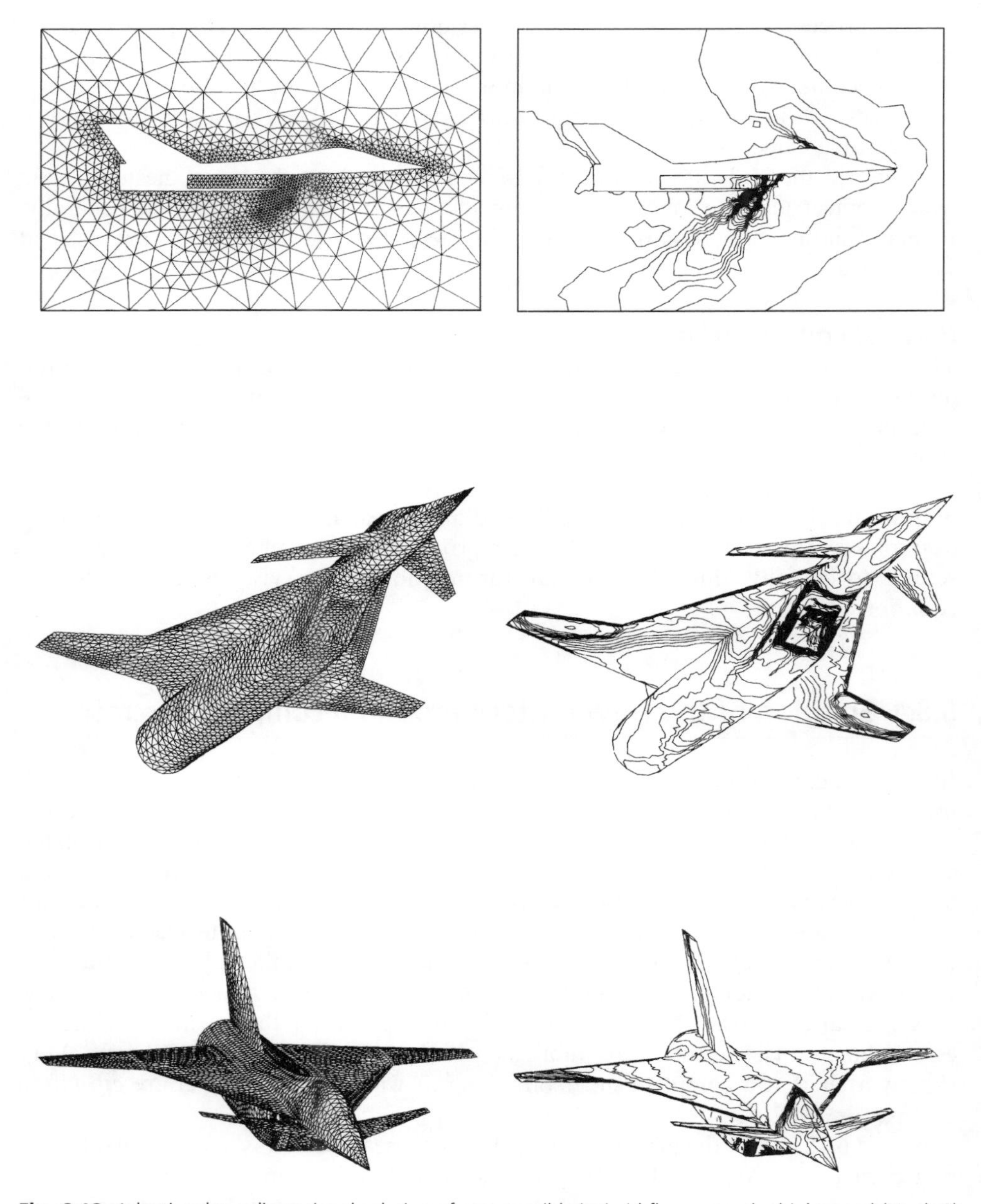

Fig. 6.12 Adaptive three-dimensional solution of compressible inviscid flow around a high speed (Mach 2) aircraft.[16] Nodes: 70 000, elements: 125 000.

methods. In this particular multigrid applications three meshes of different refinement were used and the iteration is fully described in reference 26. In this example the total number of unknown quantities was 1 616 000 in the finest mesh and indeed the details of the subdivision are given in the legend of the plate.

(a)

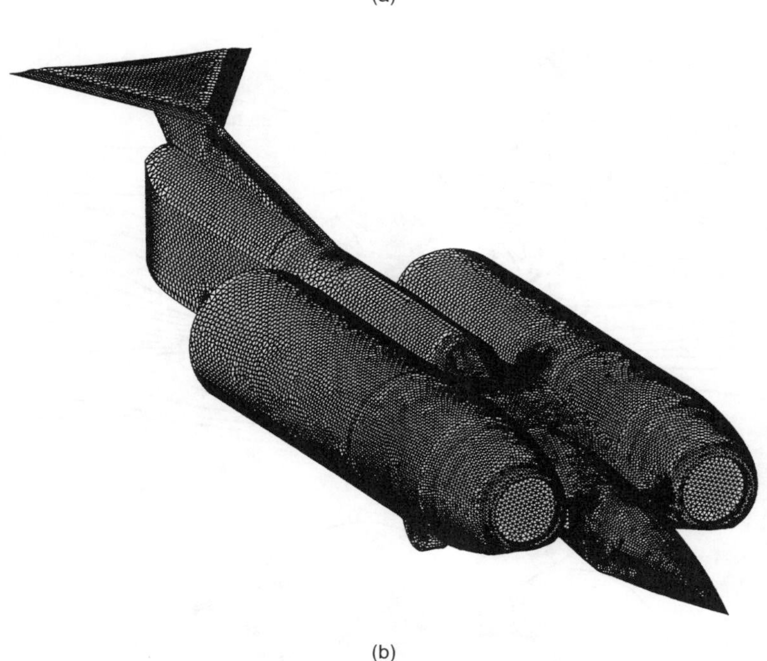

(b)

Fig. 6.13 Supersonic car, THRUST SSC.[27] (a) car and (b) finite element surface mesh. (Image used in (a) courtesy of SSC Programme Ltd. Photographer Jeremy C.R. Davey.)

6.8.2 THRUST – the supersonic car[27,28,72]

A very similar problem to that posed by the analysis of the whole aircraft was given much more recently by the team led by Professor Morgan. This was the analysis of a car which was attempting to create the world speed record by establishing this in the supersonic range. This attempt was indeed successfully made on 15 October 1997. Unlike in the problem of the aircraft, the alternative of wind tunnel tests was not available. Whilst in aircraft design, wind tunnels which are supersonic and subsonic are well used in practice (though at a cost which is considerably more than that of a numerical analysis) the possibility of doing such a test on a motor car was virtually non-existent. The reason for this is the fact that the speed of the air flow past the body of the car and the speed of the ground relative to the car are identical. Any test would therefore require the bed of the wind tunnel to move at a speed in excess of 750+ miles an hour. For this reason calculations were therefore preferable.

The moving ground will of course create a very important boundary layer such as that which we will discuss in later sections. However the simple omission of viscosity permitted the inviscid solution by a standard Euler-type program to be used. It is

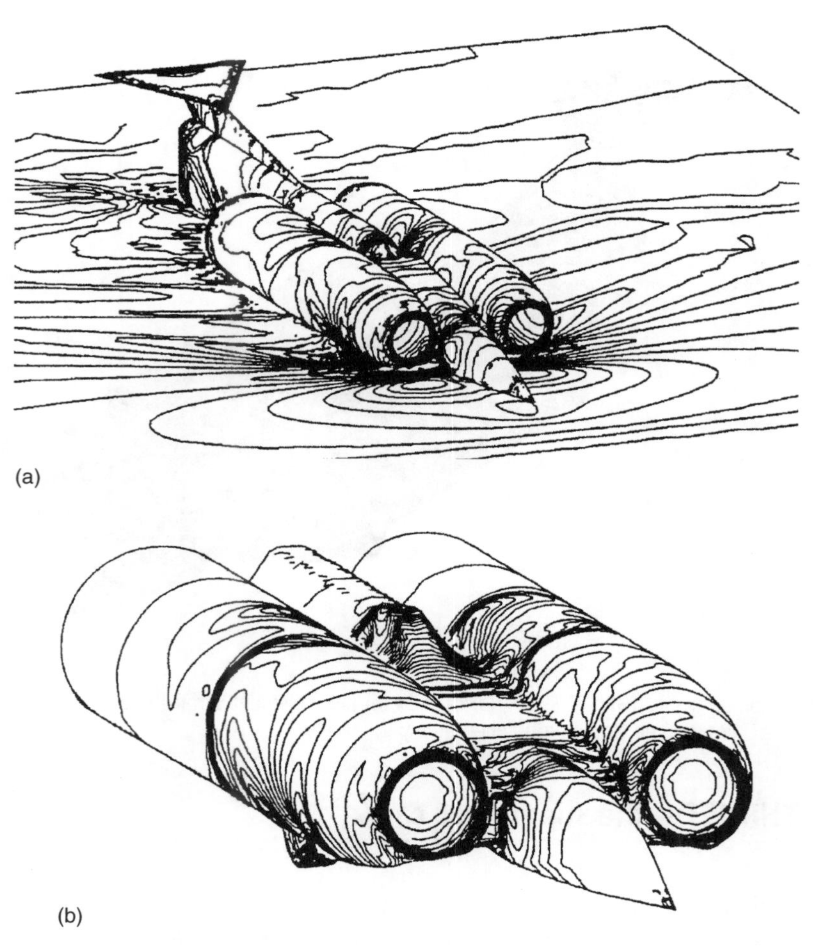

Fig. 6.14 Supersonic car, THRUST SSC[27] pressure contours (a) full configuration, (b) front portion.

well known that the Euler solution is perfectly capable of simulating all shocks very adequately and indeed results in very well defined pressure distributions over the bodies whether it is over an aircraft or a car. The object of the analysis was indeed that of determining such pressure distributions and the lift caused by these pressures. It was essential that the car should remain on the ground, indeed this is one of the conditions of the ground speed record and any design which would result in substantial lift overcoming the gravity on the car would be disastrous for obvious reasons.

The complete design of the supersonic car was thus made with several alternative geometries until the computer results were satisfactory. Here it is interesting however to have some experimental data and the preliminary configuration was tested by a rocket driven sled. This was available for testing rocket projectiles at Pendine Sands, South Wales, UK. Here a 1:25 scale model of the car was attached to such a rocket and 13 supersonic and transonic runs were undertaken.

In Fig. 6.13(a), we show a photograph[27] of the car concerned after winning the speed record in the Nevada desert. In Fig. 6.13(b) a surface mesh is presented from which the full three-dimensional mesh at the surrounding atmosphere was generated (surface mesh, nodes: 39 528, elements: 79 060; volume mesh, nodes: 134 272, elements: 887 634). We do not show the complete mesh as of course in a three-dimensional problem this is counterproductive.

In Fig. 6.14, pressure contours[27] on the surface of the car body are given and somewhat similar contours are shown on the covers of this book.

In Fig. 6.15, a detailed comparison of CFD results[27] with experiments is shown. The results of this analysis show a remarkable correlation with experiments. The data points which do not appear close to the straight line are the result of the sampling point being close to, but the wrong side of, a shock wave. If conventional correlation techniques for inviscid flow (viscous correction) are applied, these data points also lie on the straight line. In total, nine pressure points were used situated on the upper and

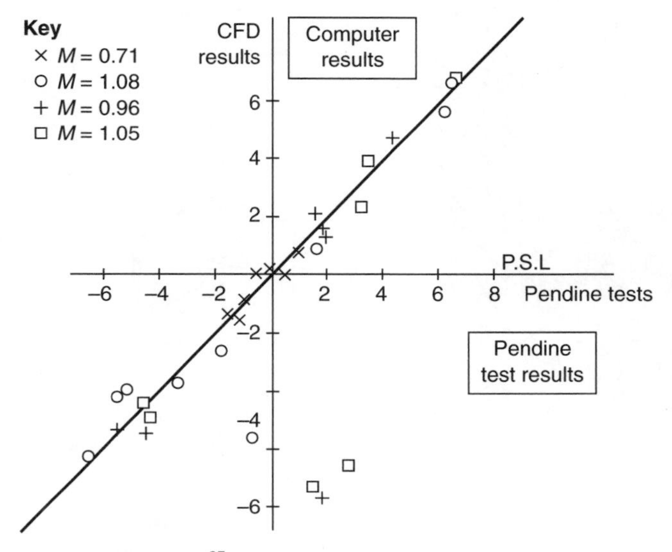

Fig. 6.15 Supersonic car, THRUST SSC[27] comparison of finite element and experimental results.

lower surfaces of the car. The plot shows the comparison of pressures at specific positions on the car for Mach numbers of 0.71, 0.96, 1.05 and 1.08.

6.8.3 Other examples

There are many other three-dimensional examples which could at this stage be quoted but we only show here a three-dimensional analysis of an engine intake[16] at Mach 2. This is given in Fig. 6.16.

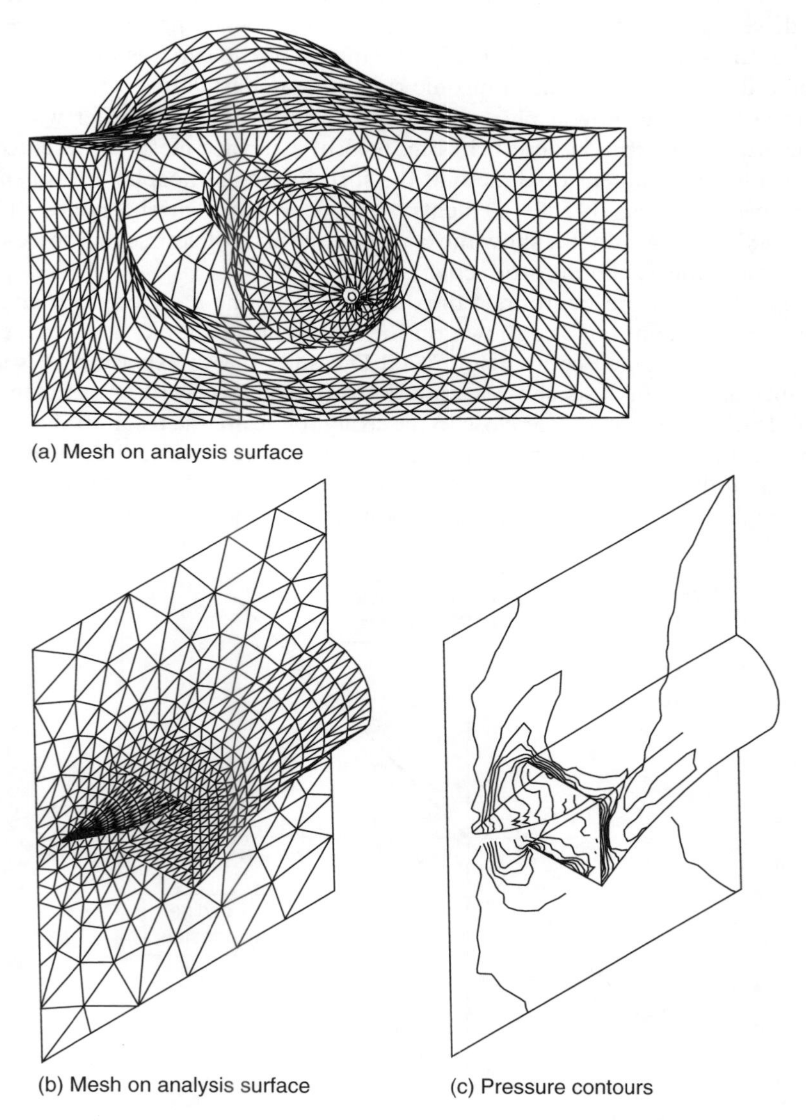

Fig. 6.16 Three-dimensional analysis of an engine intake[16] at Mach 2 (14 000 elements).

6.9 Transient two and three-dimensional problems

In all of the previous problems the time stepping was used simply as an iterative device for reaching the steady-state solutions. However this can be used in real time and the transient situation can be studied effectively. Many such transient problems have been

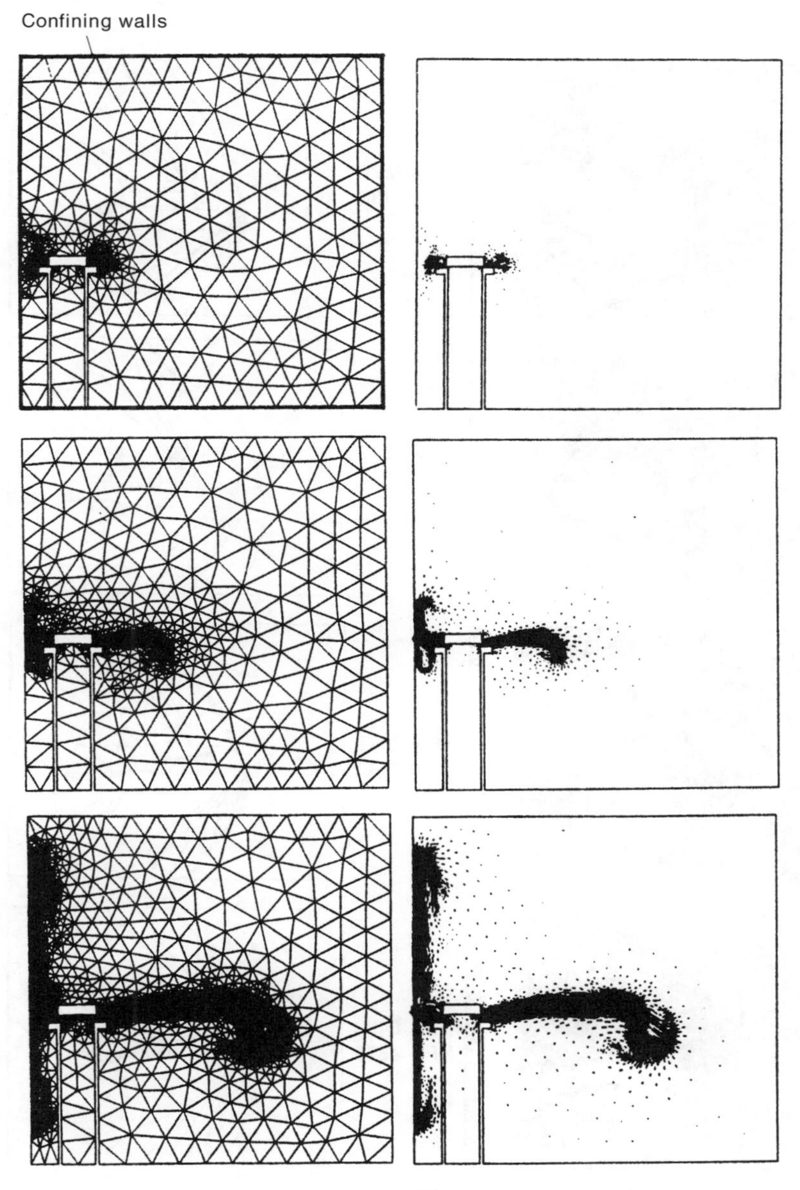

Fig. 6.17 A transient problem with adaptive remeshing.[73] Simulation of a sudden failure of a pressure vessel. Progression of refinement and velocity patterns shown. Initial mesh 518 nodes.

dealt with from time to time and here we illustrate the process on three examples. The first one concerns an exploding pressure vessel[73] as a two-dimensional model as shown in Fig. 6.17. Here of course adaptivity had to be used and the mesh is regenerated every few steps to reproduce the transient motion of the shock front.

A similar computation is shown in Fig. 6.18 where a diagrammatic form of a shuttle launch is modelled again as a two-dimensional problem.[73] Of course this two-dimensional model is purely imaginary but it is useful for showing the general

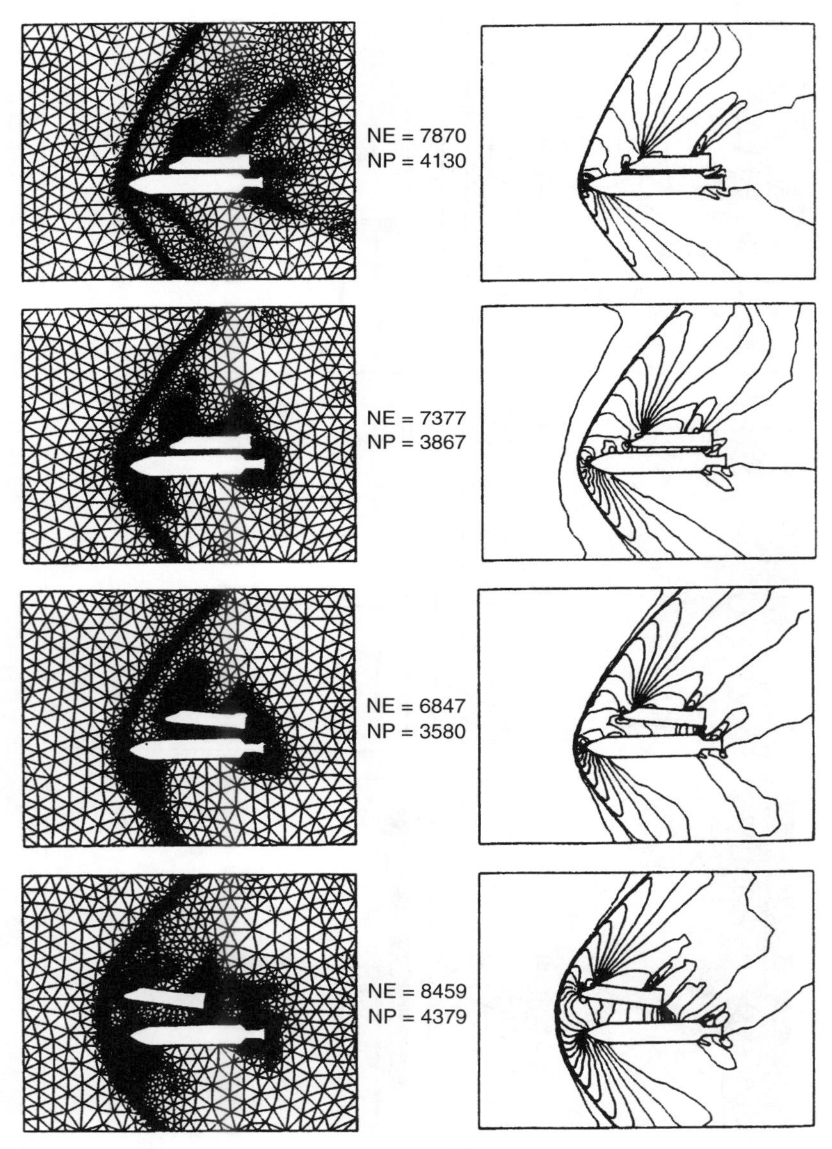

Fig. 6.18 A transient problem with adaptive remeshing.[73] Model of the separation of shuttle and rocket. Mach 2, angle of attack −4°, initial mesh 4130 nodes.

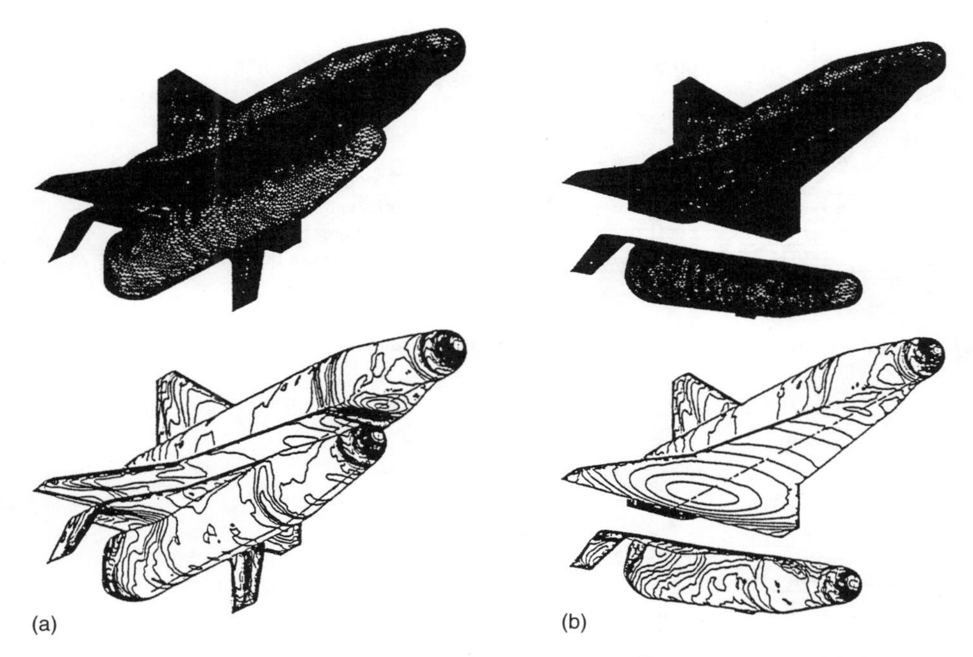

Fig. 6.19 Separation of a generic shuttle vehicle and rocket booster.[29] (a) Initial surface mesh and surface pressure; (b) final surface mesh and surface pressure.

configuration. In Fig. 6.19 however, we show a three-dimensional shuttle approximating closely to reality.[29] The picture shows the initial configuration and the separation from the rocket.

6.10 Viscous problems in two dimensions

Clearly the same procedures which we have discussed previously could be used for the full Navier–Stokes equations by the introduction of viscous and other heat diffusion terms. Although this is possible we will note immediately that very rapid gradients of velocity will develop in the boundary layers (we have remarked on this already in Chapter 4) and thus special refinement will be needed there. In the first example we illustrate a viscous solution by using meshes designed *a priori* with fine subdivision near the boundary. However, in general the refinement must be done adaptively and here various methodologies of doing so exist. The simplest of course is the direct use of mesh refinement with elongated elements which we have also discussed in Chapter 4. This will be dealt with by a few examples in Sec. 6.10.2. However in Sec. 6.10.3 we shall address the question of much finer refinement with very elongated elements in the boundary layer. Generally we shall do such a refinement with such a structured grid near the solid surfaces merging into the general unstructured meshing outside. In that section we shall introduce methods which can automatically separate structured and unstructured regions both in the boundary layer and in the shock regions.

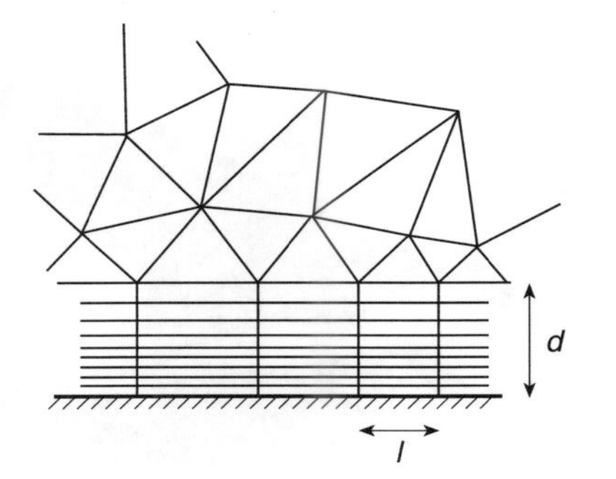

Unstructured, adaptively refined triangles

10–15 'body' layers of quadrilateral elements with length l, corresponding adaptive layer thickness d and number of layers decided by user

(a) A two-dimensional sublayer of structured quadrilaterals

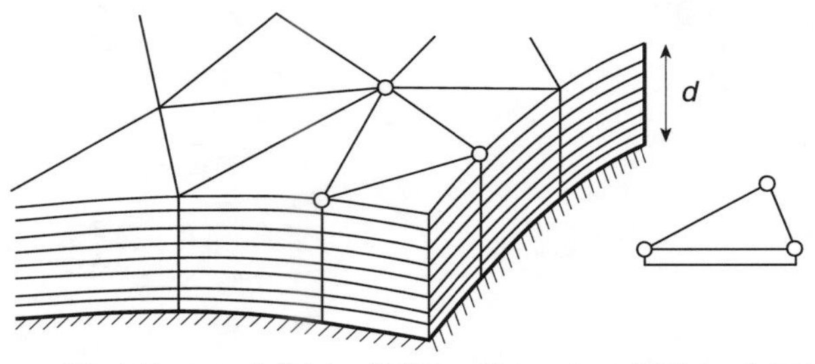

'Body' layer subdivision in three dimensions joining a tetrahedral mesh

(b) A three-dimensional sublayer of prismatic elements

Fig. 6.20 Refinement in the boundary layer.

The methodology is of course particularly important in problems of three dimensions. In Sec. 6.11 we show some realistic applications of boundary layer refinement and here we shall again refer to turbulence.

The special refinement which we mentioned above is well illustrated in Fig. 6.20. In this we show the possibility of using a structured mesh with quadrilaterals in the boundary layer domain (for two-dimensional problems) and a three-dimensional equivalent of such a structured mesh using prismatic elements. Indeed such elements have been used as a general tool by some investigators.[74–76]

6.10.1 A preliminary example

The example given here is that in which both shock and boundary layer development occur simultaneously in high-speed flow over a flat plate.[77] This problem was studied extensively by Carter.[78] His finite difference solution is often used for comparison purposes although some oscillations can be seen even there despite a very high refinement.

A fixed mesh which is graded from a rather fine subdivision near the boundary to a coarser one elsewhere is shown in Fig. 6.21. We obtained the solution using as usual the CBS algorithm. In Fig. 6.22, comparisons with Carter's[78] solution are presented

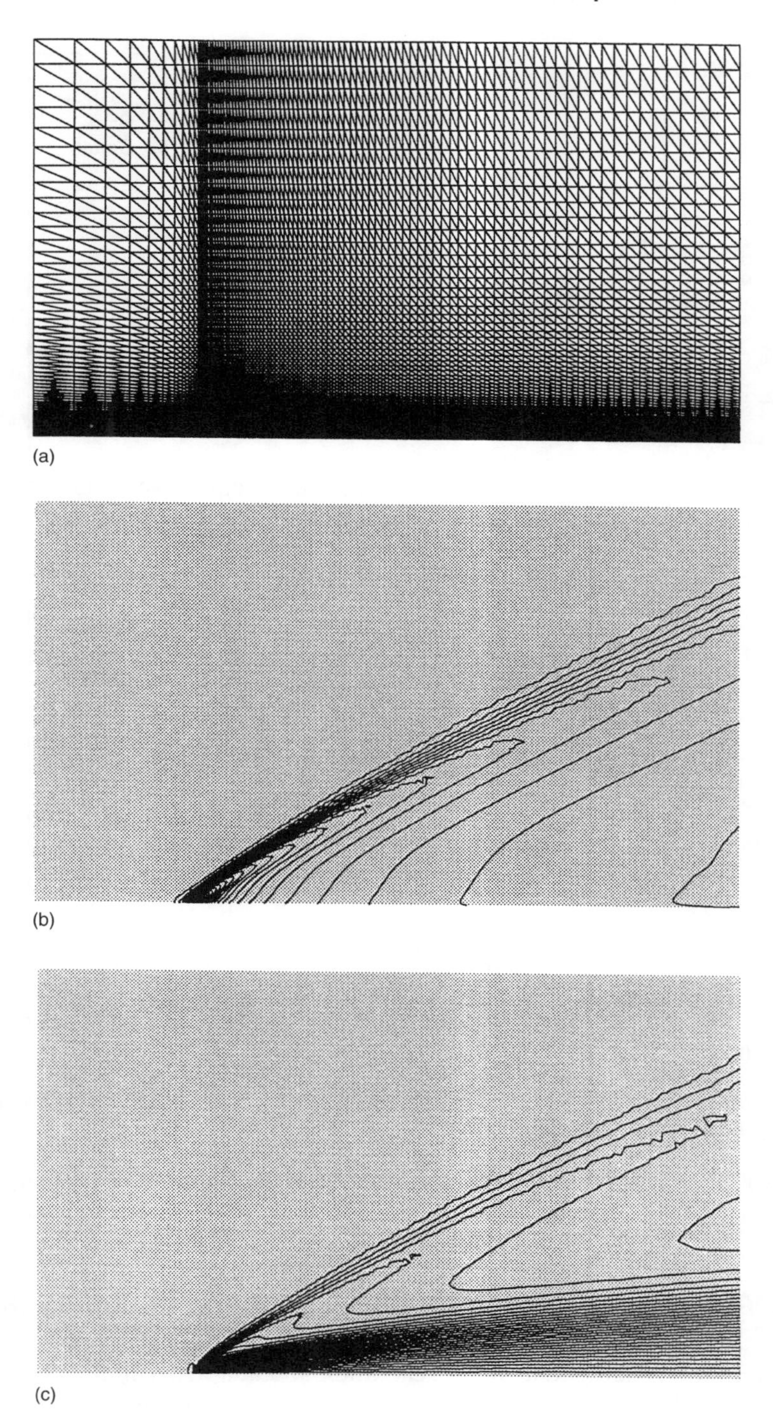

Fig. 6.21 Viscous flow past a flat plate (Carter problem).[77] Mach 3, $Re = 1000$ (a) mesh, nodes: 6750, elements: 13, 172 contours of (b) pressure and (c) Mach number.

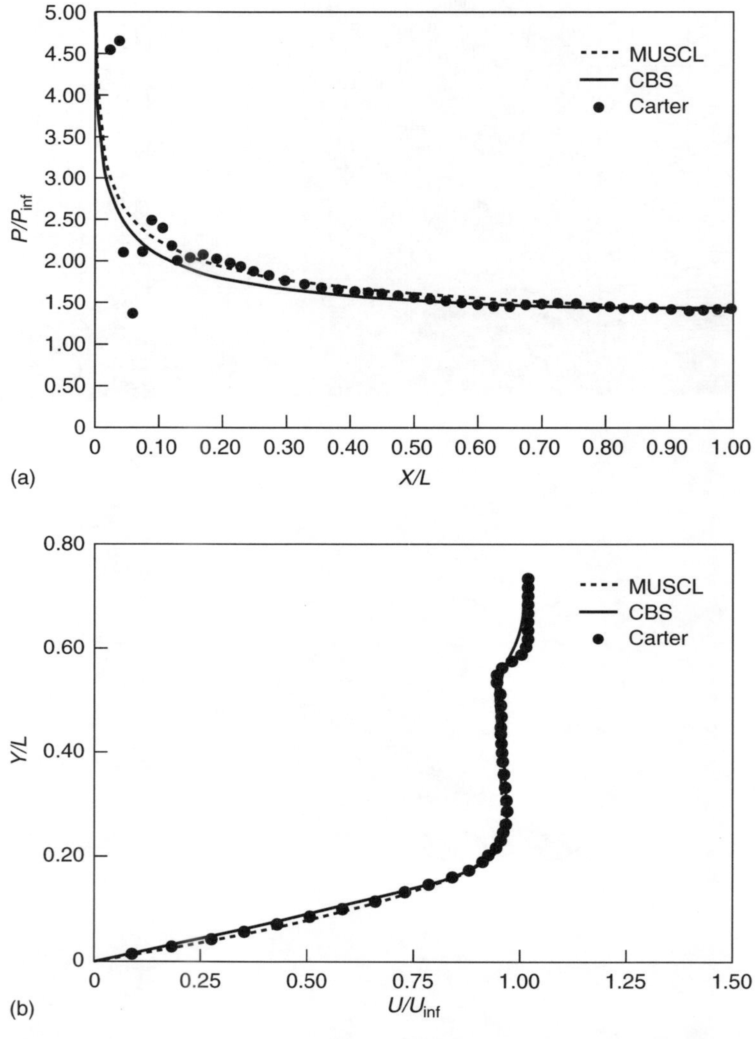

Fig. 6.22 Viscous flow past a flat plate (Carter problem).[77] Mach 3, $Re = 1000$. (a) Pressure distribution along the plate surface, (b) exit velocity profile.

and it will be noted that the CBS solution appears to be more consistent, avoiding oscillations near the leading edge.

6.10.2 Adaptive refinement in both shock and boundary layer

In this section we shall pursue mesh generation and adaptivity in precisely the same manner as we have done in Chapter 4 and previously in this chapter, i.e. using elongated finite elements in the zones where rapid variation of curvature occurs. An example of this application is given in Fig. 6.23. Here now a problem of the

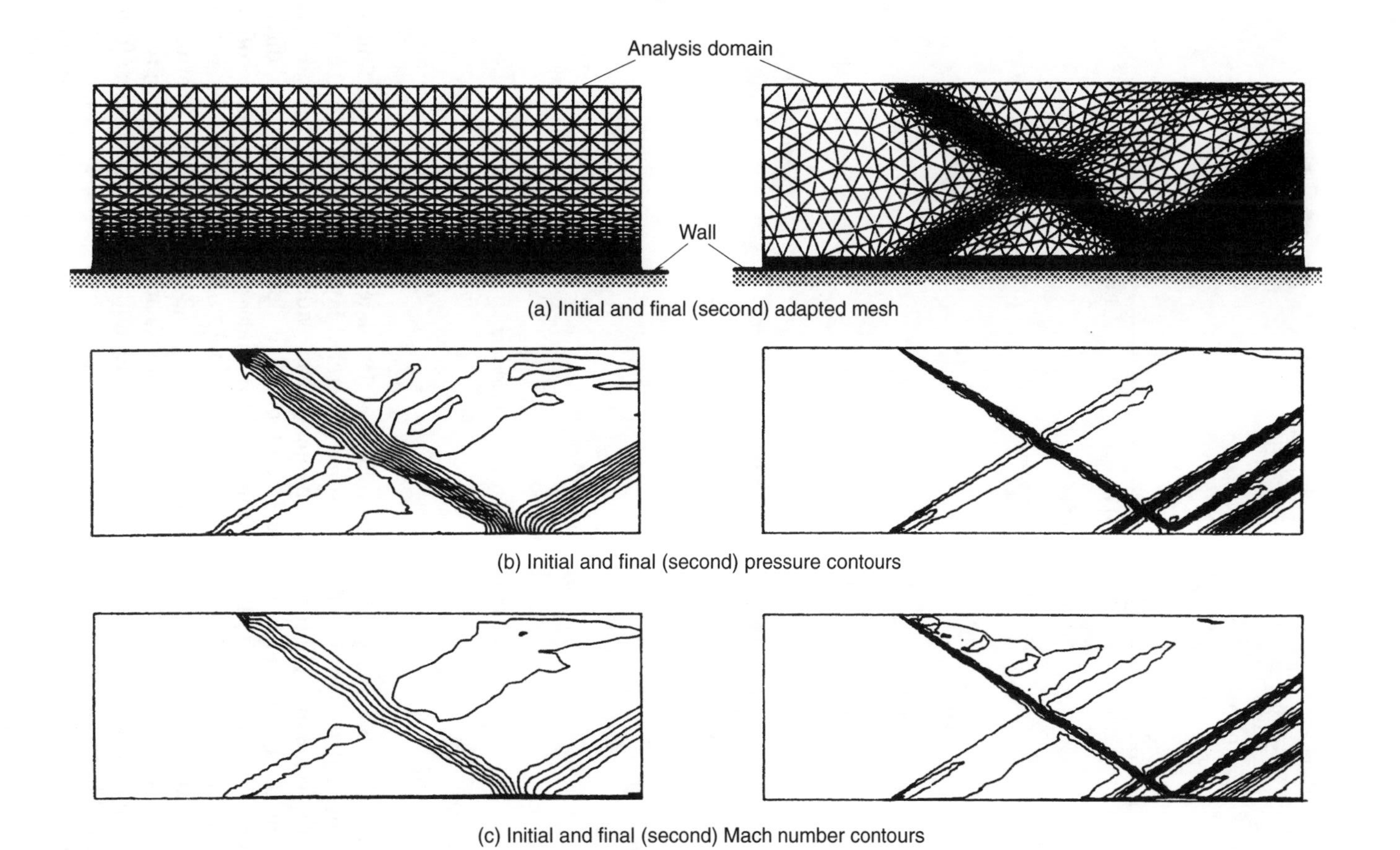

Fig. 6.23 Shock and boundary layer interaction.[79] Final mesh, nodes: 4198.

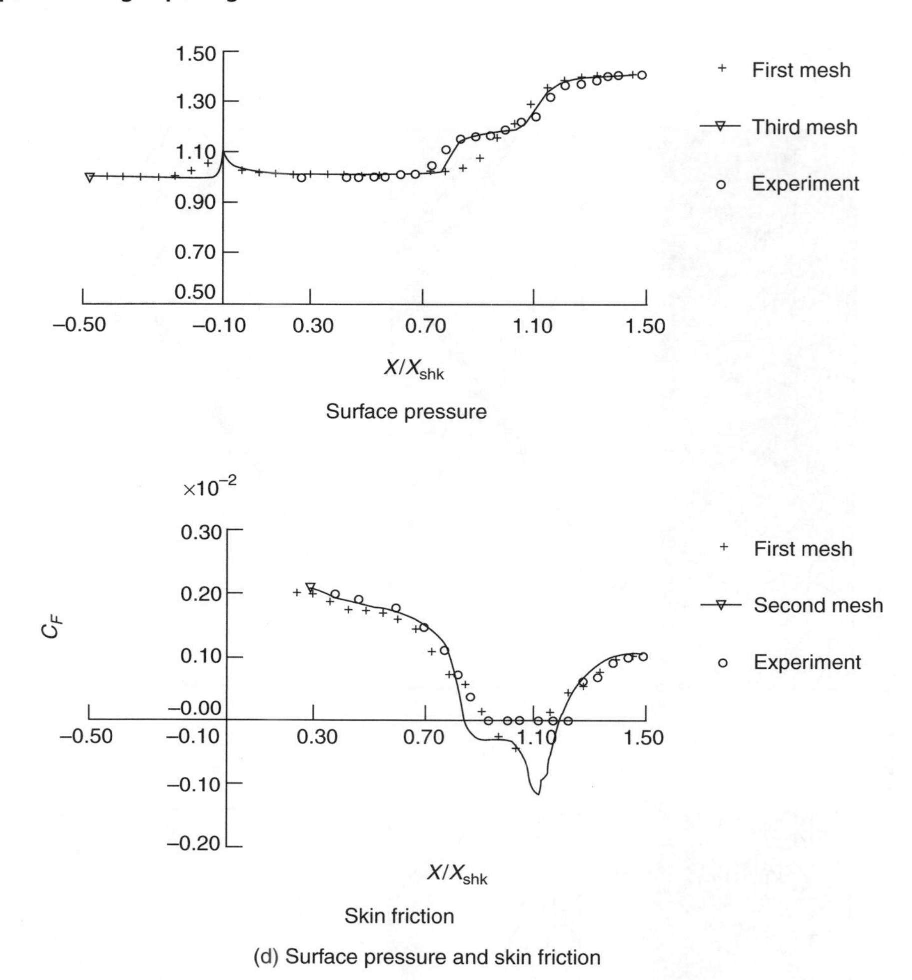

Surface pressure

Skin friction

(d) Surface pressure and skin friction

Fig. 6.23 Continued.

interaction of a boundary layer generated by a flat plate and externally impinging shock is presented.[79] In this problem, some structured layers are used near the wall in addition to the direct approach of Chapter 4. The reader will note the progressive refinement in the critical area. The second problem dealt with by such a direct approach again using the CBS algorithm is that of high-speed flow over an aerofoil. The flow is transonic and is again over a NACA0012 aerofoil. This problem was extensively studied by many researchers.[80,81] In Fig. 6.24, we show the final mesh after three iterations as well as contours of density. The density contours present some instability which are indeed observed by many authors at large distance from the aerofoil in the wake.[82]

In such a problem it would be simpler to refine near the boundary or indeed at the shock using structured meshes and the idea of introducing such refinement is explored in the next section.

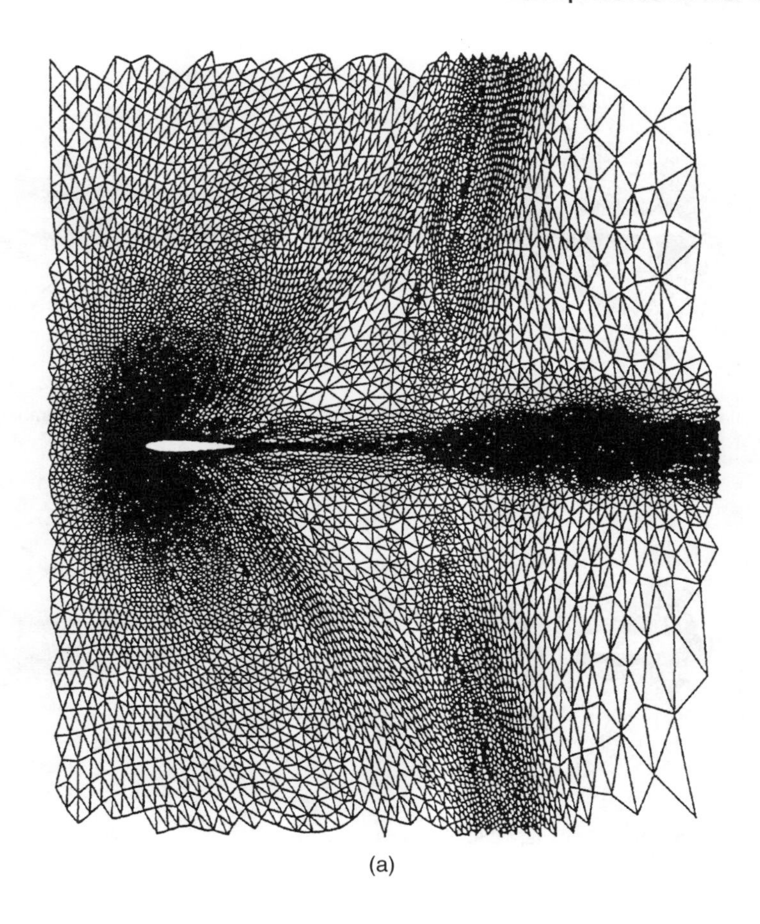

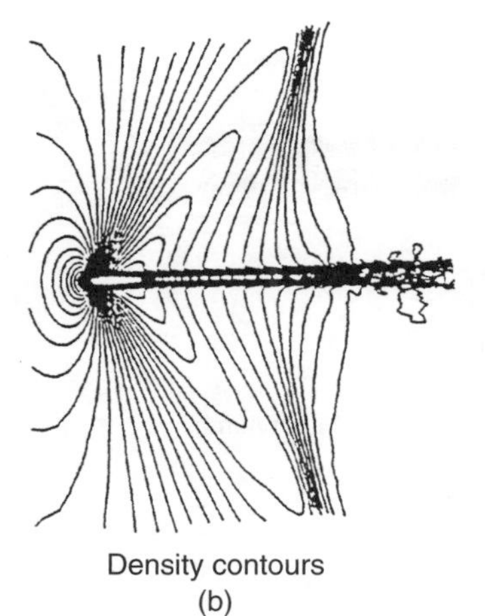

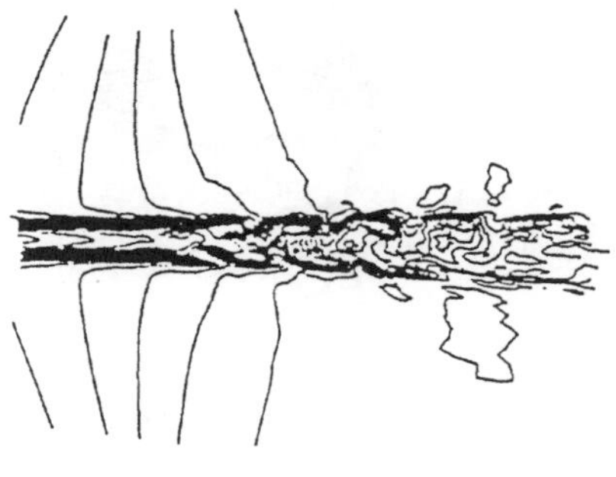

Fig. 6.24 Transonic viscous flow past a NACA0012 aerofoil,[82] Mach 0.95, (a) adapted mesh nodes: 16 388, elements: 32 522, (b) density contours, (c) density contours in the wake.

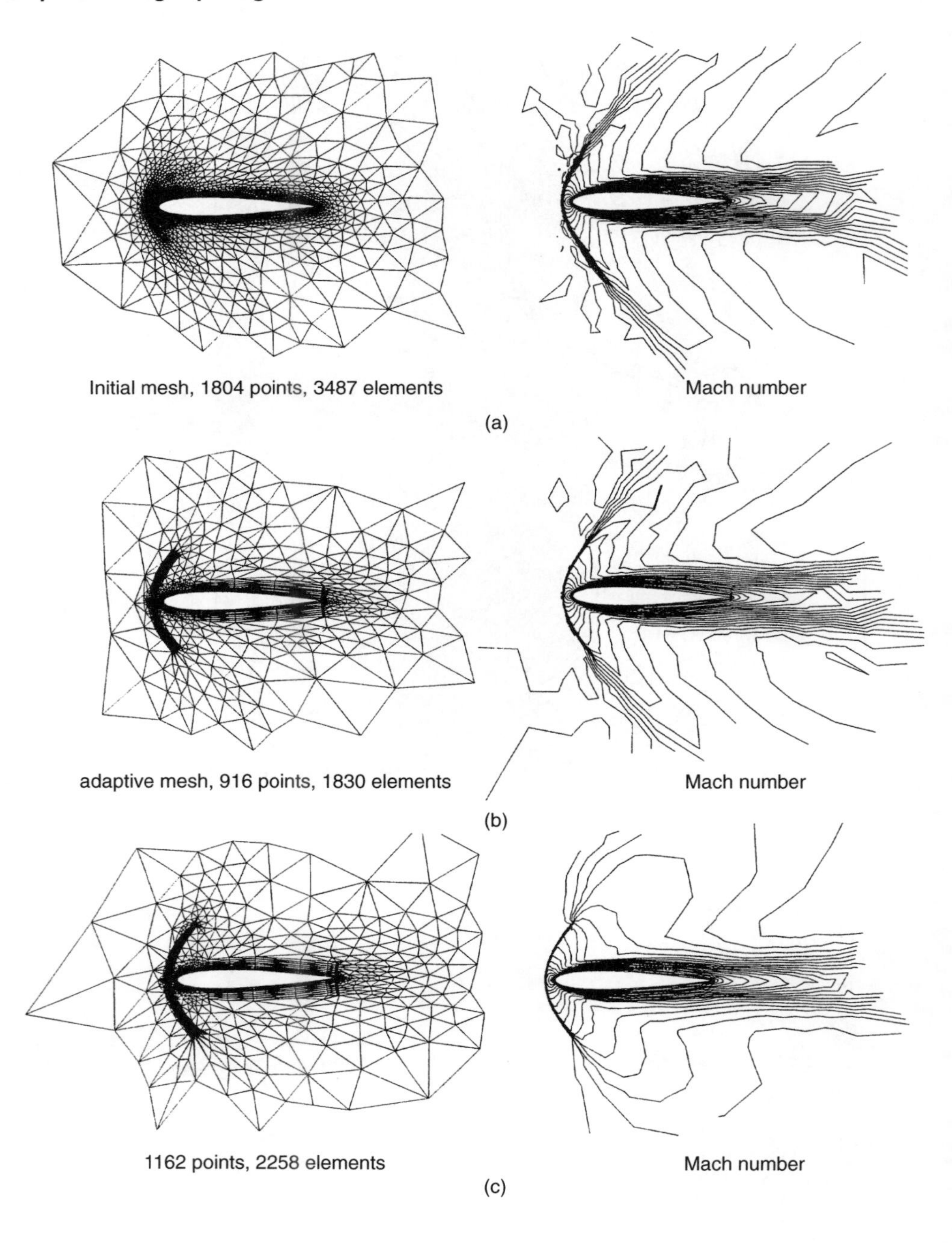

Fig. 6.25 Hybrid mesh for supersonic viscous flow past a NACA0012 aerofoil,[80] Mach 2, and contours of Mach number, (a) initial mesh, (b) first adapted mesh, (c) final mesh, (d) mesh near stagnation point (shown opposite).

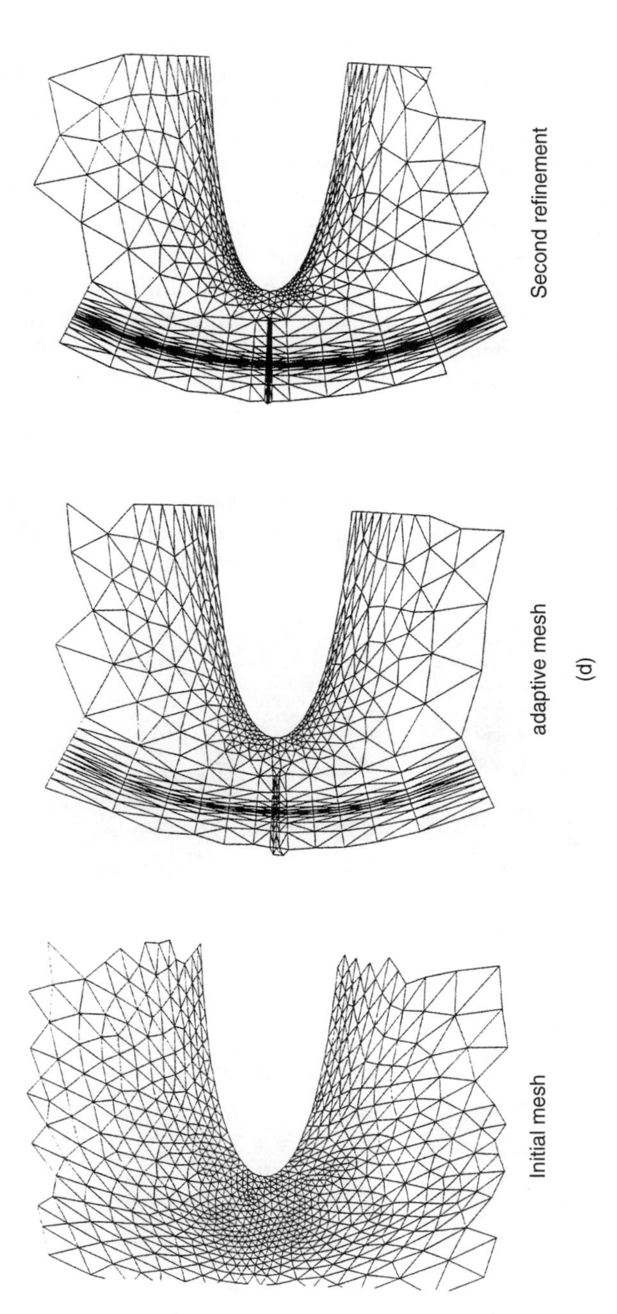

Fig. 6.25 Continued.

6.10.3 Special adaptive refinement for boundary layers and shocks

As with the direct iterative approach, it is difficult to arrive at large elongations during mesh generation, and the procedures just described tend to be inaccurate. For this reason it is useful to introduce a structured layer within the vicinity of solid boundaries to model the boundary layers and indeed it is possible to do the same in the shocks once these are defined. Within the boundary layer this can be done readily as shown in Fig. 6.20 using a layer of structured triangles or indeed quadrilaterals. On many occasions triangles have been used here to avoid the use of two kinds of elements in the same code. However if possible it is better to use directly quadrilaterals. The same problem can of course be done three dimensionally and we shall in Sec. 6.11 discuss applications of such layers. Again in the structured layer we can use either prismatic elements or simply tetrahedra though if the latter are used many more elements are necessary for the same accuracy. It is clear that unless the structured meshes near the boundary are specified *a priori*, an adaptive procedure will be somewhat complicated and on several occasions fixed boundary meshes have been used. However alternatives exist and here two possibilities should be mentioned. The first possibility, and that which has not yet been fully exploited, is that of refinement in which structured meshes are used in both shocks and boundary layers and the width of the domains is determined after some iterations. The

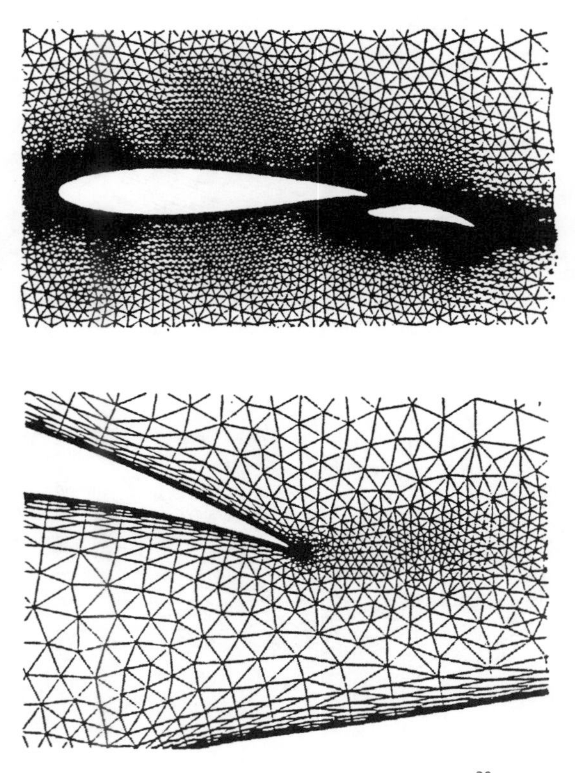

Fig. 6.26 Structured grid in boundary layer for a two-component aerofoil.[30] Advancing boundary normals.

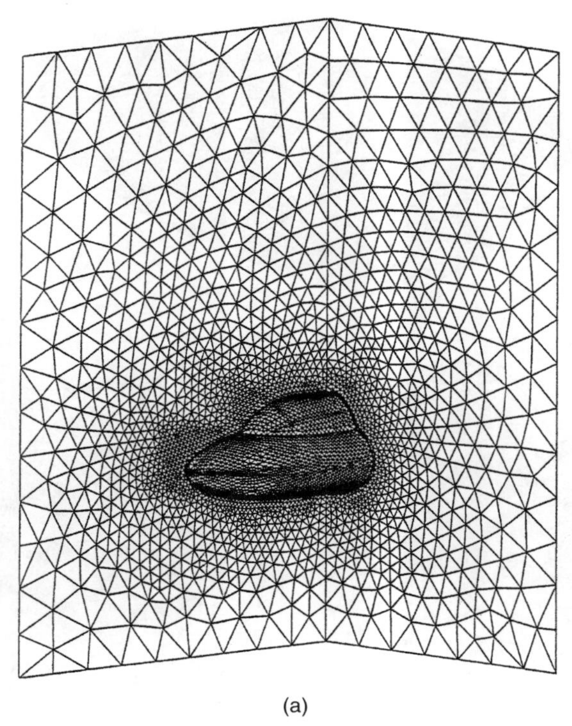

(a)

Fig. 6.27 Hypersonic viscous flow at Mach 8.15 over a double ellipsoid.[31] (a) Initial surface mesh total nodes: 25 990 and elements: 139 808,

procedure is somewhat involved and has been used with success in many trial problems as shown by Zienkiewicz and Wu.[80] We shall not describe the method in detail here but essentially structured meshes again composed of triangles or at least quadrilaterals divided into two triangles were used near the boundary and in the shock regions. The subdivision and accuracy obtained was excellent. In the second method we could imagine that normals are created on the boundaries, and a boundary layer thickness is predicted using some form of boundary layer analytical computation.[30–33] Within this layer structured meshes are adopted using a geometrical progression of thickness. The structured boundary layer meshing can of course be terminated where its need is less apparent and unstructured meshes continued outside. In this procedure we shall use the simple direct refinement of the type discussed in the previous section.

Figure 6.25 illustrates supersonic flow around an NACA0012 aerofoil using the automatic generation of structured and unstructured domains taken from reference 80. The second method is illustrated in Fig. 6.26 on a two-component aerofoil.

6.11 Three-dimensional viscous problems

The same procedures which we have described in the previous section can of course be used in three dimensions. Quite realistic high Reynolds number boundary layers were

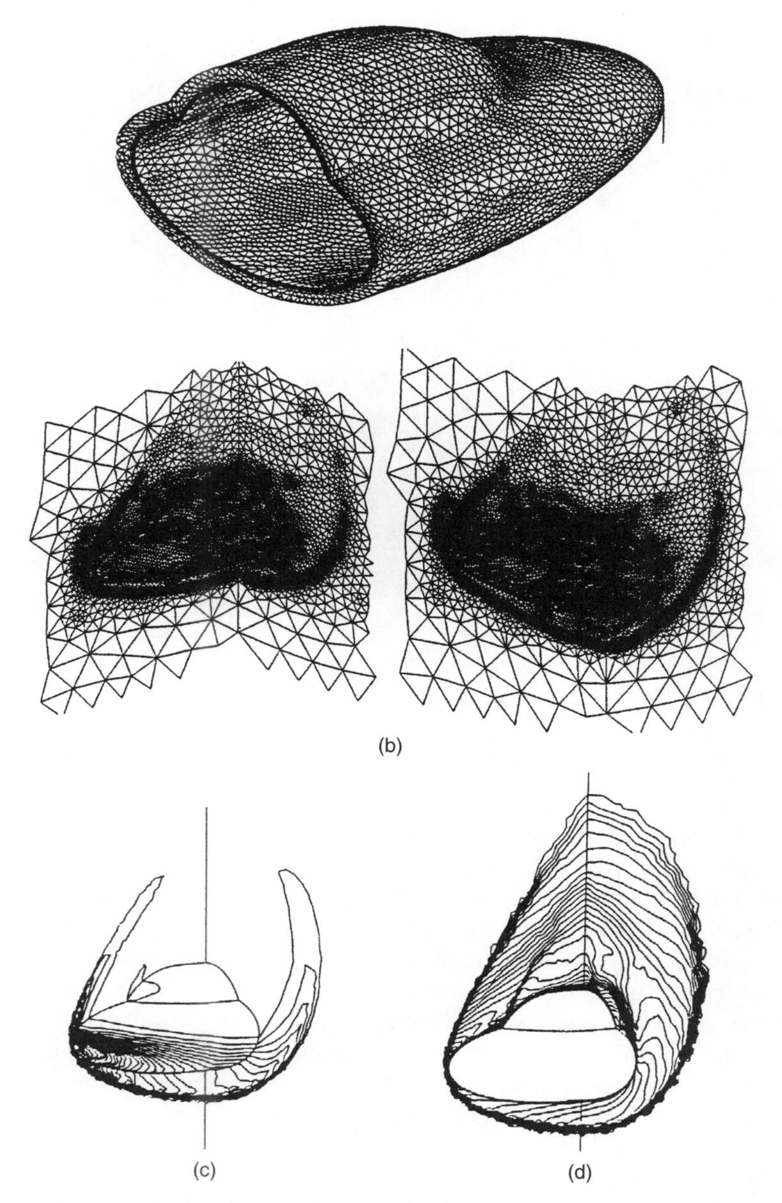

Fig. 6.27 (Continued) (b) adapted mesh total nodes: 79 023 and elements: 441 760, (c) pressure contours, (d) Mach contours.

so modelled. Figure 6.27 shows the viscous flow at a very high Reynolds number around a forebody of a double ellipsoid form.[31] In this example a structured boundary layer is assumed *a priori*.

The second example concerns a more sophisticated use of a structured subgrid procedure using local normals executed for a turbulent flow around an ONERA M6 wing attached to an aircraft body (Fig. 6.28).[34]

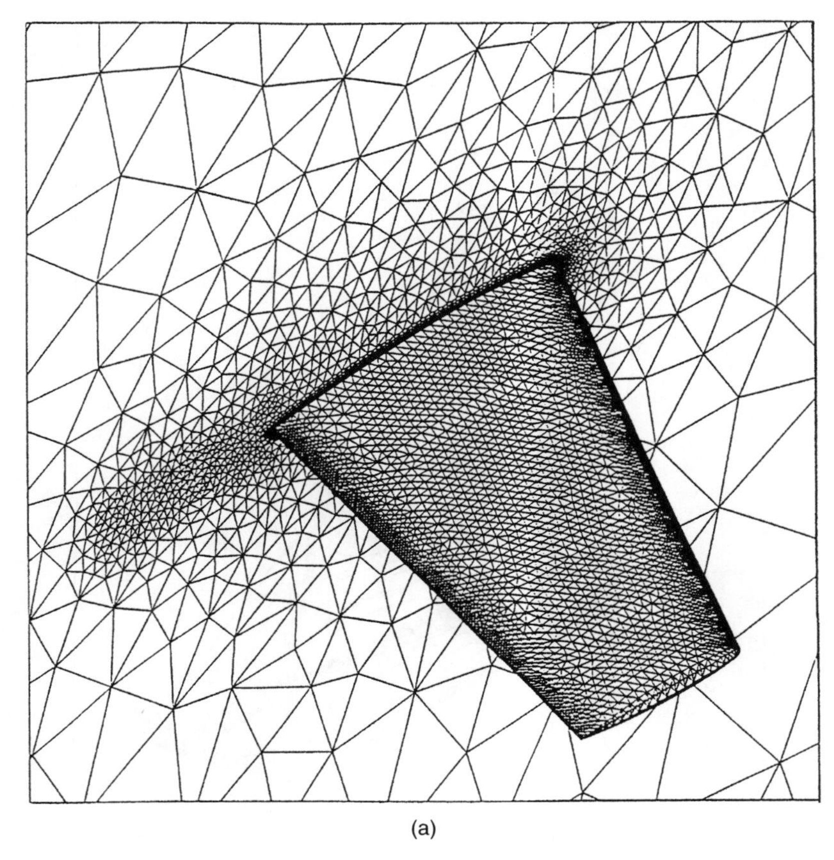

(a)

Fig. 6.28 Turbulent, viscous, compressible flow past a ONERA M6 wing.[34] (a) Surface mesh, elements: 48 056.

In this example a turbulent κ-ω model was used similar to the κ-ε model. As we described in Chapter 5 an additional solution for two parameters is required. Figure 6.28(c) also shows the comparison of the coefficient of pressure values with experimental data.[83]

6.12 Boundary layer–inviscid Euler solution coupling

It is well known that high-speed flows which exist without substantial flow separation develop a fairly thin boundary layer to which all the viscous effects are confined. The flow outside this boundary layer is purely inviscid. Such problems have for some years been solved approximately by using pure Euler solutions from which the pressure distribution is obtained. Coupling these solutions with a boundary layer approximation written for a very small thickness near the solid body provides the complete solution. The theory by which the separation between inviscid and viscous domains

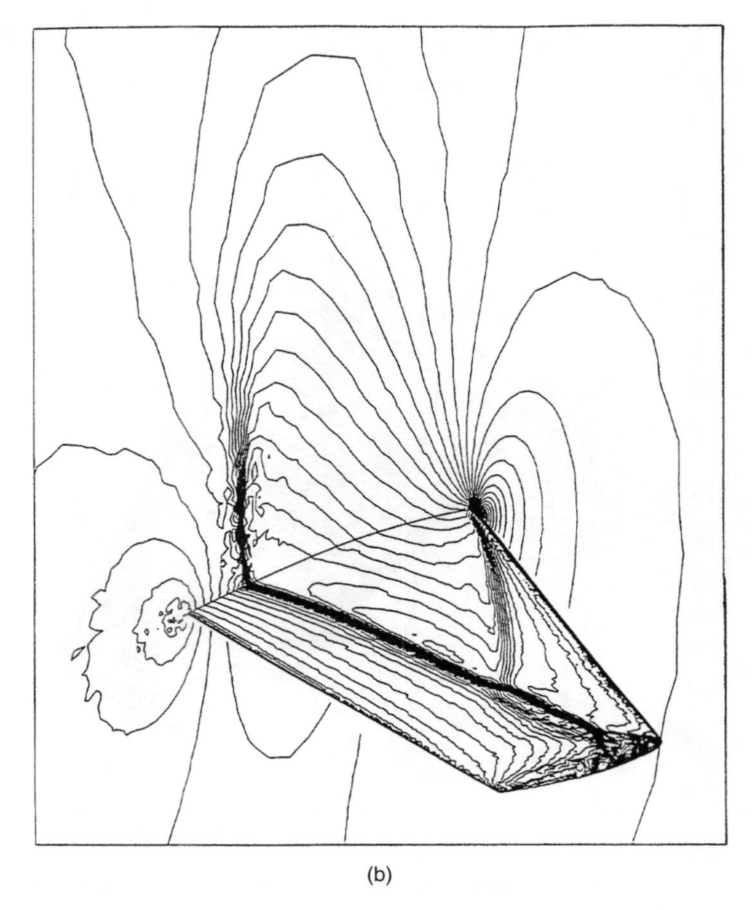

(b)

Fig. 6.28 (Continued) (b) pressure contours.

is predicted is that based on the work of Prandtl and for which much development has taken place since his original work. Clearly various methods of solving boundary layer problems can be used and many different techniques of inviscid solution can be implemented.

In the boundary layer full Navier–Stokes equations are used and generally these equations are specialized by introducing the assumptions of a boundary layer in which no pressure variation across the thickness occurs.

An alternative to solving the equations in the whole boundary layer is the integral approach in which the boundary layer equations need to be solved only on the solid surface. Here the 'transpiration velocity model' for laminar flows[84] and the 'lag-entrainment' method[85] for turbulent flows are notable approaches. Further extensions of these procedures can be found in many available research articles.[86–90]

Many recent studies illustrate the latest developments and implementation procedures of viscous–inviscid coupling.[91–93] Although the use of such viscous–inviscid

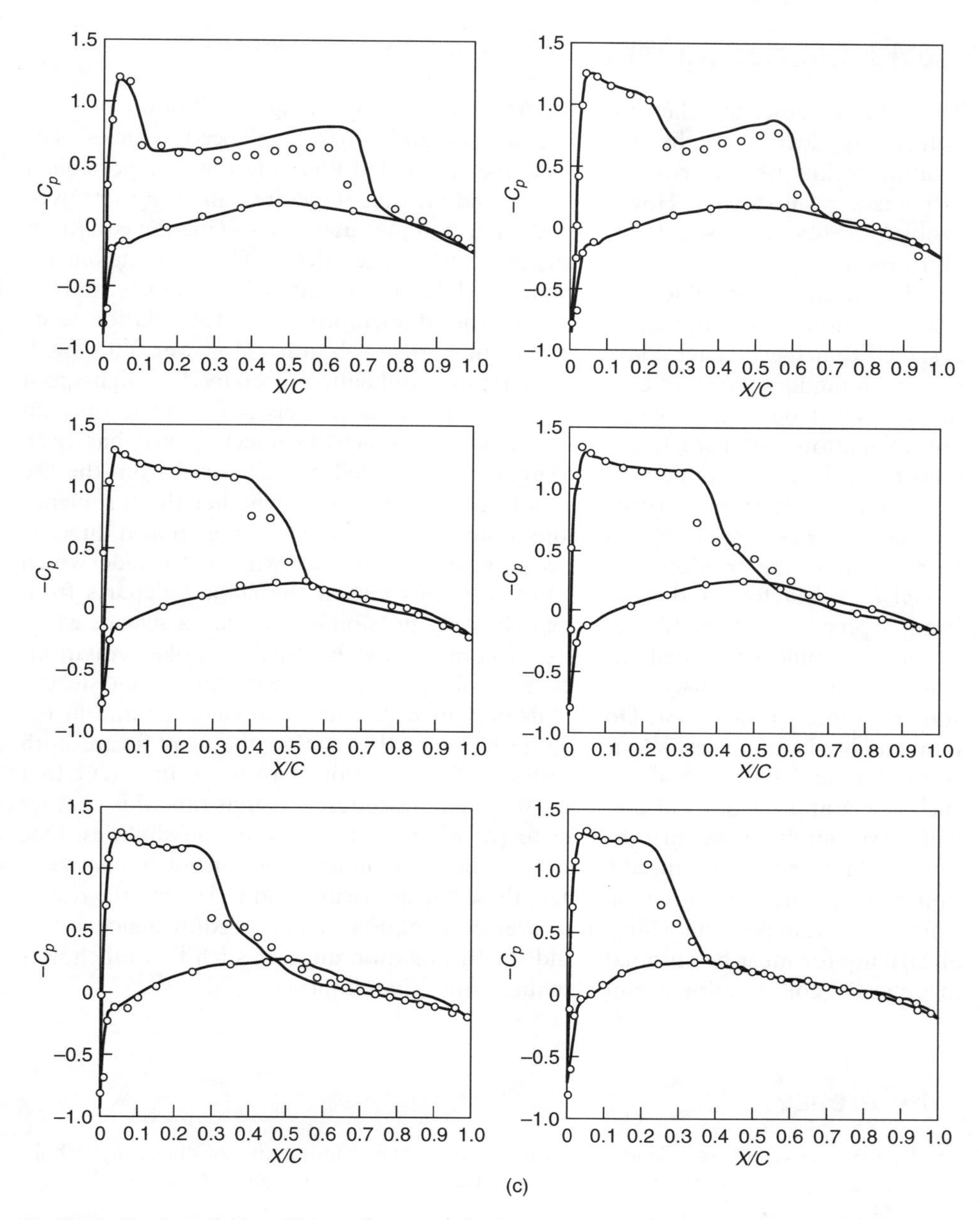

Fig. 6.28 (Continued) (c) coefficient of pressure distribution at 20%, 44%, 65%, 80%, 90% and 95% of wing span, line–numerical[34] and circle–experimental.[83]

coupling is not directly applicable in problems where boundary layer separation occurs, many studies are available to deal with separated flows.[94–96] We do not give any further details of viscous–inviscid coupling here and the reader can refer to the quoted references and Appendix E.

6.13 Concluding remarks

This chapter describes the most important and far-reaching possibilities of finite element application in the design of aircraft and other high-speed vehicles. The solution techniques described and examples presented illustrate that the possibility of realistic results exists. However, we do admit that there are still many unsolved problems. Most of these refer to either the techniques used for solving the equations or to modelling satisfactorily viscous and turbulence effects. The paths taken for simplifying and more efficient calculations have been outlined previously and we have mentioned possibilities such as multigrid methods, edge formulation, etc., designed to achieve faster convergence of numerical solutions. However full modelling of boundary layer effects is much more difficult, especially for high-speed flows. Use of boundary layer theory and turbulence models is of course only an approximation and here it must be stated that much 'engineering art' has been used to achieve acceptable results. This inside knowledge is acquired from the use of data available from experiments and becomes necessary whether the turbulence models of any type are used or whether boundary layer theories are applied directly. In either case the freedom of choice is given to the user who will decide which model is satisfactory and which is not. For this reason the subject departs from being a precise mathematical science. The only possibility for such a science exists in direct turbulence modelling. Here of course only the Navier–Stokes equations which we have previously described are solved in a transient state when steady-state solutions do not exist. Doing this may involve billions of elements and at the moment is out of reach. We anticipate however that within the near future both computers and the methods of solution will be developed to such an extent that such direct approaches will become a standard procedure. At that time this chapter will serve purely as an introduction to the essential formulation possibilities. One aspect which can be visualized is that realistic three-dimensional turbulent computations will only be used in regions where these effects are important, leaving the rest to simpler Eulerian flow modelling. However the computational procedure which we are all striving for must be automatic and the formulation must be such that all choices made in the computation are predictable rather than imposed.

References

1. R. Löhner, K. Morgan and O.C. Zienkiewicz. The solution of non-linear hyperbolic equation systems by the finite element method. *Int. J. Num. Meth. Fluids*, **4**, 1043–63, 1984.
2. R. Löhner, K. Morgan and O.C. Zienkiewicz. Domain splitting for an explicit hyperbolic solver. *Comp. Meth. Appl. Mech. Eng.*, **45**, 313–29, 1984.
3. O.C. Zienkiewicz, R. Löhner, K. Morgan and J. Peraire. High speed compressible flow and other advection dominated problems of fluid mechanics, in *Finite Elements in Fluids* (eds R.H. Gallagher, G.F. Carey, J.T. Oden and O.C. Zienkiewicz), Vol. 6, chap. 2, pp. 41–88, Wiley, Chichester, 1985.
4. R. Löhner, K. Morgan and O.C. Zienkiewicz. An adaptive finite element procedure for compressible high speed flows. *Comp. Meth. Appl. Mech. Eng.*, **51**, 441–65, 1985.

5. R. Löhner, K. Morgan, J. Peraire, O.C. Zienkiewicz and L. Kong. Finite element methods for compressible flow, in *ICFD Conf. on Numerical Methods in Fluid Dynamics* (ed. K.W. Morton and M.J. Baines), Vol. II, pp. 27–53, Clarendon Press, Oxford, 1986.
6. R. Löhner, K. Morgan, J. Peraire and M. Vahdati. Finite element, flux corrected transport (FEM–FCT) for the Euler and Navier–Stokes equations. *Int. J. Num. Meth. Eng.*, **7**, 1093–109, 1987.
7. R. Löhner, K. Morgan and O.C. Zienkiewicz. Adaptive grid refinement for the Euler and compressible Navier–Stokes equation, in *Proc. Int. Conf. on Accuracy Estimates and Adaptive Refinement in Finite Element Computations*, Lisbon, 1984.
8. R. Löhner, K. Morgan, J. Peraire and O.C. Zienkiewicz. Finite element methods for high speed flows. *AIAA paper 85-1531-CP*, 1985.
9. O.C. Zienkiewicz, K. Morgan, J. Peraire, M. Vahdati and R. Löhner. Finite elements for compressible gas flow and similar systems, in *7th Int. Conf. in Computational Methods in Applied Sciences and Engineering*, Versailles, December 1985.
10. R. Löhner, K. Morgan and O.C. Zienkiewicz. Adaptive grid refinement for the Euler and compressible Navier–Stokes equations, in *Accuracy Estimates and Adaptive Refinements in Finite Element Computations* (eds I. Babuska, O.C. Zienkiewicz, J. Gago and E.R. de A. Oliveira), chap. 15, pp. 281–98, Wiley, Chichester, 1986.
11. J. Peraire, M. Vahdati, K. Morgan and O.C. Zienkiewicz. Adaptive remeshing for compressible flow computations. *J. Comp. Phys.*, **72**, 449–66, 1987.
12. J. Peraire, K. Morgan, J. Peiro and O.C. Zienkiewicz. An adaptive finite element method for high speed flows, in *AIAA 25th Aerospace Sciences Meeting*, Reno, Nevada, AIAA paper 87-0558, 1987.
13. O.C. Zienkiewicz, J.Z. Zhu, Y.C. Liu, K. Morgan and J. Peraire. Error estimates and adaptivity: from elasticity to high speed compressible flow, in *The Mathematics of Finite Elements and Application* (*MAFELAP 87*) (ed. J.R. Whiteman), pp. 483–512, Academic Press, London, 1988.
14. L. Formaggia, J. Peraire and K. Morgan. Simulation of state separation using the finite element method. *Appl. Math. Modelling*, **12**, 175–81, 1988.
15. O.C. Zienkiewicz, K. Morgan, J. Peraire, J. Peiro and L. Formaggia. Finite elements in fluid mechanics. Compressible flow, shallow water equations and transport, in *ASME Conf. on Recent Development in Fluid Dynamics*, AMD 95, American Society of Mechanical Engineers, December 1988.
16. J. Peraire, J. Peiro, L. Formaggia, K. Morgan and O.C. Zienkiewicz. Finite element Euler computations in 3-dimensions. *Int. J. Num. Meth. Eng.*, **26**, 2135–59, 1989. (See also same title: *AIAA 26th Aerospace Sciences Meeting*, Reno, AIAA paper 87-0032, 1988.)
17. J.R. Stewart, R.R. Thareja, A.R. Wieting and K. Morgan. Application of finite elements and remeshing techniques to shock interference on a cylindrical leading edge. Reno, Nevada, AIAA paper 88-0368, 1988.
18. R.R. Thareja, J.R. Stewart, O. Hassan, K. Morgan and J. Peraire. A point implicit unstructured grid solver for the Euler and Navier–Stokes equation. *Int. J. Num. Meth. Fluids*, **9**, 405–25, 1989.
19. R. Löhner. Adaptive remeshing for transient problems with moving bodies, in *National Fluid Dynamics Congress*, Ohio, AIAA paper 88-3736, 1988.
20. R. Löhner. The efficient simulation of strongly unsteady flows by the finite element method, in *25th Aerospace Sci. Meeting*, Reno, Nevada, AIAA paper 87-0555, 1987.
21. R. Löhner. Adaptive remeshing for transient problems. *Comp. Meth. Appl. Mech. Eng.*, **75**, 195–214, 1989.
22. J. Peraire. A finite element method for convection dominated flows. Ph.D. thesis, University of Wales, Swansea, 1986.

23. O. Hassan, K. Morgan and J. Peraire. An implicit–explicit scheme for compressible viscous high speed flows. *Comp. Meth. Appl. Mech. Eng.*, **76**, 245–58, 1989.
24. N.P. Weatherill, E.A. Turner-Smith, J. Jones, K. Morgan and O. Hassan. An integrated software environment for multi-disciplinary computational engineering. *Engng. Comp.*, **16**, 913–33, 1999.
25. P.M.R. Lyra and K. Morgan. A review and comparative study of upwind biased schemes for compressible flow computation. Part I: 1-D first-order schemes. *Arch. Comp. Meth. Engng.*, **7**, 19–55, 2000.
26. K. Morgan and J. Peraire. Unstructured grid finite element methods for fluid mechanics. *Rep. Prog. Phys.*, **61**, 569–638, 1998.
27. R. Ayers, O. Hassan, K. Morgan and N.P. Weatherill. The role of computational fluid dynamics in the design of the Thrust supersonic car. *Design Optim. Int. J. Prod. & Proc. Improvement*, **1**, 79–99, 1999.
28. K. Morgan, O. Hassan and N.P. Weatherill. Why didn't the supersonic car fly? *Mathematics Today, Bulletin of the Institute of Mathematics and its Applications*, **35**, 110–14, Aug. 1999.
29. O. Hassan, L.B. Bayne, K. Morgan and N.P. Weatherill. An adaptive unstructured mesh method for transient flows involving moving boundaries. *ECCOMAS '98*, Wiley, New York.
30. O. Hassan, E.J. Probert, N.P. Weatherill, M.J. Marchant and K. Morgan. The numerical simulation of viscous transonic flow using unstructured grids, *AIAA-94-2346*, June 20–23, Colorado Springs, USA.
31. O. Hassan, E.J. Probert, K. Morgan and J. Peraire. Mesh generation and adaptivity for the solution of compressible viscous high speed flows. *Int. J. Num. Meth. Eng.*, **38**, 1123–48, 1995.
32. O. Hassan, K. Morgan, E.J. Probert and J. Peraire. Unstructured tetrahedral mesh generation for three dimensional viscous flows. *Int. J. Num. Meth. Eng.*, **39**, 549–67, 1996.
33. O. Hassan, E.J. Probert and K. Morgan. Unstructured mesh procedures for the simulation of three dimensional transient compressible inviscid flows with moving boundary components. *Int. J. Num. Meth. Fluids*, **27**, 41–55, 1998.
34. M.T. Manzari, O. Hassan, K. Morgan and N.P. Weatherill. Turbulent flow computations on 3D unstructured grids. *Finite Elements in Analysis and Design*, **30**, 353–63, 1998.
35. E.J. Probert, O. Hassan and K. Morgan. An adaptive finite element method for transient compressible flows with moving boundaries. *Int. J. Num. Meth. Eng.*, **32**, 751–65, 1991.
36. K. Morgan, N.P. Weatherill, O. Hassan, P.J. Brookes, R. Said and J. Jones. A parallel framework for multidisciplinary aerospace engineering simulations using unstructured meshes. *Int. J. Num. Meth. Fluids*, **31**, 159–73, 1999.
37. K. Morgan, P.J. Brookes, O. Hassan and N.P. Weatherill. Parallel processing for the simulation of problems involving scattering of electromagnetic waves. *Comp. Meth. Appl. Mech. Eng.*, **152**, 157–74, 1998.
38. R. Said, N.P. Weatherill, K. Morgan and N.A. Verhoeven. Distributed parallel Delaunay mesh generation. *Comp. Meth. Appl. Mech. Eng.*, **177**, 109–25, 1999.
39. C. Hirsch. *Numerical Computation of Internal and External Flows*. Vol. I, Wiley, Chichester, 1988.
40. L. Demkowicz, J.T. Oden and W. Rachowicz. A new finite element method for solving compressible Navier–Stokes equations based on an operator splitting method and h-p adaptivity. *Comp. Meth. Appl. Mech. Eng.*, **84**, 275–326, 1990.
41. J. Vadyak, J.D. Hoffman and A.R. Bishop. Flow computations in inlets at incidence using a shock fitting Bicharacteristic method. *AIAA Journal*, **18**, 1495–502, 1980.
42. K.W. Morton and M.F. Paisley. A finite volume scheme with shock fitting for steady Euler equations. *J. Comp. Phys.*, **80**, 168–203, 1989.
43. J. von Neumann and R.D. Richtmyer. A method for the numerical calculations of hydrodynamical shocks. *J. Math. Phys.*, **21**, 232–7, 1950.

44. A. Lapidus. A detached shock calculation by second order finite differences. *J. Comp. Phys.*, **2**, 154–77, 1967.
45. J.L. Steger. Implicit finite difference simulation of flow about two dimensional geometries. *AIAA J.*, **16**, 679–86, 1978.
46. R.W. MacCormack and B.S. Baldwin. A numerical method for solving the Navier–Stokes equations with application to shock boundary layer interaction. *AIAA paper 75-1*, 1975.
47. A. Jameson and W. Schmidt. Some recent developments in numerical methods in transonic flows. *Comp. Meth. Appl. Mech. Eng.*, **51**, 467–93, 1985.
48. K. Morgan, J. Peraire, J. Peiro and O.C. Zienkiewicz. Adaptive remeshing applied to the solution of a shock interaction problem on a cylindrical leading edge, in P. Stow (Ed.), *Computational Methods in Aeronautical Fluid Dynamics*, Clarenden Press, Oxford, 1990, pp. 327–44.
49. R. Codina. A discontinuity capturing crosswind dissipation for the finite element solution of the convection diffusion equation. *Comp. Meth. Appl. Mech. Eng.*, **110**, 325–42, 1993.
50. T.J.R. Hughes and M. Malett. A new finite element formulation for fluid dynamics IV: A discontinuity capturing operator for multidimensional advective–diffusive problems. *Comp. Mech. Appl. Mech. Eng.*, **58**, 329–36, 1986.
51. C. Johnson and A. Szepessy. On convergence of a finite element method for a nonlinear hyperbolic conservation law. *Math. Comput.*, **49**, 427–44, 1987.
52. A.C. Galeão and E.G. Dutra Do Carmo. A consistent approximate upwind Petrov–Galerkin method for convection dominate problems. *Comp. Meth. Appl. Mech. Eng.*, **68**, 83–95, 1988.
53. P. Hansbo and C. Johnson. Adaptive streamline diffusion methods for compressible flow using conservation variables. *Comp. Meth. Appl. Mech. Eng.*, **87**, 267–80, 1991.
54. F. Shakib, T.J.R. Hughes and Z. Johan. A new finite element formulation for computational fluid dynamics X. The compressible Euler and Navier–Stokes equations. *Comp. Mech. Appl. Mech. Eng.*, **89**, 141–219, 1991.
55. P. Nithiarasu, O.C. Zienkiewicz, B.V.K.S. Sai, K. Morgan, R. Codina and M. Vázquez. Shock capturing viscosities for the general algorithm. *10th Int. Conf. on Finite Elements in Fluids*, Ed. M. Hafez and J.C. Heinrich, 350–6, Jan 5–8, 1998, Tucson, Arizona, USA.
56. P. Nithiarasu, O.C. Zienkiewicz, B.V.K.S. Sai, K. Morgan, R. Codina and M. Vázquez. Shock capturing viscosities for the general fluid mechanics algorithm. *Int. J. Num. Meth. Fluids*, **28**, 1325–53, 1998.
57. G. Sod. A survey of several finite difference methods for systems of non-linear hyperbolic conservation laws. *J. Comp. Phys.*, **27**, 1–31, 1978.
58. T.E. Tezduyar and T.J.R. Hughes. Development of time accurate finite element techniques for first order hyperbolic systems with particular emphasis on Euler equation. Stanford University paper, 1983.
59. P. Woodward and P. Colella. The numerical simulation of two dimensional flow with strong shocks. *J. Comp. Phys.*, **54**, 115–73, 1984.
60. O.C. Zienkiewicz and J.Z. Zhu. A simple error estimator and adaptive procedure for practical engineering analysis. *Int. J. Num. Meth. Eng.*, **24**, 337–57, 1987.
61. J.T. Oden and L. Demkowicz. Advance in adaptive improvements: a survey of adaptive methods in computational fluid mechanics, in *State of the Art Survey in Computational Fluid Mechanics* (eds A.K. Noor and J.T. Oden), American Society of Mechanical Engineers, 1988.
62. M.T. Manzari, P.R.M. Lyra, K. Morgan and J. Peraire. An unstructured grid FEM/MUSCL algorithm for the compressible Euler equations. *Proc. VIII Int. Conf. on Finite Elements in Fluids: New Trends and Applications*, Swansea 1993, pp. 379–88, Pineridge Press.

63. P.R.M. Lyra, K. Morgan, J. Peraire and J. Peiro. TVD algorithms for the solution of compressible Euler equations on unstructured meshes. *Int. J. Num. Meth. Fluids*, **19**, 827–47, 1994.
64. P.R.M. Lyra, M.T. Manzari, K. Morgan, O. Hassan and J. Peraire. Upwind side based unstructured grid algorithms for compressible viscous flow computations. *Int. J. Eng. Anal. Des.*, **2**, 197–211, 1995.
65. R.A. Nicolaides. On finite element multigrid algorithms and their use, in J.R. Whiteman (ed.), *The Mathematics of Finite Elements and Applications III, MAFELAP 1978*, Academic Press, London, 1979, pp. 459–66.
66. W. Hackbusch and U. Trottenberg (Eds). *Multigrid Methods.* Lecture Notes in Mathematics 960, Springer-Verlag, Berlin, 1982.
67. R. Löhner and K. Morgan. An unstructured multigrid method for elliptic problems. *Int. J. Num. Meth. Eng.*, **24**, 101–15, 1987.
68. M.C. Rivara. Local modification of meshes for adaptive and or multigrid finite element methods. *J. Comp. Appl. Math.*, **36**, 79–89, 1991.
69. S. Lopez and R. Casciaro. Algorithmic aspects of adaptive multigrid finite element analysis. *Int. J. Num. Meth. Eng.*, **40**, 919–36, 1997.
70. A. Jameson, T.J. Baker and N.P. Weatherill. Calculation of inviscid transonic flow over a complete aircraft. *AIAA 24th Aerospace Sci. Meeting*. Reno, Nevada, AIAA paper 86-0103, 1986.
71. V. Billey, J. Periaux, P. Perrier and B. Stoufflet. 2D and 3D Euler computations with finite element methods in aerodynamics. *Lecture Notes in Math.*, **1270**, 64–81, 1987.
72. R. Noble, A. Green, D. Tremayne and SSC Program Ltd. *THRUST*, Transworld, London, 1998.
73. E.J. Probert. Finite element method for convection dominated flows. Ph.D. thesis, University of Wales, Swansea, 1986.
74. Y. Kallinderis and S. Ward. Prismatic grid generation for 3-dimensional complex geometries. *AIAA J.*, **31**, 1850–6, 1993.
75. Y. Kallinderis. Adaptive hybrid prismatic tetrahedral grids. *Int. J. Num. Meth. Fluids*, **20**, 1023–37, 1995.
76. A.J. Chen and Y. Kallinderis. Adaptive hybrid (prismatic-tetrahedral) grids for incompressible flows. *Int. J. Num. Meth. Fluids*, **26**, 1085–105, 1998.
77. O.C. Zienkiewicz, P. Nithiarasu, R. Codina, M. Vázquez and P. Ortiz. The Characteristic-Based-Split procedure: An efficient and accurate algorithm for fluid problems. *Int. J. Num. Meth. Fluids*, **31**, 359–92, 1999.
78. J.E. Carter. Numerical solutions of the Navier–Stokes equations for the supersonic laminar flow over a two-dimensional compression corner. *NASA TR-R-385*, 1972.
79. O. Hassan. Finite element computations of high speed viscous compressible flows. Ph.D. thesis, University of Wales, Swansea, 1990.
80. O.C. Zienkiewicz and J. Wu. Automatic directional refinement in adaptive analysis of compressible flows. *Int. J. Num. Meth. Eng.*, **37**, 2189–210, 1994.
81. M.J. Castro-Diaz, H. Borouchaki, P.L. George, F. Hecht and B. Mohammadi. Anisotropic mesh adaptation: theory, validation and applications. *Computational: Fluid Dynamics '96 – Proc. 3rd ECCOMAS Conf.*, Ed. J-A. Désidéri *et al.*, Wiley, Chichester, pp. 181–86.
82. P. Nithiarasu and O.C. Zienkiewicz. Adaptive mesh generation for fluid mechanics problems. *Int. J. Num. Meth. Eng.*, **47** (to appear, 2000).
83. V. Schmitt and F. Charpin. Pressure distributions on the ONERA M6 wing at transonic Mach numbers. *AGARD Report AR-138*, Paris, 1979.
84. M.J. Lighthill. On displacement thickness. *J. Fluid Mech.*, **4**, 383, 1958.
85. J.E. Green, D.J. Weeks and J.W.F. Brooman. Prediction of turbulent boundary layers and wakes in compressible flow by a lag-entrainment method. *Aeronautical Research Council Repo. and Memo Rept. No. 3791*, 1973.

86. P. Bradshaw. The analogy between streamline curvature and buoyancy in turbulent shear flow, *J. Fluid Mech.*, **36**, 177–91, 1969.
87. J.E. Green. The prediction of turbulent boundary layer development in compressible flow. *J. Fluid Mech.*, **31**, 753–78, 1969.
88. H.B. Squire and A.D. Young. The calculation of the profile drag of aerofoils. *Aeronautical Research Council Repo. and Memo 1838*, 1937.
89. R.E. Melnok, R.R. Chow and H.R. Mead. Theory of viscous transonic flow over airfoils at high Reynolds Number. *AIAA Paper 77-680*, June 1977.
90. J.C. Le Balleur. Calcul par copulage fort des écoluements visqueux transsoniques incluant sillages et décollemants. Profils d'aile portant. *La Recherche Aerospatiale*, May–June 1981.
91. J. Szmelter and A. Pagano. Viscous flow modelling using unstructured meshes for aeronautical applications. *Lecture Notes in Physics 453*, ed. S.M. Deshpande *et al.*, Springer-Verlag, Berlin (1994).
92. J. Szmelter. Viscous coupling techniques using unstructured and multiblock meshes. *ICAS Paper ICAS-96-1.7.5*, Sorrento, 1996.
93. J. Szmelter. Aerodynamic wing optimisation. *AIAA Paper 99-0550*, January, 1999.
94. J.C. Le Balleur. Viscous–inviscid calculation of high lift separated compressible flows over airfoils and wings. *Proceedings AGARD/FDP High Lift Aerodynamics, AGARD-CP515*, Banff, Canada, October 1992.
95. J.C. Le Balleur. Calculation of fully three-dimensional separated flows with an unsteady viscous–inviscid interaction method. *5th Int. Symp. on Numerical and Physical Aspects of Aerodynamic Flows*, Long Beach CA (USA), January, 1992.
96. J.C. Le Balleur and P. Girodroux-Lavigne. Calculation of dynamic stall by viscous–inviscid interaction over airfoils and helicopter-blade sections. *AHS 51st Annual Forum and Technology Display*, Fort Worth, TX USA, May, 1995.

7

Shallow-water problems

7.1 Introduction

The flow of water in shallow layers such as occur in coastal estuaries, oceans, rivers, etc., is of obvious practical importance. The prediction of tidal currents and elevations is vital for navigation and for the determination of pollutant dispersal which, unfortunately, is still frequently deposited there. The transport of sediments associated wth such flows is yet another field of interest.

In free surface flow in relatively thin layers the horizontal velocities are of primary importance and the problem can be reasonably approximated in two dimensions. Here we find that the resulting equations, which include in addition to the horizontal velocities the free surface elevation, can once again be written in the same conservation form as the Euler equations studied in previous chapters:

$$\frac{\partial \mathbf{\Phi}}{\partial t} + \frac{\partial \mathbf{F}_i}{\partial x_i} + \frac{\partial \mathbf{G}_i}{\partial x_i} + \mathbf{Q} = \mathbf{0} \qquad \text{for } i = 1, 2 \tag{7.1}$$

Indeed, the detailed form of these equations bears a striking similarity to those of compressible gas flow – despite the fact that now a purely incompressible fluid (water) is considered. It follows therefore that:

1. The methods developed in the previous chapters are in general applicable.
2. The type of phenomena (e.g. shocks, etc.) which we have encountered in compressible gas flows will occur again.

It will of course be found that practical interest focuses on different aspects. The objective of this chapter is therefore to introduce the basis of the derivation of the equation and to illustrate the numerical approximation techniques by a series of examples.

The approximations made in the formulation of the flow in shallow-water bodies are similar in essence to those describing the flow of air in the earth's environment and hence are widely used in meteorology. Here the vital subject of weather prediction involves their daily solution and a very large amount of computation. The interested reader will find much of the background in standard texts dealing with the subject, e.g. references 1 and 2.

A particular area of interest occurs in the linearized version of the shallow-water equations which, in periodic response, are similar to those describing acoustic phenomena. In the next chapter we shall therefore discuss some of these periodic phenomena involved in the action and forces due to waves.[3]

7.2 The basis of the shallow-water equations

In previous chapters we have introduced the essential Navier–Stokes equations and presented their incompressible, isothermal form, which we repeat below assuming full incompressibility. We now have the equations of mass conservation:

$$\frac{\partial u_i}{\partial x_i} = 0 \tag{7.2a}$$

and momentum conservation:

$$\frac{\partial u_j}{\partial t} + \frac{\partial}{\partial x_i}(u_i u_j) + \frac{1}{\rho}\frac{\partial p}{\partial x_j} - \frac{1}{\rho}\frac{\partial}{\partial x_i}\tau_{ij} - g_j = 0 \tag{7.2b}$$

with i, j being 1, 2, 3.

In the case of shallow water flow which we illustrate in Fig. 7.1 and where the direction x_3 is vertical, the vertical velocity u_3 is small and the corresponding accelerations negligible. The momentum equation in the vertical direction can therefore be

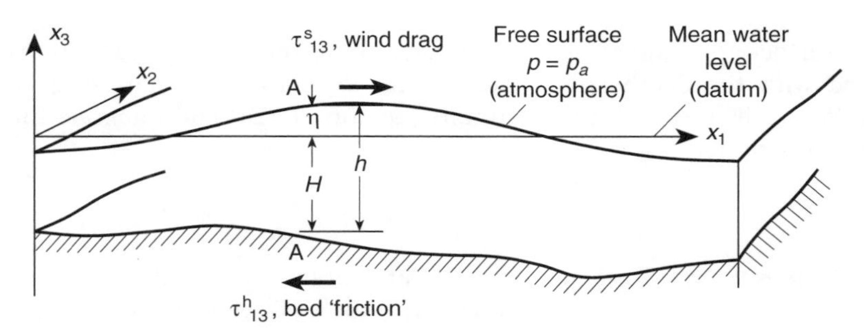

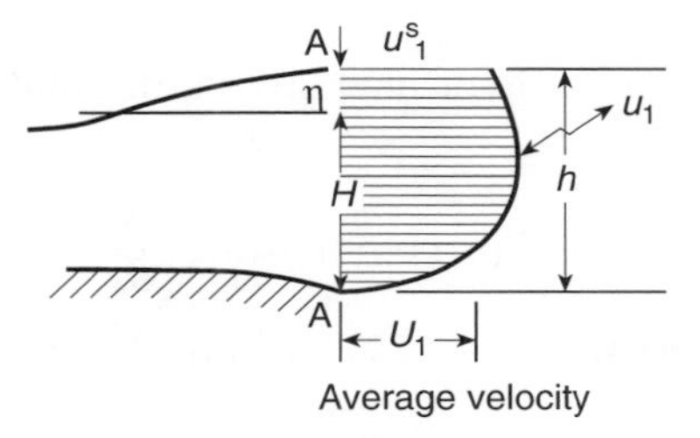

Fig. 7.1 The shallow-water problem. Notation.

reduced to

$$\frac{1}{\rho}\frac{\partial p}{\partial x_3} + g = 0 \tag{7.3}$$

where $g_3 = -g$ is the gravity acceleration. After integration this yields

$$p = \rho g(\eta - x_3) + p_a \tag{7.4}$$

as, when $x_3 = \eta$, the pressure is atmospheric (p_a) (which may on occasion not be constant over the body of the water and can thus influence its motion).

On the free surface the vertical velocity u_3 can of course be related to the total time derivative of the surface elevation, i.e. (see Sec. 5.3 of Chapter 5)

$$u_3^{\mathrm{s}} = \frac{D\eta}{\mathrm{d}t} \equiv \frac{\partial \eta}{\partial t} + u_1^{\mathrm{s}}\frac{\partial \eta}{\partial x_1} + u_2^{\mathrm{s}}\frac{\partial \eta}{\partial x_2} \tag{7.5a}$$

Similarly, at the bottom,

$$u_3^{\mathrm{b}} = \frac{DH}{\mathrm{d}t} \equiv u_1^{\mathrm{b}}\frac{\partial \eta}{\partial x_1} + u_2^{\mathrm{b}}\frac{\partial \eta}{\partial x_2} \tag{7.5b}$$

assuming that the total depth H does not vary with time. Further, if we assume that for viscous flow no slip occurs then

$$u_1^{\mathrm{b}} = u_2^{\mathrm{b}} = 0 \tag{7.6}$$

and also by continuity

$$u_3^{\mathrm{b}} = 0$$

Now a further approximation will be made. In this the governing equations will be integrated with the depth coordinate x_3 and *depth-averaged governing equations* derived. We shall start with the continuity equation (7.2a) and integrate this in the x_3 direction, writing

$$\int_{-H}^{\eta} \frac{\partial u_3}{\partial x_3}\,\mathrm{d}x_3 + \int_{-H}^{\eta} \frac{\partial u_1}{\partial x_1}\,\mathrm{d}x_3 + \int_{-H}^{\eta} \frac{\partial u_2}{\partial x_2}\,\mathrm{d}x_3 = 0 \tag{7.7}$$

As the velocities u_1 and u_2 are unknown and are not uniform, as shown in Fig. 7.1(b), it is convenient at this stage to introduce the notion of average velocities defined so that

$$\int_{-H}^{\eta} u_i\,\mathrm{d}x_3 = U_i(H + \eta) \equiv U_i h \tag{7.8}$$

with $i = 1, 2$. We shall now recall the Leibnitz rule of integrals stating that for any function $F(r, s)$ we can write

$$\int_{-a}^{b} \frac{\partial}{\partial s} F(r, s)\,\mathrm{d}r \equiv \frac{\partial}{\partial s}\int_{-a}^{b} F(r, s)\,\mathrm{d}r - F(b, s)\frac{\partial b}{\partial s} + F(a, s)\frac{\partial a}{\partial s} \tag{7.9}$$

With the above we can rewrite the last two terms of Eq. (7.7) and introducing Eq. (7.6) we obtain

$$\int_{-H}^{\eta} \frac{\partial u_i}{\partial x_i}\,\mathrm{d}x_3 = \frac{\partial}{\partial x_i}(U_i h) - u_i^{\mathrm{s}}\frac{\partial \eta}{\partial x_i} \tag{7.10}$$

with $i = 1, 2$. The first term of Eq. (7.7) is, by simple integration, given as

$$\int_{-H}^{\eta} \frac{\partial u_3}{\partial x_3} \, dx_3 = u_3^s \tag{7.11}$$

which, on using (7.5a), becomes

$$\int_{-H}^{\eta} \frac{\partial u_3}{\partial x_3} \, dx_3 = \frac{\partial \eta}{\partial t} + u_i^s \frac{\partial \eta}{\partial x_i} \tag{7.12}$$

Addition of Eqs (7.10) and (7.12) gives the depth-averaged continuity conservation finally as

$$\frac{\partial \eta}{\partial t} + \frac{\partial (hU_i)}{\partial x_i} \equiv \frac{\partial h}{\partial t} + \frac{\partial (hU_i)}{\partial x_i} = 0 \tag{7.13}$$

Now we shall perform similar depth integration on the momentum equations in the horizontal directions. We have thus

$$\int_{-H}^{\eta} \left[\frac{\partial u_i}{\partial t} + \frac{\partial}{\partial x_j}(u_i u_j) + \frac{1}{\rho}\frac{\partial p}{\partial x_i} - \frac{1}{\rho}\frac{\partial \tau_{ij}}{\partial x_i} - g_i \right] dx_3 = 0 \tag{7.14}$$

with $i = 1, 2$.

Proceeding as before we shall find after some algebraic manipulation that a conservative form of depth-averaged equations becomes

$$\frac{\partial (hU_i)}{\partial t} + \frac{\partial}{\partial x_j}\left[hU_i U_j + \delta_{ij}\frac{1}{2}g(h^2 - H^2) - \frac{1}{\rho}\int_{-H}^{\eta} \tau_{ij} \, dx_3 \right]$$
$$- \frac{1}{\rho}(\tau_{3i}^s - \tau_{3i}^b) - hg_i - g(h - H)\frac{\partial H}{\partial x_i} + \frac{h}{\rho}\frac{\partial p_a}{\partial x_i} = 0 \tag{7.15}$$

In the above the shear stresses on the surface can be prescribed externally, given, say, the wind drag. The bottom shear is frequently expressed by suitable hydraulic resistance formulae, e.g. the Chézy expression, giving

$$\tau_{3i}^b = \frac{\rho g |\mathbf{U}| U_i}{Ch^2} \tag{7.16}$$

where

$$|\mathbf{U}| = \sqrt{U_i U_i}; \qquad i = 1, 2$$

and C is the Chézy coefficient.

In Eq. (7.15) g_i stands for the Coriolis accelerations, important in large-scale problems and defined as

$$g_1 = \hat{g} U_2 \qquad g_2 = -\hat{g} U_1 \tag{7.17}$$

where $\hat{g}$ is the Coriolis parameter.

The τ_{ij} stresses require the definition of a viscosity coefficient, μ_H, generally of the averaged turbulent kind, and we have

$$\tau_{ij} = \mu_H \left(\frac{\partial u_i}{\partial x_j} + \frac{\partial u_j}{\partial x_i} - \frac{2}{3}\delta_{ij}\frac{\partial u_k}{\partial x_k} \right) \tag{7.18}$$

Approximating in terms of average velocities, the remaining integral of Eq. (7.15) can be written as

$$\frac{1}{\rho}\int_{-H}^{\eta} \tau_{ij}\,\mathrm{d}x_3 \approx \frac{1}{\rho}\mu_H h\left(\frac{\partial U_i}{\partial x_j}+\frac{\partial U_j}{\partial x_i}-\frac{2}{3}\delta_{ij}\frac{\partial U_k}{\partial x_k}\right)=\frac{h}{\rho}\bar{\tau}_{ij} \tag{7.19}$$

Equations (7.13) and (7.15) cast the shallow-water problem in the general form of Eq. (7.1), where the appropriate vectors are defined below.

Thus, with $i = 1, 2$,

$$\mathbf{U}=\begin{Bmatrix} h \\ hU_1 \\ hU_2 \end{Bmatrix} \tag{7.20a}$$

$$\mathbf{F}_i=\begin{Bmatrix} hU_i \\ hU_1U_i+\delta_{1i}\frac{1}{2}g(h^2-H^2) \\ hU_2U_i+\delta_{2i}\frac{1}{2}g(h^2-H^2) \end{Bmatrix} \tag{7.20b}$$

$$\mathbf{G}_i=\begin{Bmatrix} 0 \\ -(h/\rho)\bar{\tau}_{1i} \\ -(h/\rho)\bar{\tau}_{2i} \end{Bmatrix} \tag{7.20c}$$

in which the relation (7.19) is used to give the internal average $\bar{\tau}$ in terms of the average velocity gradients and

$$\mathbf{Q}=\begin{Bmatrix} 0 \\ -h\hat{g}U_2-g(h-H)\dfrac{\partial H}{\partial x_1}+\dfrac{h}{\rho}\dfrac{\partial p_a}{\partial x_1}-\dfrac{1}{\rho}\tau_{31}^{s}+\dfrac{gU_1|\mathbf{U}|}{Ch^2} \\ h\hat{g}U_1-g(h-H)\dfrac{\partial H}{\partial x_2}+\dfrac{h}{\rho}\dfrac{\partial p_a}{\partial x_2}-\dfrac{1}{\rho}\tau_{32}^{s}+\dfrac{gU_2|\mathbf{U}|}{Ch^2} \end{Bmatrix} \tag{7.20d}$$

The above, conservative, form of shallow-water equations was first presented in references 4 and 5 and is generally applicable. However, many variants of the general shallow-water equations exist in the literature, introducing various approximations.

In the following sections of this chapter we shall discuss time-stepping solutions of the full set of the above equations in transient situations and in corresponding steady-state applications. Here non-linear behaviour will of course be included but for simplicity some terms will be dropped. In particular, we shall in most of the examples omit consideration of viscous stresses $\bar{\tau}_{ij}$, whose influence is small compared with the bottom drag stresses. This will, incidentally, help in the solution, as second-order derivatives now disappear and boundary layers can be eliminated.

If we deal with the linearized form of Eqs (7.13) and (7.15), we see immediately that on omission of all non-linear terms, bottom drag, etc., and approximately $h \sim H$, we can write these equations as

$$\frac{\partial h}{\partial t}+\frac{\partial}{\partial x_i}(HU_i)=0 \tag{7.21a}$$

$$\frac{\partial (HU_i)}{\partial t}+gH\frac{\partial}{\partial x_i}(h-H)=0 \tag{7.21b}$$

Noting that

$$\eta = h - H \qquad \text{and} \qquad \frac{\partial h}{\partial t} = \frac{\partial \eta}{\partial t}$$

the above becomes

$$\frac{\partial \eta}{\partial t} + \frac{\partial}{\partial x_i}(HU_i) = 0 \tag{7.22a}$$

$$\frac{\partial (HU_i)}{\partial t} + gH\frac{\partial \eta}{\partial x_i} = 0 \tag{7.22b}$$

Elimination of HU_i immediately yields

$$\frac{\partial^2 \eta}{\partial t^2} - \frac{\partial}{\partial x_i}\left(gH\frac{\partial \eta}{\partial x_i}\right) = 0 \tag{7.23}$$

or the standard Helmholtz wave equation. For this, many special solutions are analysed in the next chapter.

The shallow-water equations derived in this section consider only the depth-averaged flows and hence cannot reproduce certain phenomena that occur in nature and in which some velocity variation with depth has to be allowed for. In many such problems the basic assumption of a vertically hydrostatic pressure distribution is still valid and a form of shallow-water behaviour can be assumed.

The extension of the formulation can be achieved by an *a priori* division of the flow into strata in each of which different velocities occur. The final set of discretized equations consists then of several, coupled, two-dimensional approximations. Alternatively, the same effect can be introduced by using several different velocity 'trial functions' for the vertical distribution, as was suggested by Zienkiewicz and Heinrich.[6] Such generalizations are useful but outside the scope of the present text.

7.3 Numerical approximation

Both finite difference and finite element procedures have for many years been used widely in solving the shallow-water equations. The latter approximation has been applied relatively recently and Kawahara[7] and Navon[8] survey the early applications to coastal and oceanographic engineering. In most of these the standard procedures of spatial discretization followed by suitable time-stepping schemes are adopted.[9–16] In meteorology the first application of the finite element method dates back to 1972, as reported in the survey given in reference 17, and the range of applications has been increasing steadily.[4,5,18–41]

At this stage the reader may well observe that with the exception of source terms, the isothermal compressible flow equations can be transformed into the depth-integrated shallow-water equations with the variables being changed as follows:

$$\rho \text{ (density)} \rightarrow h \text{ (depth)}$$

$$u_i \text{ (velocity)} \rightarrow U_i \text{ (mean velocity)}$$

$$p \text{ (pressure)} \rightarrow \tfrac{1}{2}g(h^2 - H^2)$$

These similarities suggest that the characteristic-based-split algorithm adopted in the previous chapters for compressible flows be used for the shallow-water equations.[42,43]

The extension of effective finite element solutions of high-speed flows to shallow-water problems has already been successful in the case of the Taylor–Galerkin method.[4,5] However, the semi-implicit form of the general CBS formulation provides a critical time step dependent only on the current velocity of the flow **U** (for pure convection):

$$\Delta t \leqslant \frac{h}{|\mathbf{U}|} \tag{7.24}$$

where h is the element size, instead of a critical time step in terms of the wave celerity $c = \sqrt{gh}$:

$$\Delta t \leqslant \frac{h}{c + |\mathbf{U}|} \tag{7.25}$$

which places a severe contraint on fully explicit methods such as the Taylor–Galerkin approximation and others,[4,5,32] particularly for the analysis of long-wave propagation in shallow waters and in general for low Froude number problems.

Important savings in computation can be reached in these situations obtaining for some practical cases up to 20 times the critical (explicit) time step, without seriously affecting the accuracy of the results. When nearly critical to supercritical flows must be studied, the fully explicit form is recovered, and the results observed for these cases are also excellent.[43,44]

In the examples that follow we shall illustrate several problems solved by the CBS procedure, and also with the Taylor–Galerkin method.

7.4 Examples of application

7.4.1 Transient one-dimensional problems – a performance assessment

In this section we present some relatively simple examples in one space dimension to illustrate the applicability of the algorithms.

The first, illustrated in Fig. 7.2, shows the progress of a solitary wave[45] onto a shelving beach. This frequently studied situation[46,47] shows well the progressive steepening of the wave often obscured by schemes that are very dissipative.

The second example, of Fig. 7.3, illustrates the so-called 'dam break' problem diagrammatically. Here a dam separating two stationary water levels is suddenly removed and the almost vertical waves progress into the two domains. This problem, somewhat similar to those of a shock tube in compressible flow, has been solved quite successfully even without artificial diffusivity.

The final example of this section, Fig. 7.4, shows the formation of an idealized 'bore' or a steep wave progressing into a channel carrying water at a uniform speed caused by a gradual increase of the downstream water level. Despite the fact that

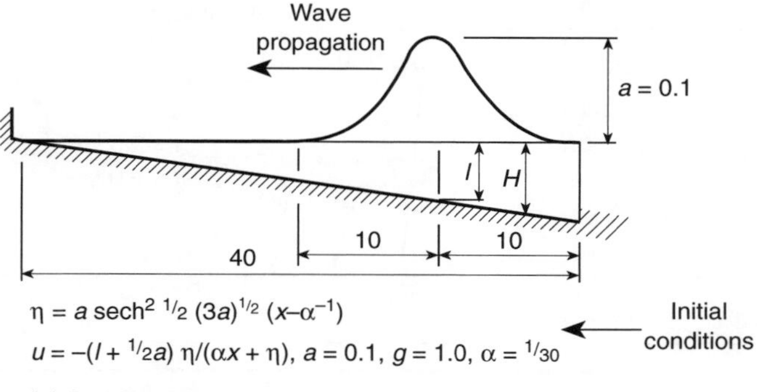

(a) Problem statement

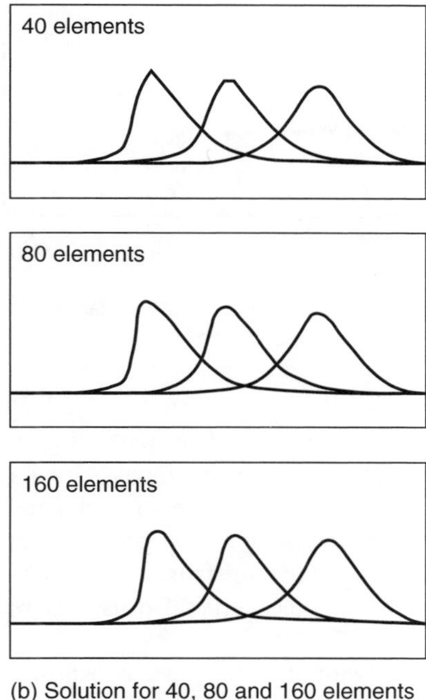

(b) Solution for 40, 80 and 160 elements at various times

Fig. 7.2 Shoaling of a wave.

the flow speed is 'subcritical' (i.e. velocity $< \sqrt{gh}$), a progressively steepening, travelling shock clearly develops.

7.4.2 Two-dimensional periodic tidal motions

The extension of the computation into two space dimensions follows the same pattern as that described in compressible formulations. Again linear triangles are

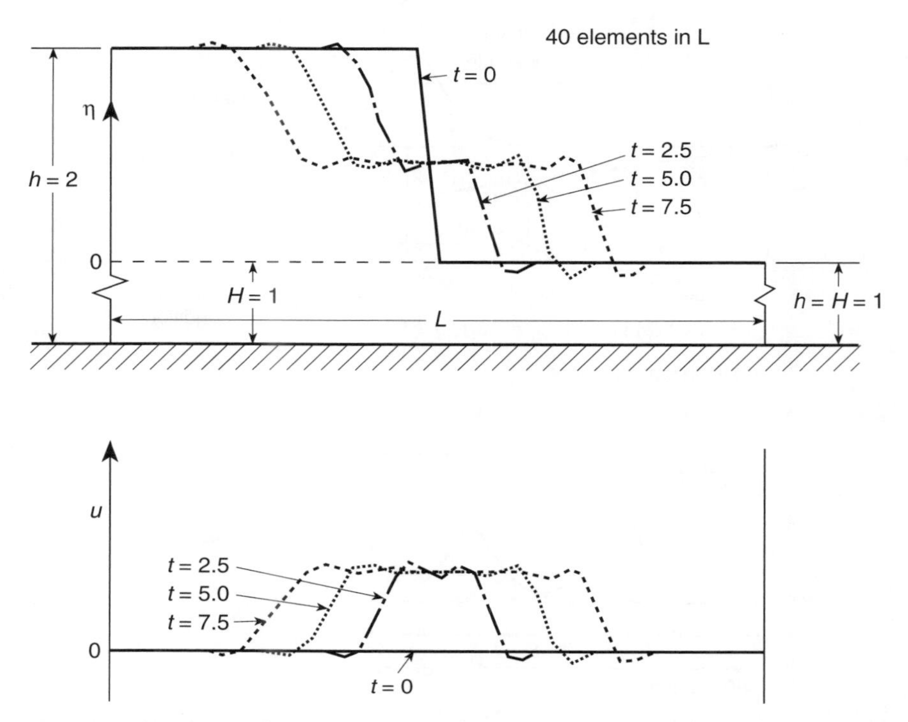

Fig. 7.3 Propagation of waves due to dam break ($C_{Lap} = 0$). 40 elements in analysis domain. $C = \sqrt{gH} = 1$, $\Delta t = 0.25$.

used to interpolate the values of h, hU_1 and hU_2. The main difference in the solutions is that of emphasis. In the shallow-water problem, shocks either do not develop or are sufficiently dissipated by the existence of bed friction so that the need for artificial viscosity and local refinement is not generally present. For this reason we have not introduced here the error measures and adaptivity – finding that meshes sufficiently fine to describe the geometry also usually prove sufficiently accurate.

The first example of Fig. 7.5 is presented merely as a test problem. Here the frictional resistance is linearized and an exact solution known for a periodic response[48] is used for comparison. This periodic response is obtained numerically by performing some five cycles with the input boundary conditions. Although the problem is essentially one dimensional, a two-dimensional uniform mesh was used and the agreement with analytical results is found to be quite remarkable.

In the second example we enter the domain of more realistic applications.[4,5,42–44,49] Here the 'test bed' is provided by the Bristol Channel and the Severn Estuary, known for some of the highest tidal motions in the world. Figure 7.6 shows the location and the scale of the problem.

The objective is here to determine tidal elevations and currents currently existing (as a possible preliminary to a subsequent study of the influence of a barrage which some day may be built to harness the tidal energy). Before commencement of the

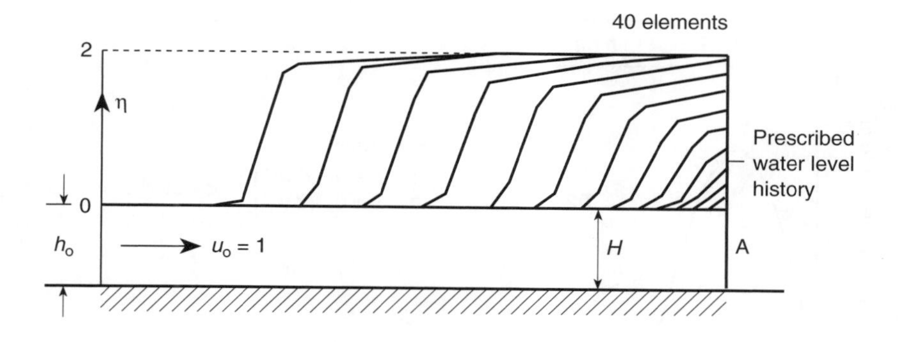

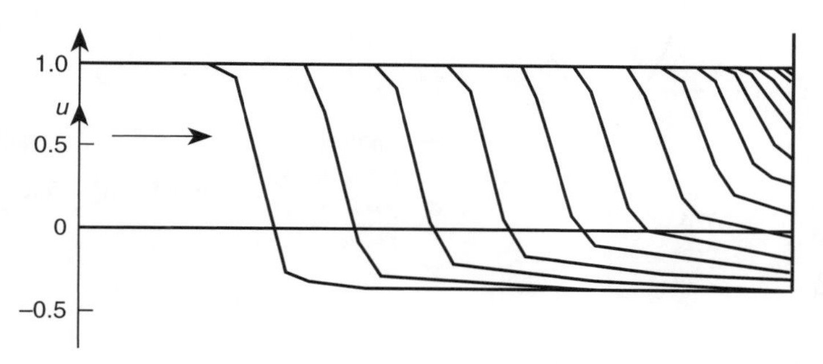

Fig. 7.4 A 'bore' created in a stream due to water level rise downstream (A). Level at A, $\eta = 1 - \cos \pi t/30$ $(0 \leqslant t \leqslant 30)$, 2 $(30 \leqslant t)$. Levels and velocities at intervals of 5 time units, $\Delta t = 0.5$.

analysis the extent of the analysis domain must be determined by an arbitrary, seaward, boundary. On this the measured tidal heights will be imposed.

This height-prescribed boundary condition is not globally conservative and also can produce undesired reflections. These effects sometimes lead to considerable errors in the calculations, particularly if long-term computations are to be carried out (like, for instance, in some pollutant dispersion analysis). For these cases, more general open boundary conditions can be applied, as, for example, those described in references 35 and 36.

The analysis was carried out on four meshes of linear triangles shown in Fig. 7.7. These meshes encompass two positions of the external boundary and it was found that the differences in the results obtained by four separate analyses were insignificant.

The mesh sizes ranged from 2 to 5 km in minimum size for the fine and coarse subdivisions. The average depth is approximately 50 m but of course full bathygraphy information was used with depths assigned to each nodal point.

The numerical study of the Bristol Channel was completed by a comparison of performance between the explicit and semi-explicit algorithms.[43] The results for the coarse mesh were compared with measurements obtained by the I[illegible] of Oceanographic Science (IOS) for the M_2 tide,[49] with time steps corres[illegible] o

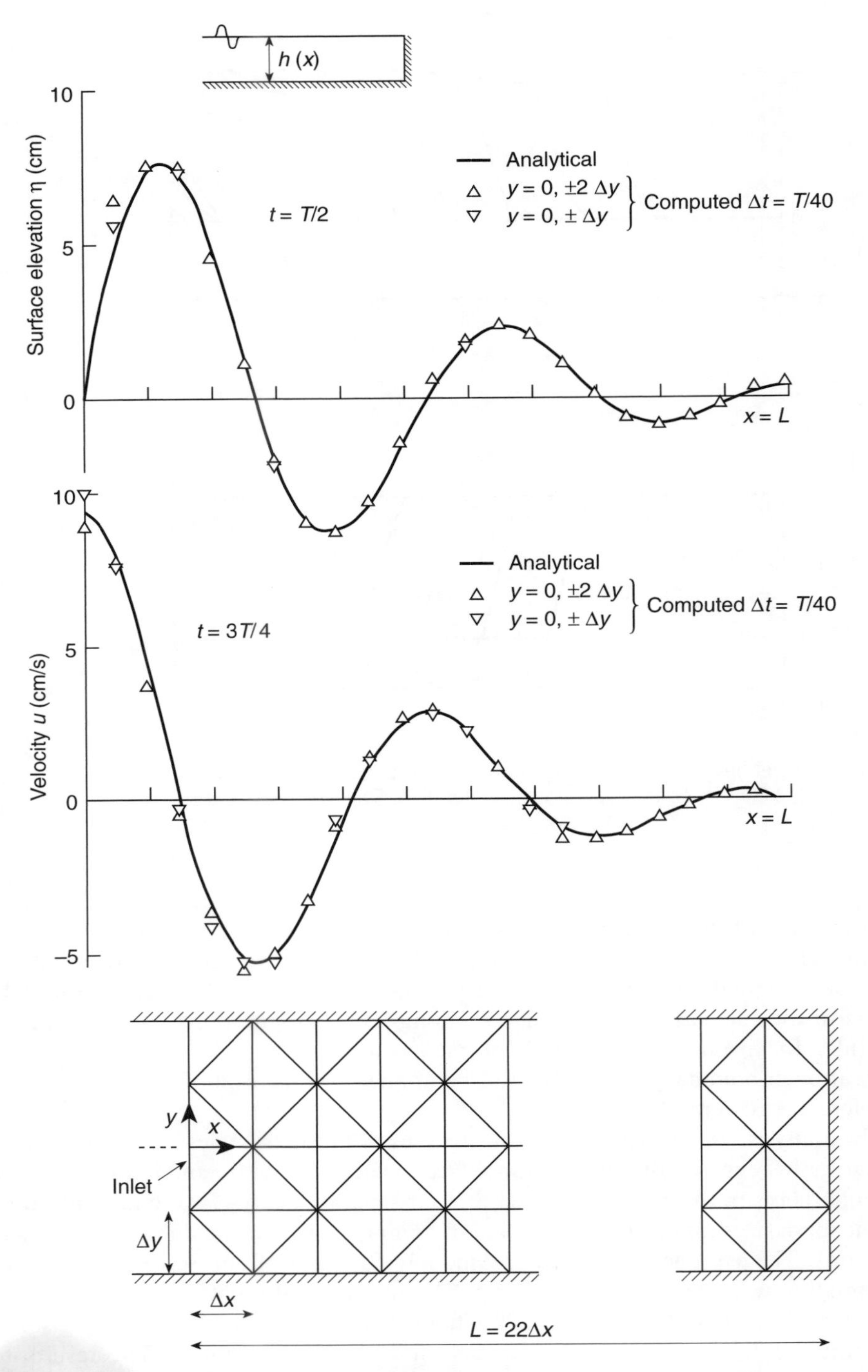

Fig. 7.5 Steady-state oscillation in a rectangular channel due to periodic forcing of surface elevation at an inlet. Linear frictional dissipation.[32]

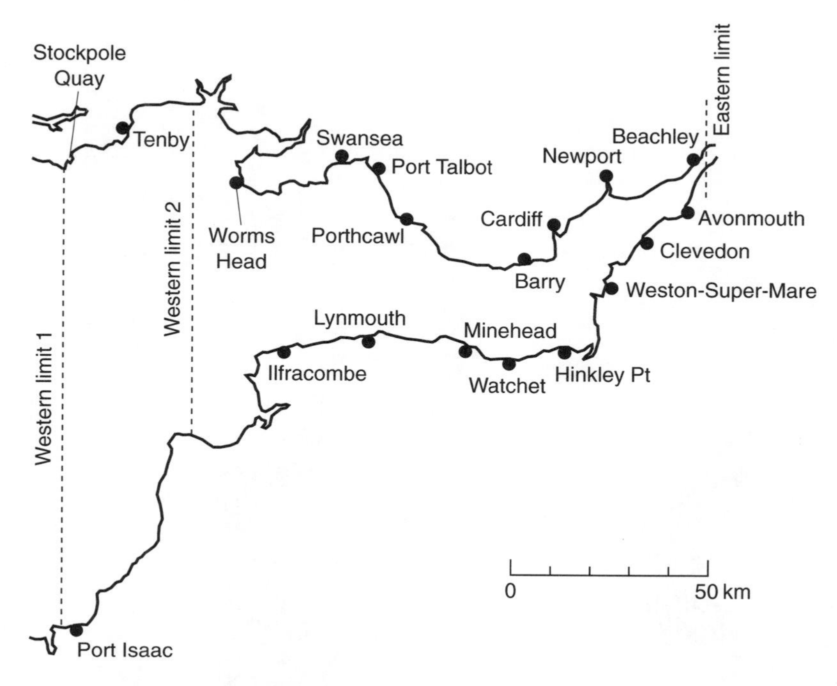

Fig. 7.6 Location map. Bristol Channel and Severn Estuary.

the critical (explicit) time step (50 s), 4 times (200 s) and 8 times (400 s) the critical time step. A constant real friction coefficient (Manning) of 0.038 was adopted for all of the estuary. Coriolis forces were included. The analysis proved that the Coriolis effect was very important in terms of phase errors. Table 7.1 represents a comparison between observations and computations in terms of amplitudes and phases for seven different points which are represented in the location map (Fig. 7.6), for the three different time steps described above. The maximum error in amplitude only increases by 1.4% when the time step of 400 s is used with respect to the time step of 50 s, while the absolute error in phases (−13°) is two degrees more than the case of 50 s (−11°). These bounds show a remarkable accuracy for the semi-explicit model. In Fig. 7.8 the distribution of velocities at different times of the tide is illustrated (explicit model).

In the analysis presented we have omitted details of the River Severn upstream of the eastern limit (see Figs 7.6 and 7.9(a)), where a 'bore' moving up the river can be observed. An approach to this phenomenon is made by a simplified straight extension of the mesh used previously, preserving an approximate variation of the bottom and width until the point G (Gloucester) (77.5 km from Avonmouth), but obviously neglecting the dissipation and inertia effects of the bends. Measurement points are located at B and E, and the results (elevations) are presented in Fig. 7.9(d) for the points A, B, E in time, along with a steady river flow. A typical shape for a tidal bore can be observed for the point E, with fast flooding and a

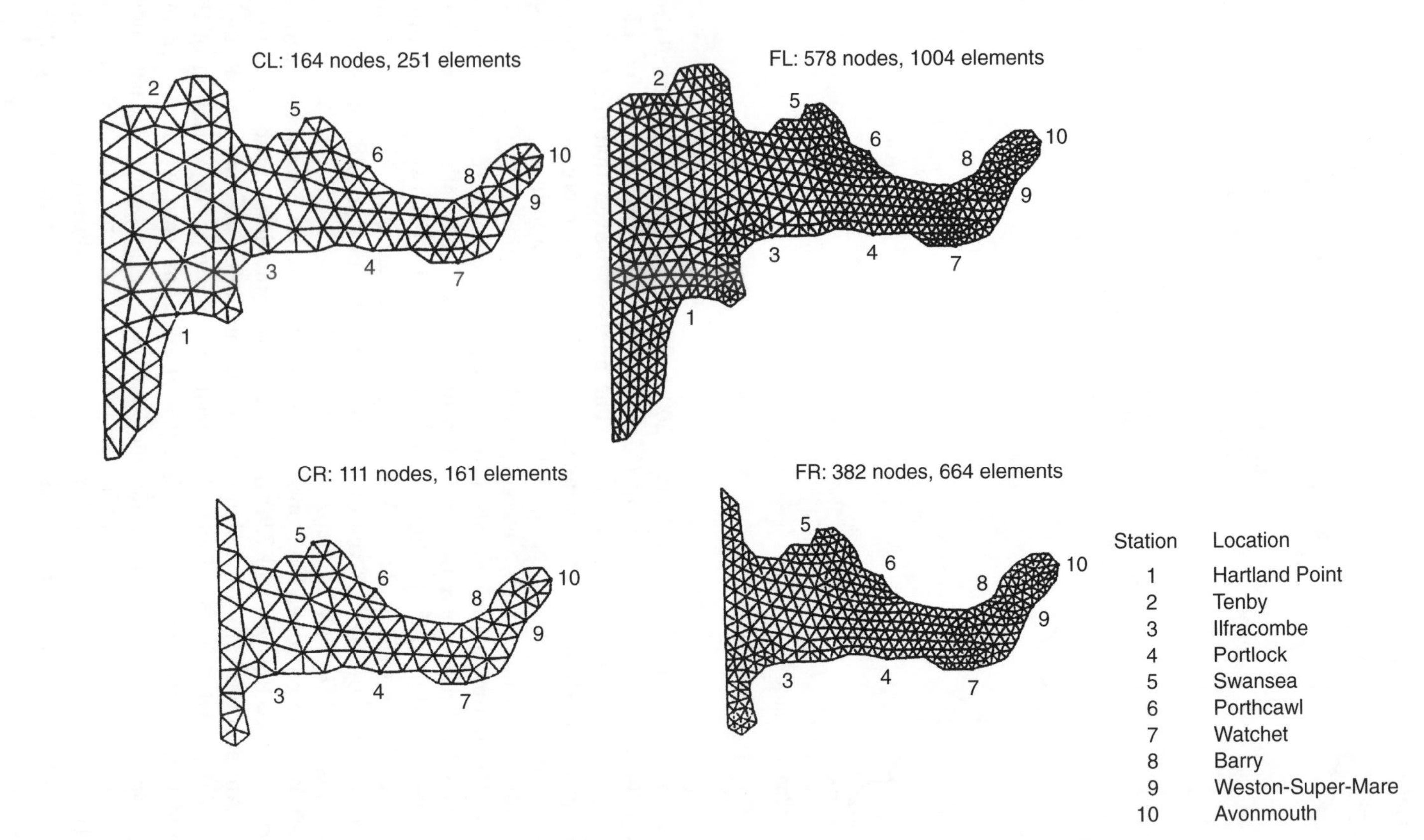

Fig. 7.7 Finite element meshes. Bristol Channel and Severn Estuary.

Table 7.1 Bristol Channel and Severn Estuary – observed results and FEM computation (FL mesh) of tidal half-amplitude (m × 10^2)

Location	Observed	FEM
Tenby	262	260 (−1%)
Swansea	315	305 (−3%)
Cardiff	409	411 (0%)
Porthcawl	317	327 (+3%)
Barry	382	394 (+3%)
Port Talbot	316	316 (−1%)
Newport	413	420 (+2%)
Ilfracombe	308	288 (−6%)
Minehead	358	362 (+1%)

smooth ebbing of water. (The flooding from the minimum to maximum level is in less than 25 minutes.)

7.4.3 Tsunami waves

A problem of some considerable interest in earthquake zones is that of so-called tidal waves or tsunamis. These are caused by sudden movements in the earth's

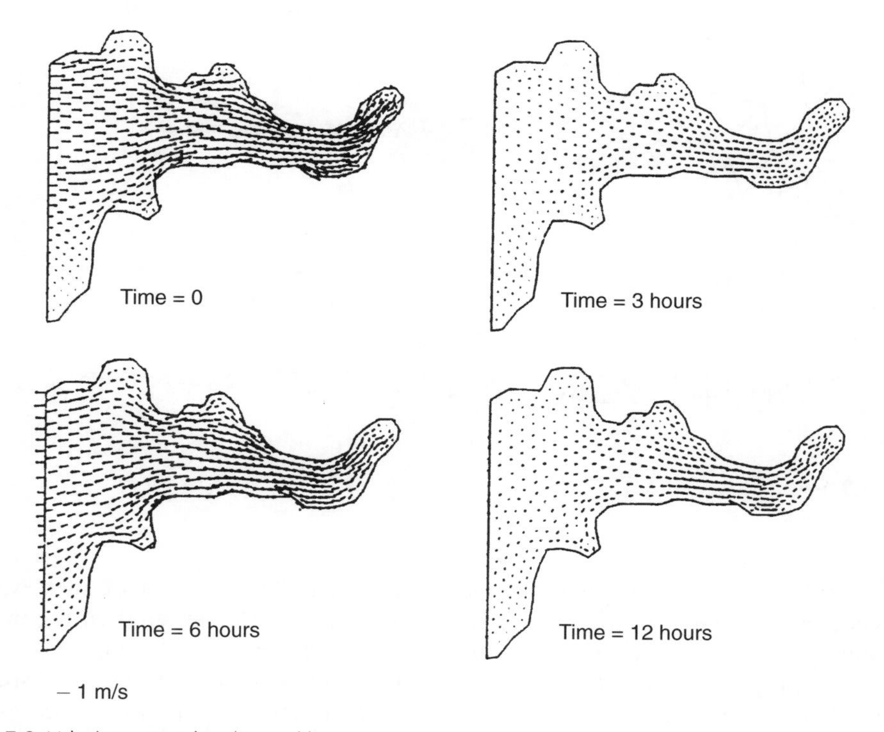

Fig. 7.8 Velocity vector plots (FL mesh).

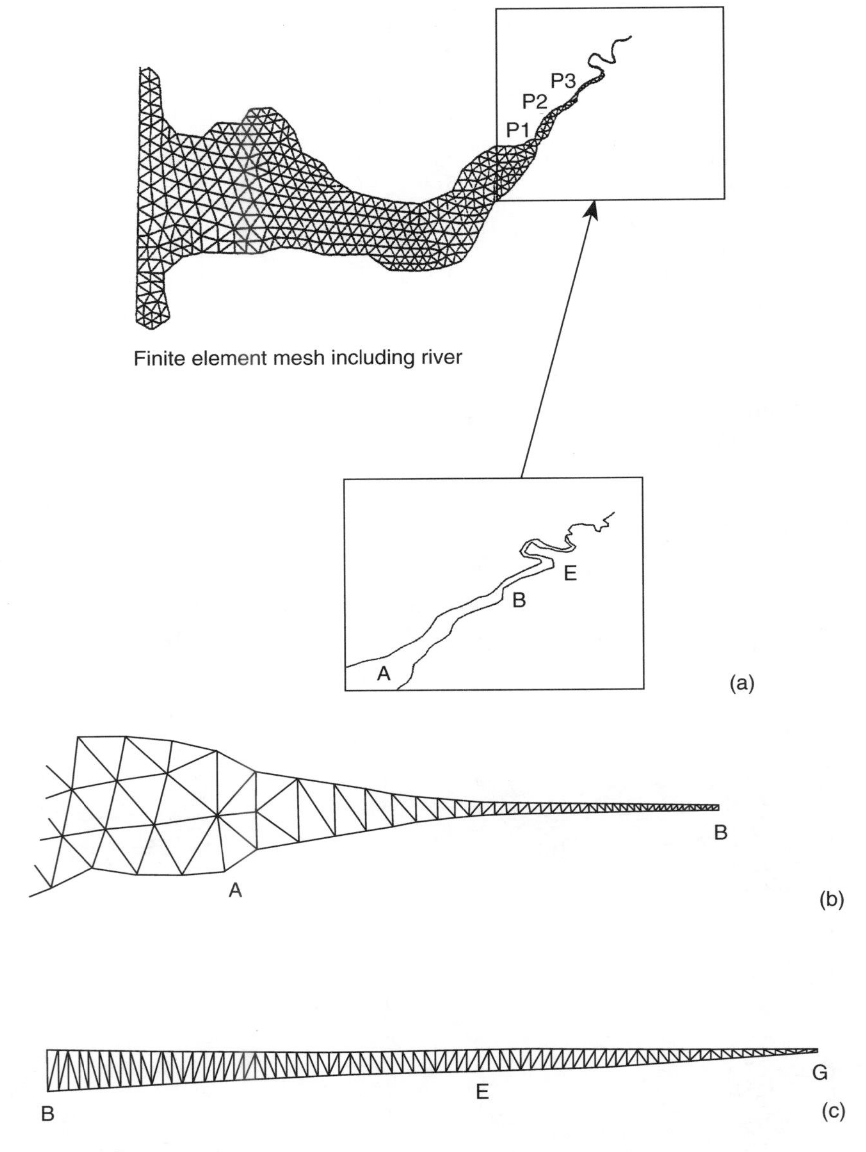

Fig. 7.9 Severn bore.

crust and can on occasion be extremely destructive. The analysis of such waves presents no difficulties in the general procedure demonstrated and indeed is computationally cheaper as only relatively short periods of time need be considered. To illustrate a typical possible tsunami wave we have created one in the Severn Estuary just analysed (to save problems of mesh generation, etc., for another more likely configuration).

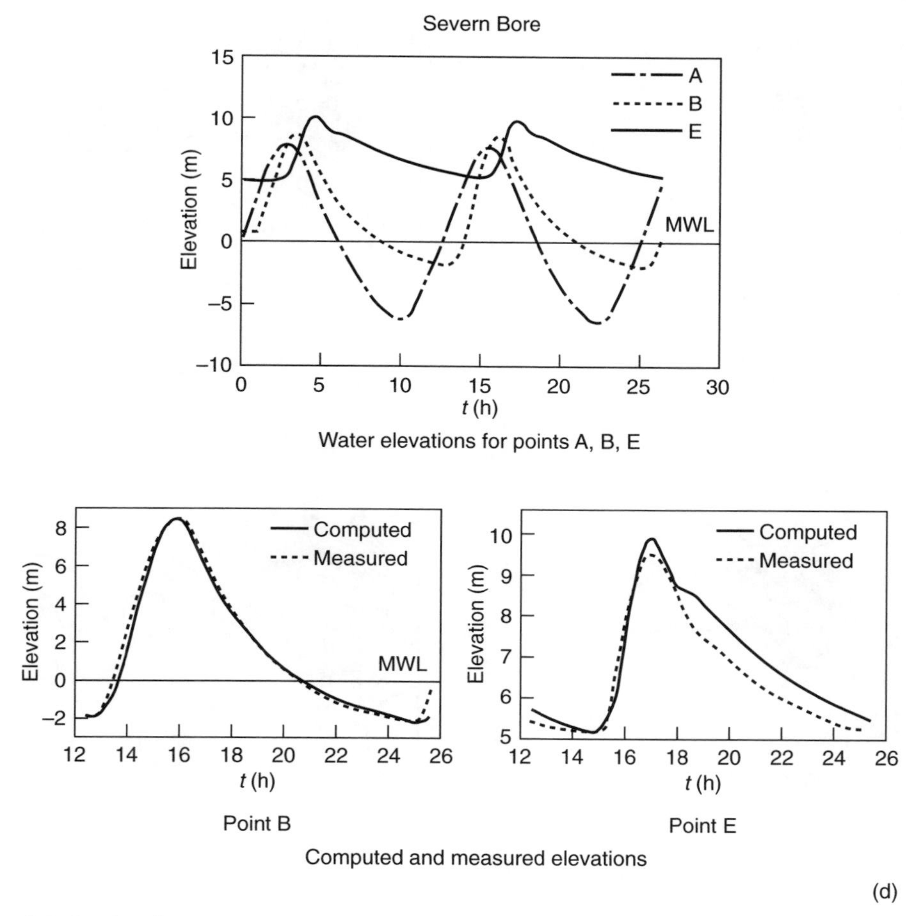

Fig. 7.9 Continued.

Here the tsunami is forced by an instantaneous raising of an element situated near the centre of the estuary by some 6 m and the previously designed mesh was used (FL). The progress of the wave is illustrated in Fig. 7.10. The tsunami wave was superimposed on the tide at its highest level – though of course the tidal motion was allowed for.

One particular point only needs to be mentioned in this calculation. This is the boundary condition on the seaward, arbitrary, limit. Here the Riemann decomposition of the type discussed earlier has to be made if tidal motion is to be incorporated and note taken of the fact that the tsunami forms only an outgoing wave. This, in the absence of tides, results simply in application of the free boundary condition there.

The clean way in which the tsunami is seen to leave the domain in Fig. 7.10 testifies to the effectiveness of this process.

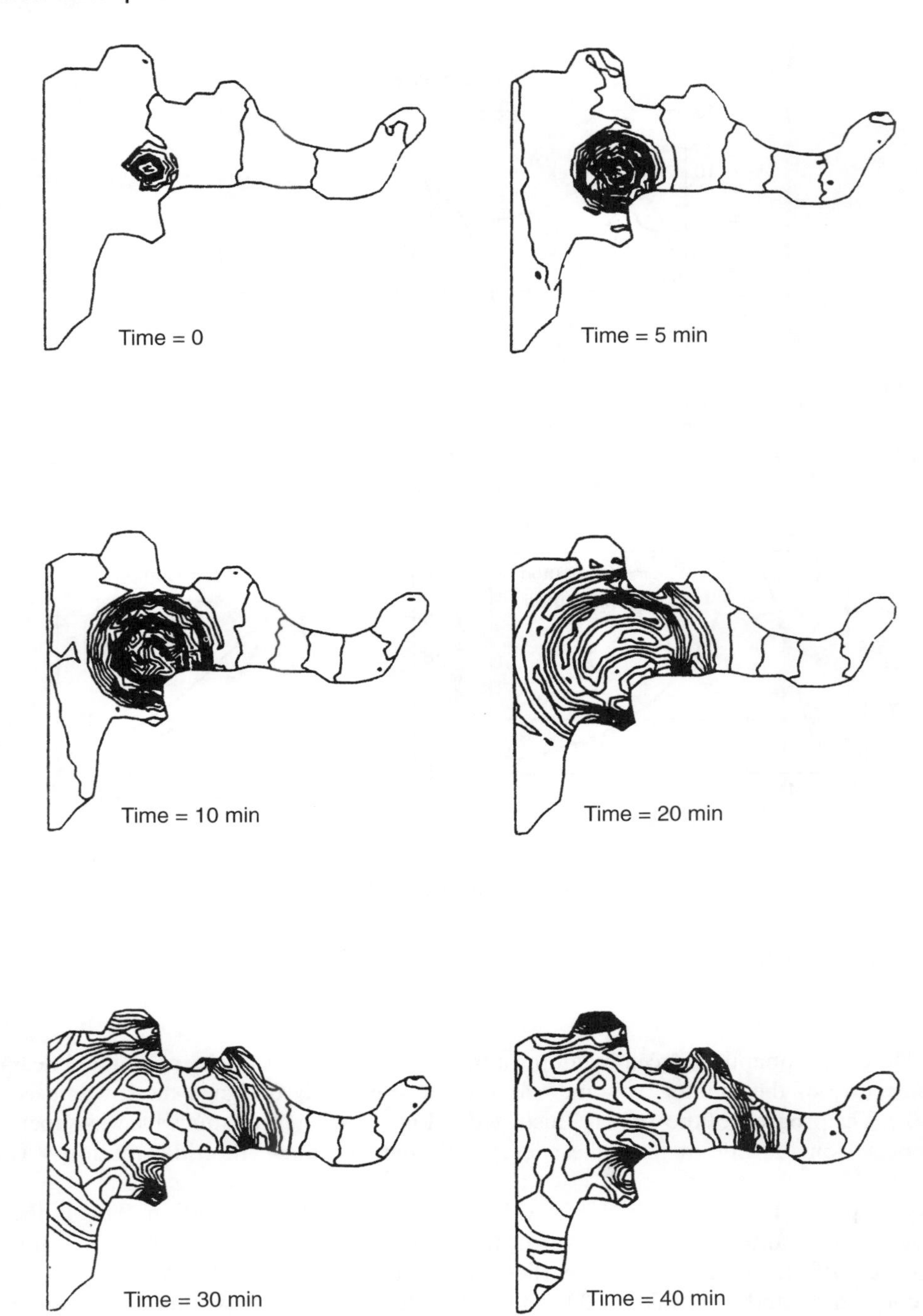

Fig. 7.10 Severn tsunami. Generation during high tide. Water height contours (times after generation).

7.4.4 Steady-state solutions

On occasion steady-state currents such as may be caused by persistent wind motion or other influences have to be considered. Here once again the transient form of explicit computation proves very effective and convergence is generally more rapid than in compressible flow as the bed friction plays a greater role.

The interested reader will find many such steady-state solutions in the literature. In Fig. 7.11 we show a typical example. Here the currents are induced by the breaking of waves which occurs when these reach small depths creating so-called radiation stresses.[6,30,50] Obviously as a preliminary the wave patterns have to be computed using procedures to be given later. The 'forces' due to breaking are the cause of longshore currents and rip currents in general. The figure illustrates this effect on a harbour.

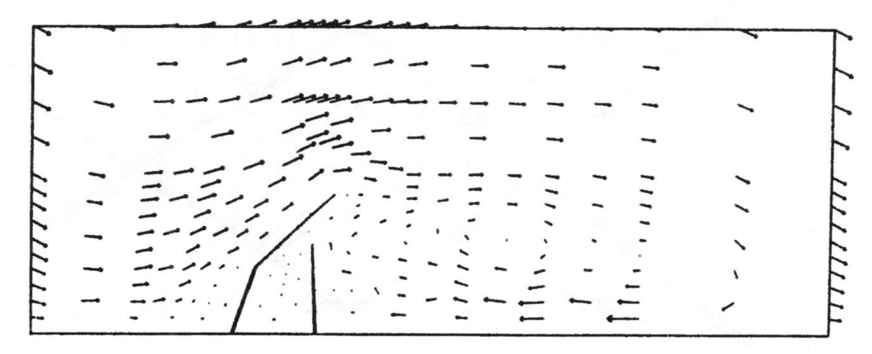

Fig. 7.11 Wave-induced steady-state flow past a harbour.[30]

It is of interest to remark that in the problem discussed, the side boundaries have been 'repeated' to model an infinite harbour series.[50]

Another type of interesting steady-state (and also transient) problem concerns supercritical flows over hydraulic structures, with shock formation similar to those present in high-speed compressible flows. To illustrate this range of flows, the problem of a symmetric channel of variable width with a supercritical inflow is shown here. For a supercritical flow in a rectangular channel with a symmetric transition on both sides, a combination of a 'positive' jump and 'negative' waves, causing a decrease in depth, appears. The profile of the negative wave is gradual and an approximate solution can be obtained by assuming no energy losses and that the flow near the wall turns without separation. The constriction and enlargement analysed here was 15°, and the final mesh used was of only 6979 nodes, after two remeshings. The supercritical flow had an inflow Froude number of 2.5 and the boundary conditions were as follows: heights and velocities prescribed in inflow (left boundary of Fig. 7.12), slip boundary on walls (upper and lower boundaries in Fig. 7.12) and free variables on the outflow boundary (right side of Fig. 7.12). The explicit version with local time step was adopted. Figure 7.12 represents contours of heights, where 'cross'-waves and 'negative' waves are contained. One can observe the 'gradual' change in the behaviour of the negative wave created at the origin of the wall enlargement.

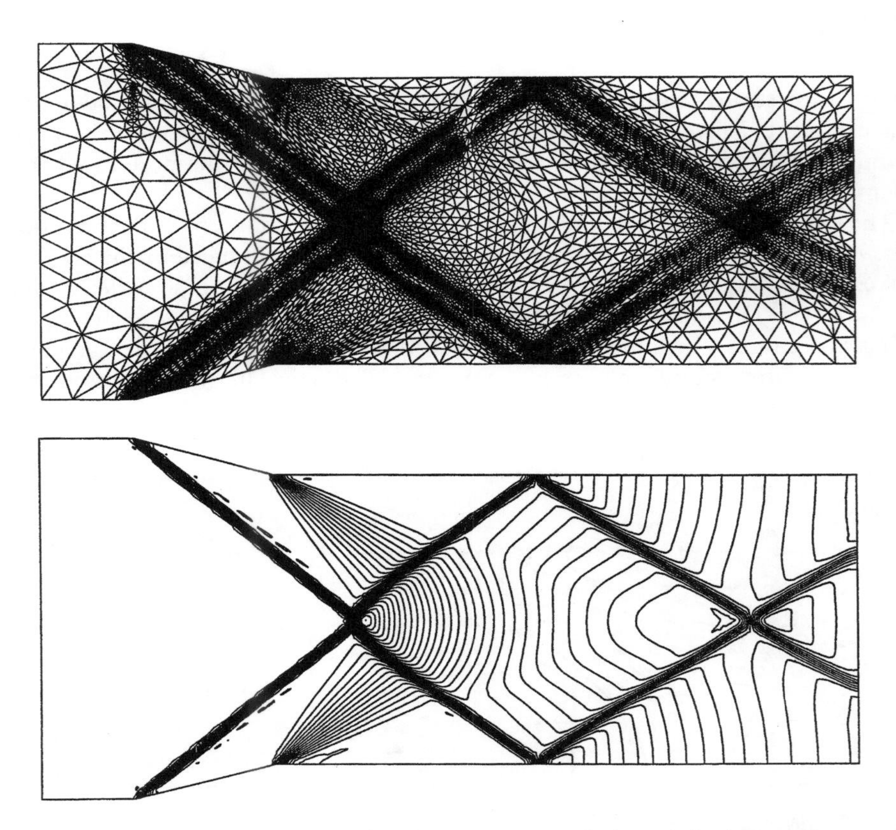

Fig. 7.12 Supercritical flow and formation of shock waves in symmetric channel of variable width contours of h. Inflow Froude number = 2.5. Constriction: 15°.

7.5 Drying areas

A special problem encountered in transient, tidal, computations is that of boundary change due to changes of water elevation. This has been ignored in the calculation presented for the Bristol Channel–Severn Estuary as the movements of the boundary are reasonably small in the scale analysed. However, in that example these may be of the order of 1 km and in tidal motions near Mont St. Michel, France, can reach 12 km. Clearly on some occasions such movements need to be considered in the analysis and many different procedures for dealing with the problem has been suggested. In Fig. 7.13 we show the simplest of these which is effective if the total movement can be confined to one element size. Here the boundary nodes are repositioned along the normal direction as required by elevation changes $\Delta\eta$.

If the variations are larger than those that can be absorbed in a single element some alternatives can be adopted, such as partial remeshing over layers surrounding the distorted elements or a general smooth displacement of the mesh.

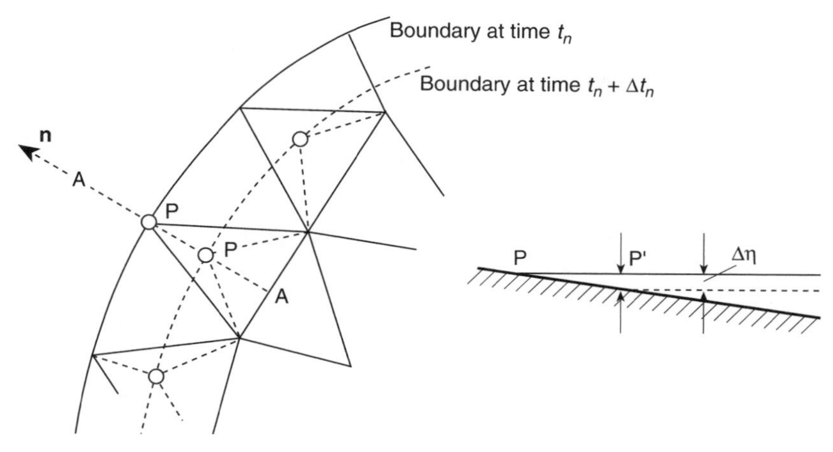

Fig. 7.13 Adjustment of boundary due to tidal variation.

7.6 Shallow-water transport

Shallow-water currents are frequently the carrier for some quantities which may disperse or decay in the process. Typical here is the transport of hot water when discharged from power stations, or of the sediment load or pollutants. The mechanism of sediment transport is quite complex[51] but in principle follows similar rules to that of the other equations. In all cases it is possible to write *depth-averaged transport equations* in which the average velocities U_i have been determined independently.

A typical averaged equation can be written – using for instance temperature (T) as the transported quantity – as

$$\frac{\partial(hT)}{\partial t} + \frac{\partial(hU_iT)}{\partial x_i} - \frac{\partial}{\partial x_i}\left(hk\frac{\partial T}{\partial x_i}\right) + R = 0 \qquad \text{for } i = 1, 2 \tag{7.26}$$

where h and U_i are the previously defined and computed quantities, k is an appropriate diffusion coefficient and R is a source term.

A quasi-implicit form of the general CBS algorithm can be obtained when diffusion terms are included. In this situation practical horizontal viscosity ranges (and diffusivity in the case of transport equations) can produce limiting time steps much lower than the convection limit. To circumvent this restraint, a quasi-implicit computation, requiring an implicit computation of the viscous terms, is recommended.

The application of the CBS method for any scalar transport equation is straightforward, because of the absence of the pressure gradient term. Then, the second and third step of the method are not necessary. The computation of the scalar hT is analogous to the intermediate momentum computation, but now a new time integration parameter θ_3 is introduced for the viscous term such that $0 \leqslant \theta_3 \leqslant 1$.

The application of the characteristic–Galerkin procedure gives the following final matrix form (neglecting terms higher than second order):

$$(\mathbf{M} + \theta_3\Delta t\mathbf{D})\Delta\tilde{\mathbf{T}} = -\Delta t[\mathbf{C}\tilde{\mathbf{T}}^n + \mathbf{MR}^n] - \frac{\Delta t^2}{2}[\mathbf{K}_u\tilde{\mathbf{T}}^n + \mathbf{f}_R] - \Delta t\mathbf{D}\tilde{\mathbf{T}}^n + bt \tag{7.27}$$

where now $\mathbf{T}$ is the vector of nodal hT values:

$$\mathbf{M} = \int_\Omega \mathbf{N}^\mathrm{T}\mathbf{N}\,\mathrm{d}\Omega$$

$$\mathbf{C} = \int_\Omega \mathbf{N}^\mathrm{T} U_j \frac{\partial \mathbf{N}}{\partial x_j}\,\mathrm{d}\Omega$$

$$\mathbf{K}_u = \int_\Omega \frac{\partial}{\partial x_k}(\mathbf{N}^\mathrm{T} U_k)\frac{\partial}{\partial x_j}(\mathbf{N}U_j)\,\mathrm{d}\Omega$$

$$\mathbf{D} = \int_\Omega \frac{\partial \mathbf{N}^\mathrm{T}}{\partial x_i} k \frac{\partial \mathbf{N}}{\partial x_i}\,\mathrm{d}\Omega$$

$$\mathbf{f}_R = \int_\Omega \frac{\partial}{\partial x_k}(\mathbf{N}^\mathrm{T} U_k)\mathbf{N}\,\mathrm{d}\Omega \cdot R_i^n$$

and

$$bt = \Delta t \int_\Gamma \mathbf{N}k\frac{\partial T}{\partial x_i} \cdot n_i\,\mathrm{d}\Gamma$$

As an illustration of a real implementation, the parameters involved in the study of transport of salinity in an industrial application for a river area are considered here.

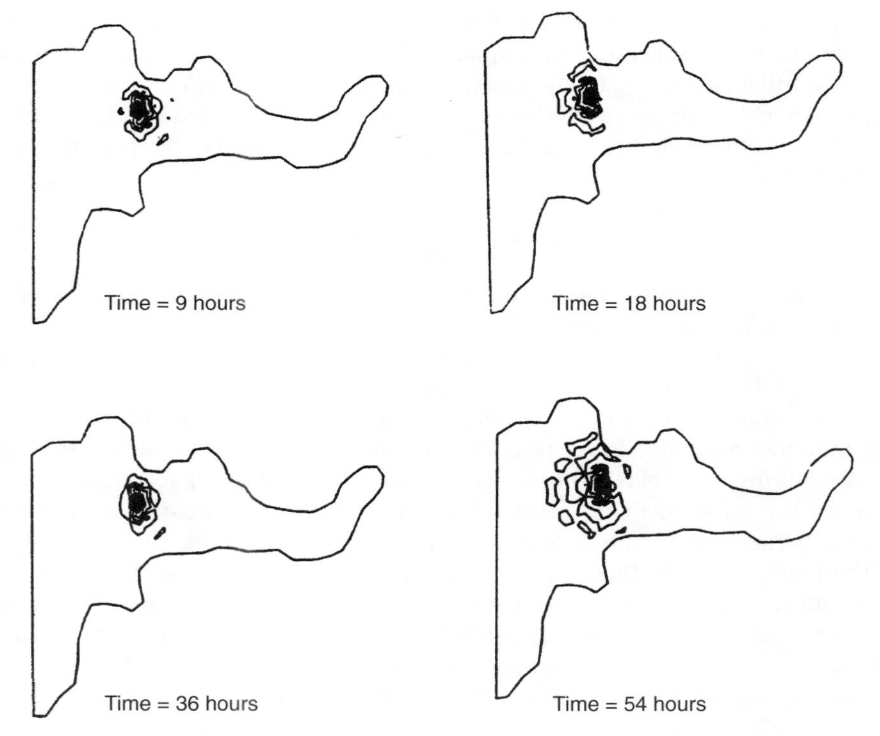

Fig. 7.14 Heat convection and diffusion in tidal currents. Temperature contours at several times after discharge of hot fluid.

The region studied was approximately 55 kilometres long and the mean value of the eddy diffusivity was of $k = 40\,\mathrm{m\,s^{-1}}$. The limiting time step for convection (considering eight components of tides) was 3.9 s. This limit was severely reduced to 0.1 s if the diffusion term was active and solved explicitly. The convective limit was recovered assuming an implicit solution with $\theta_3 = 0.5$. The comparisons of diffusion error between computations with 0.1 s and 3.9 s had a maximum diffusion error of 3.2% for the 3.9 s calculation, showing enough accuracy for engineering purposes, taking into account that the time stepping was increased 40 times, reducing dramatically the cost of computation. This reduction is fundamental when, in practical applications, the behaviour of the transported quantity must be computed for long-term periods, as was this problem, where the evolution of the salinity needed to be calculated for more than 60 periods of equivalent M_2 tides and for very different initial conditions.

In Fig. 7.14 we show by way of an example the dispersion of a *continuous* hot water discharge in an area of the Severn Estuary. Here we note not only the convection movement but also the diffusion of the temperature contours.

References

1. M.B. Abbott. *Computational Hydraulics: Elements of the Theory of Free Surface Flows*, Pitman, London, 1979.
2. G.J. Haltiner and R.T. Williams. *Numerical Prediction and Dynamic Meteorology*, Wiley, New York, 1980.
3. O.C. Zienkiewicz, R.W. Lewis and K.G. Stagg (eds). *Numerical Methods in Offshore Engineering*, Wiley, Chichester, 1978.
4. J. Peraire. A finite element method for convection dominated flows. Ph.D. thesis, University of Wales, Swansea, 1986.
5. J. Peraire, O.C. Zienkiewicz and K. Morgan. Shallow water problems: a general explicit formulation. *Int. J. Num. Meth. Eng.*, **22**, 547–74, 1986.
6. O.C. Zienkiewicz and J.C. Heinrich. A unified treatment of steady state shallow water and two dimensional Navier–Stokes equations. Finite element penalty function approach. *Comp. Meth. Appl. Mech. Eng.*, **17/18**, 673–89, 1979.
7. M. Kawahara. On finite-element methods in shallow-water long-wave flow analysis, in *Computational Methods in Nonlinear Mechanics* (ed. J.T. Oden), pp. 261–87, North-Holland, Amsterdam, 1980.
8. I.M. Navon. A review of finite element methods for solving the shallow water equations, in *Computer Modelling in Ocean Engineering*, 273–78, Balkema, Rotterdam, 1988.
9. J.J. Connor and C.A. Brebbia. *Finite-Element Techniques for Fluid Flow*, Newnes-Butterworth, London and Boston, 1976.
10. J.J. O'Brien and H.E. Hulburt. A numerical model of coastal upwelling. *J. Phys. Oceanogr.*, **2**, 14–26, 1972.
11. M. Crepon, M.C. Richez and M. Chartier. Effects of coastline geometry on upwellings. *J. Phys. Oceanogr.*, **14**, 365–82, 1984.
12. M.G.G. Foreman. An analysis of two-step time-discretisations in the solution of the linearized shallow-water equations. *J. Comp. Phys.*, **51**, 454–83, 1983.
13. W.R. Gray and D.R. Lynch. Finite-element simulation of shallow-water equations with moving boundaries, in *Proc. 2nd Conf. on Finite-Elements in Water Resources* (eds C.A. Brebbia *et al.*), pp. 2.23–2.42, 1978.

14. T.D. Malone and J.T. Kuo. Semi-implicit finite-element methods applied to the solution of the shallow-water equations. *J. Geophys. Res.*, **86**, 4029–40, 1981.
15. G.J. Fix. Finite-element models for ocean-circulation problems. *SIAM J. Appl. Math.*, **29**, 371–87, 1975.
16. C. Taylor and J. Davis. Tidal and long-wave propagation, a finite-element approach. *Computers and Fluids*, 3, 125–48, 1975.
17. M.J.P. Cullen. A simple finite element method for meteorological problems. *J. Inst. Math. Appl.*, **11**, 15–31, 1973.
18. H.H. Wang, P. Halpern, J. Douglas, Jr. and I. Dupont. Numerical solutions of the one-dimensional primitive equations using Galerkin approximations with localised basis functions. *Mon. Weekly Rev.*, **100**, 738–46, 1972.
19. I.M. Navon. Finite-element simulation of the shallow-water equations model on a limited area domain. *Appl. Math. Modelling*, **3**, 337–48, 1979.
20. M.J.P. Cullen. The finite element method, in *Numerical Methods Used in Atmosphere Models*, Vol. 2, chap. 5, pp. 330–8, WMO/GARP Publication Series 17, World Meteorological Organisation, Geneva, Switzerland, 1979.
21. M.J.P. Cullen and C.D. Hall. Forecasting and general circulation results from finite-element models. *Q. J. Roy. Met. Soc.*, **102**, 571–92, 1979.
22. D.E. Hinsman, R.T. Williams and E. Woodward. Recent advances in the Galerkin finite-element method as applied to the meteorological equations on variable resolution grids, in *Finite-Element Flow-Analysis* (ed. T. Kawai), University of Tokyo Press, Tokyo, 1982.
23. I.M. Navon. A Numerov–Galerkin technique applied to a finite-element shallow-water equations model with enforced conservation of integral invariants and selective lumping. *J. Comp. Phys.*, **52**, 313–39, 1983.
24. I.M. Navon and R. de Villiers. GUSTAF, a quasi-Newton nonlinear ADI FORTRAN IV program for solving the shallow-water equations with augmented Lagrangians. *Computers and Geosci.*, **12**, 151–73, 1986.
25. A.N. Staniforth. A review of the application of the finite-element method to meteorological flows, in *Finite-Element Flow-Analysis* (ed. T. Kawai), pp. 835–42, University of Tokyo Press, Tokyo, 1982.
26. A.N. Staniforth. The application of the finite element methods to meteorological simulations – a review. *Int. J. Num. Meth. Fluids*, **4**, 1-22, 1984.
27. R.T. Williams and O.C. Zienkiewicz. Improved finite element forms for shallow-water wave equations. *Int. J. Num. Meth. Fluids*, **1**, 91–7, 1981.
28. M.G.G. Foreman. A two-dimensional dispersion analysis of selected methods for solving the linearised shallow-water equations. *J. Comp. Phys.*, **56**, 287–323, 1984.
29. I.M. Navon. FEUDX: a two-stage, high-accuracy, finite-element FORTRAN program for solving shallow-water equations. *Computers and Geosci.*, **13**, 225–85, 1987.
30. P. Bettess, C.A. Fleming, J.C. Heinrich, O.C. Zienkiewicz and D.I. Austin. A numerical model of longshore patterns due to a surf zone barrier, in *16th Coastal Engineering Conf.*, Hamburg, West Germany, October, 1978.
31. M. Kawahara, N. Takeuchi and T. Yoshida. Two step explicit finite element method for tsunami wave propagation analysis. *Int. J. Num. Meth. Eng.*, **12**, 331–51, 1978.
32. J.H.W. Lee, J. Peraire and O.C. Zienkiewicz. The characteristic Galerkin method for advection dominated problems – an assessment. *Comp. Meth. Appl. Mech. Eng.*, **61**, 359–69, 1987.
33. D.R. Lynch and W. Gray. A wave equation model for finite element tidal computations. *Computers and Fluids*, **7**, 207–28, 1979.
34. O. Daubert, J. Hervouet and A. Jami. Description on some numerical tools for solving incompressible turbulent and free surface flows. *Int. J. Num. Meth. Eng.*, **27**, 3–20, 1989.

35. G. Labadie, S. Dalsecco and B. Latteaux. Resolution des équations de Saint-Venant par une méthode d'éléments finis. *Electricité de France*, Report HE/41, 1982.
36. T. Kodama, T. Kawasaki and M. Kawahara. Finite element method for shallow water equation including open boundary condition. *Int. J. Num. Meth. Fluids*, **13**, 939–53, 1991.
37. S. Bova and G. Carey. An entropy variable formulation and applications for the two dimensional shallow water equations. *Int. J. Num. Meth. Fluids*, **23**, 29–46, 1996.
38. K. Kashiyama, H. Ito, M. Behr and T. Tezduyar. Three-step explicit finite element computation of shallow water flows on a massively parallel computer. *Int. J. Numer. Meth. Fluids*, **21**, 885–900, 1995.
39. S. Chippada, C. Dawson, M. Martinez and M.F. Wheeler. Finite element approximations to the system of shallow water equations. Part I. Continuous time *a priori* error estimates. *TICAM Report*, *Univ. of Texas at Austin*, 1995.
40. S. Chippada, C. Dawson, M. Martinez and M.F. Wheeler. Finite element approximations to the system of shallow water equations. Part II. Discrete time *a priori* error estimates. *TICAM Report*, *Univ. of Texas at Austin*, 1996.
41. O.C. Zienkiewicz, J. Wu and J. Peraire. A new semi-implicit or explicit algorithm for shallow water equations. *Math. Mod. Sci. Comput.*, **1**, 31–49, 1993.
42. O.C. Zienkiewicz and P. Ortiz. A split-characteristic based finite element model for the shallow water equations. *Int. J. Num. Meth. Fluids*, **20**, 1061–80, 1995.
43. O.C. Zienkiewicz and P. Ortiz. The characteristic based split algorithm in hydraulic and shallow-water flows. Keynote lecture, in *2nd Int. Symposium on Environmental Hydraulics*, Hong Kong, 1998.
44. P. Ortiz and O.C. Zienkiewicz. Tide and bore propagation by a new fluid algorithm, in *Finite Elements in Fluids. 9th Int. Conference*, 1995.
45. R. Löhner, K. Morgan and O.C. Zienkiewicz. The solution of non-linear hyperbolic equation systems by the finite element method. *Int. J. Num. Meth. Fluids*, **4**, 1043–63, 1984.
46. S. Nakazawa, D.W. Kelly, O.C. Zienkiewicz, I. Christie and M. Kawahara. An analysis of explicit finite element approximations for the shallow water wave equations, in *Proc. 3rd Int. Conf. on Finite Elements in Flow Problems*, Banff, Vol. 2, pp. 1–7, 1980.
47. M. Kawahara, H. Hirano, K. Tsubota and K. Inagaki. Selective lumping finite element method for shallow water flow. *Int. J. Num. Meth. Fluids*, **2**, 89–112, 1982.
48. D.R. Lynch and W.G. Gray. Analytic solutions for computer flow model testing. *Trans. ASCE, J. Hydr. Div.*, **104**(10), 1409–28, 1978.
49. Hydraulic Research Station. Severn tidal power. *Report EX985*, 1981.
50. D.I. Austin and P. Bettess. Longshore boundary conditions for numerical wave model. *Int. J. Num. Mech. Fluids*, **2**, 263–76, 1982.
51. C.K. Ziegler and W. Lick. The transport of fine grained sediments in shallow water. *Environ. Geol. Water Sci.*, **11**, 123–32, 1988.

8

Waves

Peter Bettess*

8.1 Introduction and equations

The main developments in this chapter relate to linearized surface waves in water, but acoustic and electromagnetic waves will also be mentioned. We start from the wave equation, Eq. (7.23), which was developed from the equations of momentum balance and mass conservation in shallow water. The wave elevation, η, is small in comparison with the water depth, H. If the problem is periodic, we can write the wave elevation, η, quite generally as

$$\eta(x, y, t) = \bar{\eta}(x, y) \exp(\mathrm{i}\omega t) \tag{8.1}$$

where ω is the angular frequency and $\bar{\eta}$ may be complex. Equation (7.23) now becomes

$$\boldsymbol{\nabla}^{\mathrm{T}}(H\boldsymbol{\nabla}\bar{\eta}) + \frac{\omega^2}{g}\bar{\eta} = 0 \qquad \text{or} \qquad \frac{\partial}{\partial x_i}\left(H\frac{\partial \bar{\eta}}{\partial x_i}\right) + \frac{\omega^2}{g}\bar{\eta} = 0 \tag{8.2}$$

or, for constant depth, H,

$$\nabla^2\bar{\eta} + k^2\bar{\eta} = 0 \qquad \text{or} \qquad \frac{\partial^2 \bar{\eta}}{\partial x_i \partial x_i} + k^2\bar{\eta} = 0 \tag{8.3}$$

where the wavenumber $k = \omega/\sqrt{gH}$. The wave speed is $c = \omega/k$. Equation (8.3) is the Helmholtz equation (which was also derived in Chapter 7, in a slightly different form, as Eq. (7.23)) which models very many wave problems. This is only one form of the equation of surface waves, for which there is a very extensive literature.[1–4] From now on all problems will be taken to be periodic, and the overbar on η will be dropped. The Helmholtz equation (8.3) also describes periodic acoustic waves. The wavenumber k is now given by ω/c, where as in surface waves ω is the angular frequency and c is the wave speed. This is given by $c = \sqrt{K/\rho}$, where ρ is the density of the fluid and K is the bulk modulus. Boundary conditions need to be applied to deal with radiation and absorption of acoustic waves. The first application of finite elements to acoustics was by Gladwell.[5] This was followed in 1969 by the solution of

* Professor, Department of Civil Engineering, University of Durham, UK.

acoustic equations by Zienkiewicz and Newton,[6] and further finite element models by Craggs.[7] A more comprehensive survey of the development of the method is given by Astley.[8] Provided that the dielectric constant, ε, and the permeability, μ, are constant, then Maxwell's equations for electromagnetics can be reduced to the form

$$\nabla^2\phi - \frac{\varepsilon\mu}{c^2}\frac{\partial^2\phi}{\partial t^2} = -\frac{4\pi\rho}{\varepsilon} \quad \text{and} \quad \nabla^2\mathbf{A} - \frac{\varepsilon\mu}{c^2}\frac{\partial^2\mathbf{A}}{\partial t^2} = -\frac{4\pi\mu\mathbf{J}}{c} \tag{8.4}$$

where ρ is the charge density, $\mathbf{J}$ is the current, and ϕ and $\mathbf{A}$ are scalar and vector potentials, respectively. When ρ and $\mathbf{J}$ are zero, which is a frequent case, and the time dependence is harmonic, Eqs (8.4) reduce to the Helmholtz equations. More details are given by Morse and Feshbach.[9]

For surface waves on water when the wavelength, $\lambda = 2\pi/k$, is small relative to the depth, H, the velocities and the velocity potential vary vertically as $\cosh kz$.[1,2,10,11] The full equation can now be written as

$$\boldsymbol{\nabla}^{\mathrm{T}}(cc_g\boldsymbol{\nabla}\eta) + \frac{\omega^2}{g}\eta = 0 \quad \text{or} \quad \frac{\partial}{\partial x_i}\left(cc_g\frac{\partial\eta}{\partial x_i}\right) + \frac{\omega^2}{g}\eta = 0 \tag{8.5}$$

where the group velocity, $c_g = nc$, $n = (1 + (2kH/\sinh 2kH))/2$ and the *dispersion relation*

$$\omega^2 = gk\tanh kH \tag{8.6}$$

links the angular frequency, ω, and the water depth, H, to the wavenumber, k.

8.2 Waves in closed domains – finite element models

We now consider a closed domain of any shape. For waves on water this could be a closed basin, for acoustic or electromagnetic waves it could be a resonant cavity. In the case of surface waves we consider a two-dimensional basin, with varying depth. In plan it can be divided into two-dimensional elements, of any of the types discussed in Volume 1. The wave elevation, η, at any point (ξ, η) within the element, can be expressed in terms of nodal values, using the element shape function $\mathbf{N}$, thus

$$\eta \approx \hat{\eta} = \mathbf{N}\tilde{\boldsymbol{\eta}} \tag{8.7}$$

Next Eq. (8.2) is weighted with the shape function, and integrated by parts in the usual way, to give

$$\int_\Omega \left(\frac{\partial\mathbf{N}}{\partial x_i}H\frac{\partial\mathbf{N}}{\partial x_i} - \mathbf{N}^{\mathrm{T}}\frac{\omega^2}{g}\mathbf{N}\right)\mathrm{d}\Omega\,\tilde{\boldsymbol{\eta}} = 0 \tag{8.8}$$

The integral is taken over all the elements of the domain, and $\tilde{\boldsymbol{\eta}}$ represents all the nodal values of η.

The natural boundary condition which arises is $\partial\eta/\partial n = 0$, where n is the normal to the boundary, corresponding to zero flow normal to the boundary. Physically this corresponds to a vertical, perfectly reflecting wall. Equation (8.8) can be recast in the familiar form

$$(\mathbf{K} - \omega^2\mathbf{M})\tilde{\boldsymbol{\eta}} = \mathbf{0} \tag{8.9}$$

where

$$\mathbf{M} = \int_\Omega \mathbf{N}^\mathrm{T} \frac{1}{g} \mathbf{N} \, \mathrm{d}\Omega \qquad \mathbf{K} = \int_\Omega \mathbf{B}^\mathrm{T}\mathbf{D}\mathbf{B} \, \mathrm{d}\Omega \tag{8.10}$$

$$\mathbf{D} = \begin{bmatrix} h & 0 \\ 0 & h \end{bmatrix}$$

It is thus an *eigenvalue* problem as discussed in Chapter 17 of Volume 1. The **K** and **M** matrices are analagous to structure stiffness and mass matrices. The *eigenvalues* will give the natural frequencies of oscillation of the water in the basin and the *eigenvectors* give the mode shapes of the water surface. Such an analysis was first carried out using finite elements by Taylor *et al.*[12] and the results are shown as Fig. 17.5 of Volume 1. There are analytical solutions for harbours of regular shape and constant depth.[1,3] The reader should find it easy to modify the standard element routine given in Volume 1, Chapter 20, to generate the wave equation 'stiffness' and 'mass' matrices. In the corresponding acoustic problems, the eigenvalues give the natural resonant frequencies and the eigenvectors give the modes of vibration. The model described above will give good results for harbour and basin resonance problems, and other problems governed by the Helmholtz equation. In modelling the Helmholtz equation, it is necessary to retain a mesh which is sufficiently fine to ensure an accurate solution. A 'rule of thumb', which has been used for some time, is that there should be 10 nodes per wavelength. This has been accepted as giving results of acceptable engineering accuracy for many wave problems. However, recently more accurate error analysis of the Helmholtz equation has been carried out.[13,14] In wave problems it is not sufficient to use a fine mesh only in zones of interest. The entire domain must be discretized to a suitable element density. There are essentially two types of error:

- The wave shape may not be a good representation of the true wave, that is the local elevations or pressures may be wrong.
- The wave length may be in error.

This second case causes a poor representation of the wave in one part of the problem to cause errors in another part of the problem. This effect, where errors build up across the model, is called a *pollution error*. It has been implicitly understood since the early days of modelling of the Helmholtz equation, as can be seen from the uniform size of finite element used in meshes.

Babuška *et al.*[13,14] show some results for various finite element models, using different element types, and the error as a function of element size, h, and wave number, k. The sharper error results show that the simple rule of thumb given above is not always adequate. Since the wave number, k, and the wavelength, λ, are related by $k = 2\pi/\lambda$, the condition of 10 nodes per wavelength can be written as $kh \approx 0.6$. But keeping to this limit is not sufficient. The pollution error grows as k^3h^2. Babuška *et al.* propose *a posteriori* error indicators to asssess the pollution error. See the cited references and Chapter 14, Volume 1, for further discussion of these matters.

8.3 Difficulties in modelling surface waves

The main defects of the simple surface-wave model described above are the following:

1. inaccuracy when the wave height becomes large. The equations are no longer valid when η becomes large, and for very large η, the waves will break, which introduces energy loss.
2. lack of modelling of bed friction. This will be discussed below.
3. lack of modelling of separation at re-entrant corners. At re-entrant corners there is a singularity in the velocity of the form $1/\sqrt{r}$. The velocities become large, and physically the viscous effects, neglected above, become important. They cause retardation, flow separation and eddies. This effect can only be modelled in an approximate way.

Now the response can be determined for a given excitation frequency, as discussed in Chapter 17 of Volume 1.

8.4 Bed friction and other effects

The Chézy bed friction term is non-linear and if it is included in its original form it makes the equations very difficult to solve. The usual procedure is to assume that its main effect is to damp the system, by absorbing energy, and to introduce a linear term, which in one period absorbs the same amount of energy as the Chézy term. The linearized bed friction version of Eq. (8.2) is

$$\nabla^{T}(H\nabla\eta) + \frac{\omega^2}{g}\eta - \mathrm{i}\omega M\eta = 0 \qquad \text{or} \qquad \frac{\partial}{\partial x_i}\left(H\frac{\partial\eta}{\partial x_i}\right) + \frac{\omega^2}{g}\eta - \mathrm{i}\omega M\eta = 0 \quad (8.11)$$

where M is a linearized bed friction coefficient, which can be written as $M = 8u_{\max}/3\pi C^2 H$, C is the Chézy constant and $u_{\max}$ is the maximum velocity at the bed at that point. In general the results for η will now be complex, and iteration has to be used, since M depends upon the unknown $u_{\max}$. From the finite element point of view, there is no longer any need to separate the 'stiffness' and 'mass' matrices. Instead, Eq. (8.11) is weighted using the element shape function and the entire complex element matrix is formed. The matrix right-hand side arises from whatever exciting forces are present. The re-entrant corner effect and wave-absorbing walls and permeable breakwaters can also be modelled in a similar way, as both of these introduce a damping effect, due to viscous dissipation. The method is explained in reference 15, where an example showing flow through a perforated wall in an offshore structure is solved.

8.5 The short-wave problem

Short-wave diffraction problems are those in which the wavelength is much smaller than any of the dimensions of the problem. Such problems arise in surface waves on water, acoustics and pressure waves, electromagnetic waves and elastic waves. The methods described in this chapter will solve the problems, but the requirement of 10 nodes or thereabouts per wavelength, makes the necessary finite element meshes prohibitively fine. To take one example, radar waves of wavelength 1 mm might impinge on an aircraft of 10 m wing span. It is easy to see that the computing requirements are truly astronomical.

8.5.1 Transient solution of electromagnetic scattering problems

The penalty in using a fine mesh of conventional finite elements in solving wave problems, referred to above, is the storage and solution of the system matrix. The approach of Morgan *et al.*[16,17] is to treat the problem as transient and not to assemble and solve the system matrix. The Maxwell equations are

$$\varepsilon_0 \frac{\partial \mathbf{E}}{\partial t} = \operatorname{curl} \mathbf{H} \qquad \text{and} \qquad \mu_0 \frac{\partial \mathbf{H}}{\partial t} = -\operatorname{curl} \mathbf{E} \tag{8.12}$$

where **E** and **H** are the electric and magnetic field intensity vectors respectively. The equations are combined and expressed in the conservation form

$$\frac{\partial \mathbf{U}}{\partial t} = \sum_{j=1}^{3} \frac{\partial \mathbf{F}^j}{\partial x_j} = \mathbf{0} \qquad \text{where} \qquad \mathbf{U} = \begin{bmatrix} \mathbf{E} \\ \mathbf{H} \end{bmatrix} \tag{8.13}$$

and the flux vectors, **F**, are derived from the curl operators. That is

$$\begin{aligned} \mathbf{F}^1 &= [0 \quad H_3 \quad -H_2 \quad 0 \quad -E_3 \quad E_2]^{\mathrm{T}} \\ \mathbf{F}^2 &= [-H_3 \quad 0 \quad H_1 \quad E_3 \quad 0 \quad -E_1]^{\mathrm{T}} \\ \mathbf{F}^3 &= [H_2 \quad -H_1 \quad 0 \quad -E_2 \quad E_1 \quad 0]^{\mathrm{T}} \end{aligned} \tag{8.14}$$

The algorithm used is the characteristic–Galerkin (or Lax–Wendroff) method as described in Chapter 2. Details of the algorithm as applied to the electromagnetic problem are given by Morgan *et al.* Improved CPU efficiency and reduced storage requirements are obtained by the use of a representation in which each edge of the tetrahedral mesh is numbered and the data structure employed provides the numbers of the two nodes which are associated with each edge. Because of the massive computations needed for problems of scattering by short waves, parallel processing has also been used. The problem of radar scattering by an aircraft is shown in Fig. 8.1(a), and Fig. 8.1(b) (also in colour plate included in Volume 1) shows the radar cross-section (RCS) obtained for the aircraft using a mesh with about 20 million degrees of freedom.

It would be desirable to simulate radar scattering in the millimetre wavelength range, however even the above described scheme is computationally too intense at this time.

8.5.2 Finite elements incorporating wave shapes

Another approach is to tailor the shape functions within the elements to the known nature of the wave solution. The first attempt to do this was the infinite elements of Bettess and Zienkiewicz.[11,18] The first attempt on *finite* elements was that of Astley,[19,20] using his wave envelope, or complex conjugate weighting method. See Sec. 8.13.

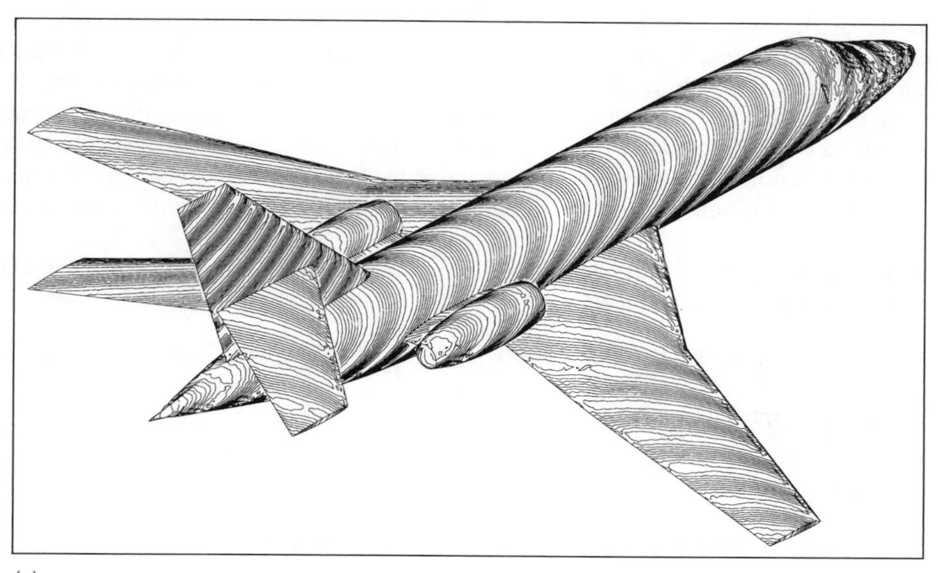
(a)

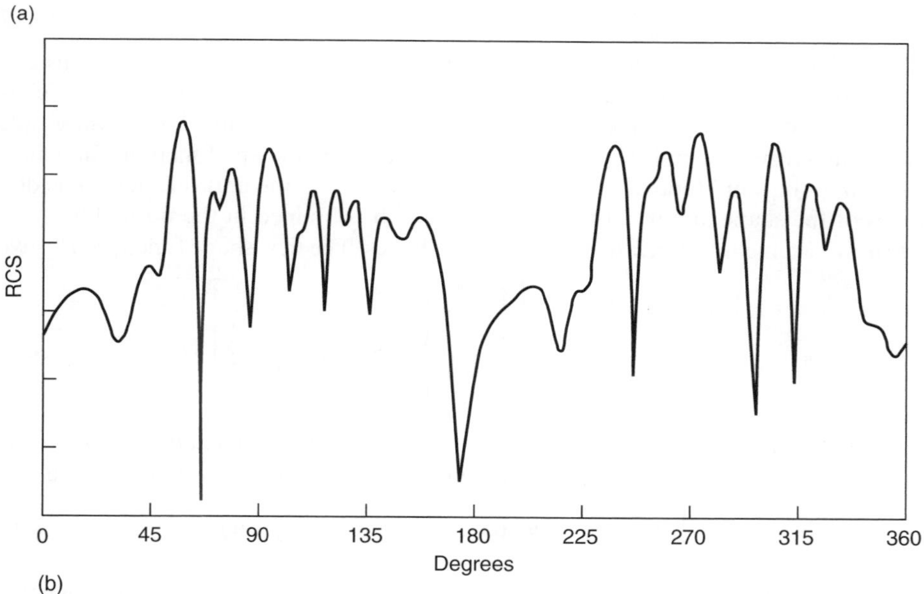

(b)

Fig. 8.1 Scattering of a plane wavelength 2 m by a perfectly conducting aircraft of length 18 m, (a) waves impacting aircraft, (b) computed distribution of RCS, Morgan.[17]

Following Astley's wave envelope technique, Chadwick, Bettess and Laghrouche[21] attempted to develop wave envelope finite elements in which the wave direction was unknown, *a priori*, and to iterate for the correct wave direction, using some type of residual. Although this method had some success, the method proposed by Melenk and Babuška[22,23] appears to be more promising. In this the element shape function incorporates the wave shape, just as in the Bettess and Zienkiewicz infinite elements and the Astley wave envelope elements. However, the innovation of Melenk and Babuška is that *multiple* wave directions are used. This is categorized as a form of the partition of unity finite element method (see Chapter 16 of Volume 1). Melenk and Babuška demonstrated that if such shape functions are used the method works for a plane wave propagated through a square mesh of square finite elements, even when the direction of the wave was not included in the nodal directions. Subsequently Bettess and Laghrouche applied the method to a range of wave problems, and enjoyed some success.[24,25]

The starting point is the standard Galerkin weighted residual form of the Helmholtz equation, which leads to

$$\int_\Omega (-\nabla^T W(\nabla\phi + k^2 W\phi)\, d\Omega + \int_\Gamma W(\nabla\phi)^T \mathbf{n}\, d\Gamma = 0 \tag{8.15}$$

The element approximation is now taken as

$$\phi = \sum_{j=1}^{n}\sum_{l=1}^{m} N_j \psi_l A_j^l \tag{8.16}$$

where N_j are the normal polynomial element shape functions,

$$\psi_l = e^{ik(x\cos\theta_l + y\sin\theta_l)} \tag{8.17}$$

and n is the number of nodes in the element, and m is the number of directions considered at each node. The shape function consists of a set of plane waves travelling in different directions, the nodal degrees of freedom corresponding to the amplitudes of the different waves and the normal polynomial element shape functions allowing a variation in the amplitude of each wave component within the finite element. The derivatives of the shape and weighting functions can be obtained in the normal way, but these now also include derivatives of the wave shapes. The new shape function, P_i, gives

$$\begin{Bmatrix} \dfrac{\partial P_{(j-1)m+l}}{\partial x} \\ \dfrac{\partial P_{(j-1)m+l}}{\partial y} \end{Bmatrix} = \left[\begin{Bmatrix} \dfrac{\partial N_j}{\partial x} \\ \dfrac{\partial N_j}{\partial y} \end{Bmatrix} + ikN_j \begin{Bmatrix} \cos\theta_l \\ \sin\theta_l \end{Bmatrix} \right] \psi_l \tag{8.18}$$

The global derivatives are obtained in the usual way from the local derivatives, using the inverse of the jacobian matrix. The element stiffness and mass matrices are

$$K_{rs} = \int_\Omega (\nabla W_r)^T \nabla P_s\, d\Omega \qquad M_{rs} = \int_\Omega W_r P_s\, d\Omega \tag{8.19}$$

where r and s are integers which vary over the range of $1, 2, \ldots, (n \times m)$. When calculating the element matrices, the integrals encountered are of the form

$$I_{jl} = \int_{-1}^{1}\int_{-1}^{1} f(\xi,\eta)\, e^{ik(x\cos\theta_j + y\sin\theta_j)}\, e^{ik(x\cos\theta_l + y\sin\theta_l)}\, d\xi\, d\eta \tag{8.20}$$

This integral has to date been performed numerically. But when the waves are short, many Gauss–Legendre integration points are needed. Typically about 10 integration points per wavelength are needed. Laghrouche and Bettess solve a range of wave diffraction problems, including that of plane waves diffracted by a cylinder. The mesh and the results are shown in Fig. 8.2. As can be seen the results are in good

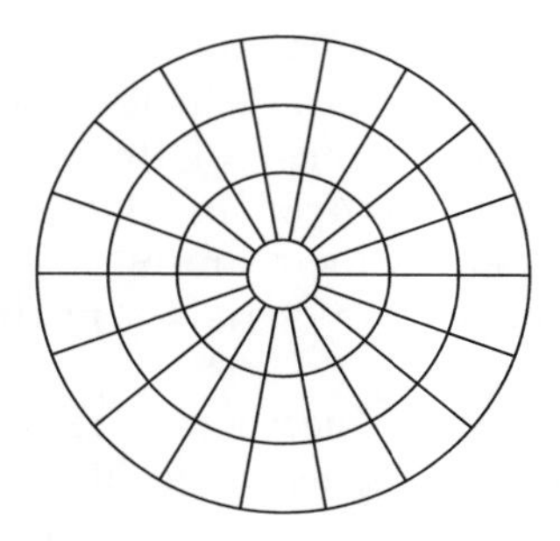

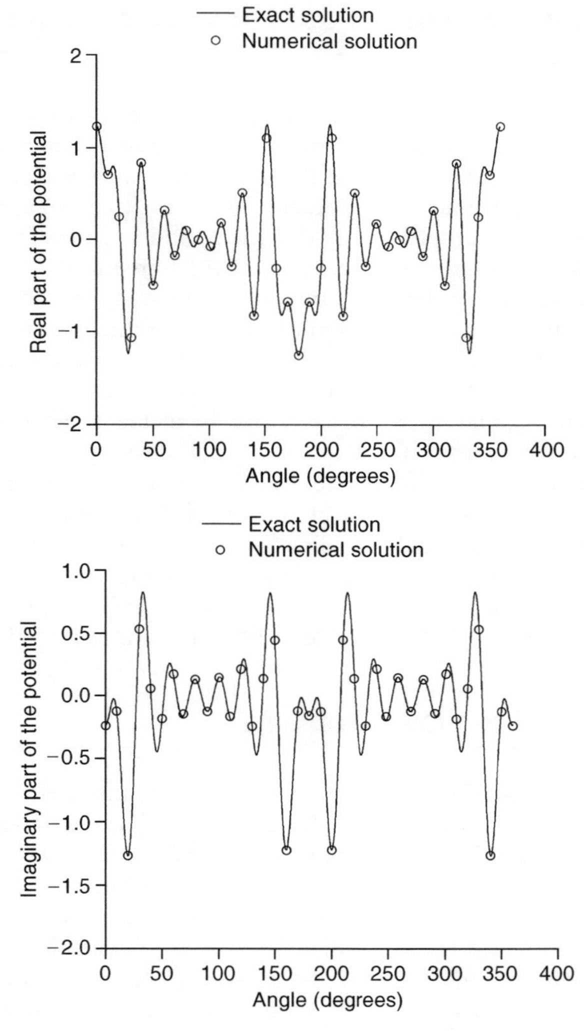

Fig. 8.2 Short waves diffracted by a cylinder, modelled using special finite elements, Laghrouche and Bettess.[24]

agreement with the analytical series solution. In this problem $ka = 8\pi$, $\lambda = 0.25a$, radius of cylinder, $a = 1$, and the mesh extends to $r = 7a$. For a conventional radial finite element mesh, the requirement of 10 nodes per wavelength would lead to a mesh with 424 160 degrees of freedom. But in the results shown, with 36 directions per node and 252 nodes there are only 9072 degrees of freedom. The dramatic reduction in the number of variables merits further investigation and development of the method.

The method still has a number of uncertainties regarding the conditioning of the system matrix and the stability of the technique and a significant problem remains in the numerical cost of integrating the element matrix.

8.6 Waves in unbounded domains (exterior surface wave problems)

Problems in this category include the diffraction and refraction of waves close to fixed and floating structures, the determination of wave forces and wave response for offshore structures and vessels, and the determination of wave patterns adjacent to coastlines, open harbours and breakwaters. In electromagnetics there are scattering problems of the type already described, and in acoustics we have various noise problems. In the interior or finite part of the domain, finite elements, exactly as described in Sec. 8.2 can be used, but special procedures must be adopted for the part of the domain extending to infinity. The main difficulty is that the problem has no outer boundary. This necessitates the use of a *radiation condition*. Such a condition was introduced in Chapter 19 of Volume 1, as Eq. (19.18), for the case of a one-dimensional wave, or a normally incident plane wave in two or more dimensions. Work by Bayliss *et al.*[26,27] has developed a suitable radiation condition, in the form of an infinite series of operators. The starting point is the representation of the outgoing wave in the form of an infinite series. Each term in the series is then annihilated by using a boundary operator. The sequence of boundary operators thus constitutes the radiation condition. In addition there is a classical form of the boundary condition for periodic problems, given by Sommerfeld. A summary of all available radiation conditions is given in Table 8.1.

8.6.1 Background to wave problems

The simplest type of exterior, or unbounded wave problem is that of some exciting device which sends out waves which do not return. This is termed the *radiation* problem. The next type of exterior wave problem is where we have a known incoming wave which encounters an object, is modified and then again radiates away to infinity. This case is known as the *scattering* problem, and is more complicated, in as much as we have to deal with both incident and radiated waves. Even when both waves are linear, this can lead to complications. Both the above cases can be complicated by wave refraction, where the wave speeds change, because of changes in the medium, for example changes in water depth. Usually this phenomenon leads to changes in the wave direction. Waves can also reflect from boundaries, both physical and computational.

Table 8.1 Radiation conditions for exterior wave problems

Dimensions		
1	2	3
	General boundary conditions	
	Transient	
$\frac{\partial\phi}{\partial x}+\frac{1}{c}\frac{\partial\phi}{\partial t}=0$	$B_m\phi=0, m\to\infty$ $B_m=\prod_{j=1}^{m}\left(\frac{\partial}{\partial r}+\frac{\partial}{\partial t}+\frac{2j-(3/2)}{r}\right)$	$B_m\phi=0, m\to\infty$ $B_m=\prod_{j=1}^{m}\left(\frac{\partial}{\partial r}+\frac{\partial}{\partial t}+\frac{2j-1}{r}\right)$
	Periodic	
$\frac{\partial\phi}{\partial x}+\frac{1}{c}\frac{\partial\phi}{\partial t}=0$	$\lim_{r\to\infty}\sqrt{r}\left(\frac{\partial\phi}{\partial r}+\mathrm{i}k\phi\right)=0$ or $B_m\phi=0, m\to\infty$ $B_m=\prod_{j=1}^{m}\left(\frac{\partial}{\partial r}+\mathrm{i}k+\frac{2j-(3/2)}{r}\right)$	$\lim_{r\to\infty} r\left(\frac{\partial\phi}{\partial r}+\mathrm{i}k\phi\right)=0$ or $B_m\phi=0, m\to\infty$ $B_m=\prod_{j=1}^{m}\left(\frac{\partial}{\partial r}+\mathrm{i}k+\frac{2j-1}{r}\right)$
	Symmetric boundary conditions	
	Transient	
$\frac{\partial\phi}{\partial r}+\frac{1}{c}\frac{\partial\phi}{\partial t}=0$	$\frac{\partial\phi}{\partial r}+\frac{\phi}{2r}+\frac{1}{c}\frac{\partial\phi}{\partial t}=0$ Axisymmetric	$\frac{\partial\phi}{\partial r}+\frac{\phi}{r}+\frac{1}{c}\frac{\partial\phi}{\partial t}=0$ Spherically symmetric
	Periodic	
$\frac{\partial\phi}{\partial r}+\mathrm{i}k\phi=0$	$\frac{\partial\phi}{\partial r}+\left(\frac{1}{2r}+\mathrm{i}k\right)\phi=0$ Axisymmetric	$\frac{\partial\phi}{\partial r}+\left(\frac{1}{r}+\mathrm{i}k\right)\phi=0$ Spherically symmetric

8.6.2 Wave diffraction

We now consider the problem of an incident wave diffracted by an object. The problem consists of an object in some medium, which diffracts the incident waves. We divide the medium as shown in Fig. 8.3, into two regions, with boundaries Γ_A, Γ_B, Γ_C and Γ_D.

These boundaries have the following meanings. Γ_A is the boundary of the body which is diffracting the waves. Γ_B is the boundary between the two computational domains, that in which the total wave elevation (or other field variable) is used, and that in which the elevation of the radiated wave is used. Γ_C is the outer boundary of the computational model, and Γ_D is the boundary at infinity. Some of these boundaries may be merged.

A variational treatment will be used, as described in Volume 1, Chapter 3. A weighted residual treatment is also possible. The elevation of the total wave, η_T, is split into those for incident and radiated waves, η_I and η_R. Hence $\eta_T=\eta_I+\eta_R$. The incident wave elevation, η_I, is assumed to be known. For the surface wave problem, the functional for the exterior can be written

$$\Pi=\iint_{\Omega_B}\frac{1}{2}\left[cc_g(\nabla\eta)^{\mathrm{T}}\nabla\eta-\frac{\omega^2c_g}{c}\eta^2\right]\mathrm{d}x\,\mathrm{d}y \tag{8.21}$$

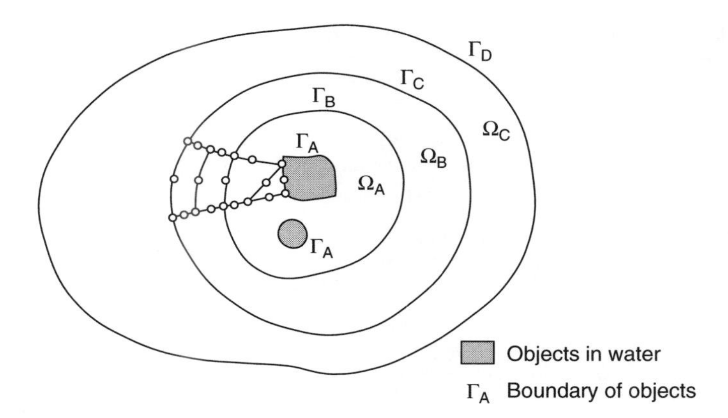

Fig. 8.3 General wave domains.

where making Π stationary with respect to variations in η corresponds to satisfying the shallow-water wave equation (8.2), with natural boundary condition $\partial\eta/\partial n = 0$, or zero velocity normal to the boundary. The functional is rewritten in terms of the incident and radiated elevations, and then Green's theorem in the plane (Volume 1, Appendix G) is applied on the domain exterior to Γ_B. But the radiation condition discussed above should be included. In order to do this the variational statement must be changed so that variations in η yield the correct boundary condition. Details are given by many authors, see for example Bettess.[28] After some manipulation the final functional for the exterior is

$$\begin{aligned}\Pi = &\iint_{\Omega_b} \frac{1}{2}\left[cc_g(\boldsymbol{\nabla}\eta^{\mathrm{s}})^{\mathrm{T}}\boldsymbol{\nabla}\eta^{\mathrm{s}} - \frac{\omega^2 cc_g}{c}(\eta^{\mathrm{s}})^2\right]\mathrm{d}x\,\mathrm{d}y \\ &+ \int_{\Gamma_b} cc_g\left[\frac{\partial\eta^i}{\partial x}\eta^{\mathrm{s}}\,\mathrm{d}y - \frac{\partial\eta^i}{\partial y}\eta^{\mathrm{s}}\,\mathrm{d}x\right] + \frac{1}{2}\int_{\Gamma_d} \mathrm{i}kcc_g(\eta^{\mathrm{s}})^2\,\mathrm{d}\Gamma\end{aligned} \tag{8.22}$$

The influence of the incident wave is thus to generate a 'forcing term' on the boundary Γ_B. For two of the most popular methods for dealing with exterior problems, linking to boundary integrals and infinite elements, the 'damping' term in Eq. (8.22), corresponding to the radiation condition, is actually irrelevant, because both methods use functions which automatically satisfy the radiation condition at infinity.

8.6.3 Incident waves, domain integrals and nodal values

It is possible to choose any known solution of the wave equation as the incident wave. Usually this is a plane monochromatic wave, for which the elevation is given by $\eta_I = a_0 \exp[ikr\cos(\theta - \gamma)]$, where γ is the angle that the incident wave makes to the positive x-axis, r and θ are the polar coordinates and a_0 is the incident wave amplitude. On the boundary Γ_B, we have two types of variables, the total elevation, η_T, on the interior, and η_R, the radiation elevation, in the exterior. Clearly the nodal values of η in the finite element model must be unique, and on this boundary, as well as the line integral, of Eq. (8.22), we must transform the nodal values, either to η_T or to η_R. This can be done simply by enforcing the change of variable, which leads to a contribution to the 'right-hand side' or 'forcing' term.[28]

8.7 Unbounded problems

There are several methods of dealing with exterior problems using finite elements in combination with other methods. Some of these methods are also applicable to finite differences. The literature in this field has grown enormously in the past 10 years, and this section will therefore be selective. The monograph by Givoli[29] is devoted exclusively to this field and gives much more detail on the competing algorithms. It is a very useful source and gives many more algorithms than can be covered here. The book edited by Geers,[30] from an IUTAM symposium, gives a very useful and up-to-date overview of the field.

The main methods include:

- boundary dampers, both plane and cylindrical (also called non-reflecting boundary conditions);
- linking to exterior solutions, both series and boundary integral (also called Dirichlet to Neumann mapping);
- infinite elements.

8.8 Boundary dampers

The nomenclature of boundary dampers comes from engineering applications. Such boundary conditions are also called *local non-reflecting boundary conditions* by mathematicians. As was seen in Chapter 19 of Volume 1, we can simply apply the plane damper at the boundary of the mesh. This was first done in fluid problems by Zienkiewicz and Newton.[6] However the more sophisticated dampers proposed by Baylisss *et al.*[26,27] can be used at little extra computational cost and a big increase in accuracy. The dampers are developed from the series given in Table 8.1. Full details are given in reference 31. For the case of two-dimensional waves the line integral which should be applied on the circular boundary of radius r is

$$A = \int_\Gamma \left[\frac{\alpha}{2}\eta^2 + \frac{\beta}{2}\left(\frac{\partial \eta}{\partial s}\right)^2 \right] \mathrm{d}s \tag{8.23}$$

where ds is an element of distance along the boundary and

$$\alpha = \frac{3/4r^2 - 2k^2 + 3\mathrm{i}k/r}{2/r + 2\mathrm{i}k} \qquad \text{and} \qquad \beta = \frac{1}{2/r + 2\mathrm{i}k} \tag{8.24}$$

For the plane damper, $\beta = 0$ and $\alpha = \mathrm{i}k$. For the cylindrical damper $\beta = 0$ and $\alpha = \mathrm{i}k - 1/2r$. The corresponding expressions for three-dimensional waves are different. Non-circular boundaries can be handled but the expressions become much more complicated. Some results are given by Bando *et al.*[31] Figure 8.4 shows the waves diffracted by a cylinder problem for which there is a solution, due to

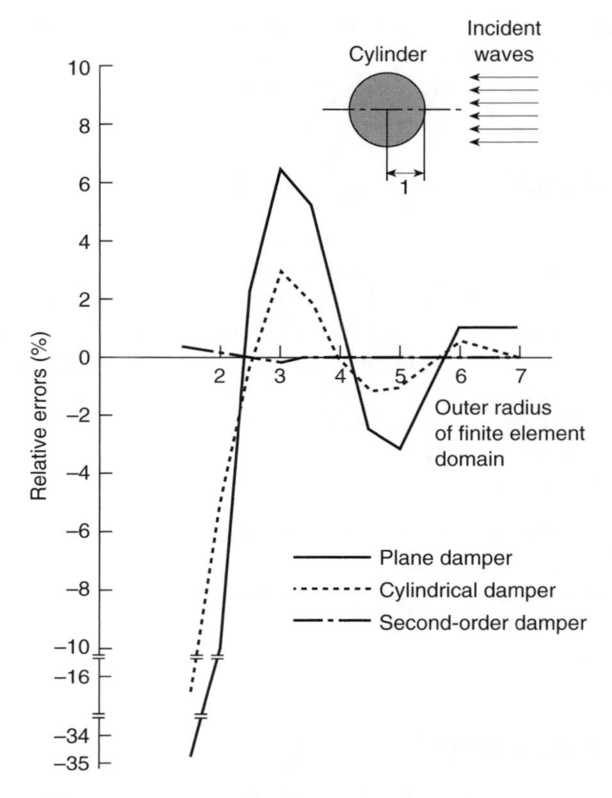

Fig. 8.4 Damper solutions for waves diffracted by circular cylinder. Comparison of relative errors for various outer radii, $(ka = 1)$. Relative error $= (abs(\eta_n) - abs(\eta_a))/abs(\eta_a)$.

Havelock. The higher-order dampers are clearly a big improvement over the plane and cylindrical dampers, for little or no extra computational cost. Engquist and Majda have also earlier proposed dampers for these problems,[32] but instead of using a hierarchy of operators like Bayliss *et al.*, they use a different method. The effect is the same, in that a hierarchy of boundary operators is defined, but the terms are different to those of Bayliss *et al.*

8.8.1 Other damper-related approaches

A great variety of methods have appeared recently based on dampers, and variants of the concept. There is not enough space to review them all in detail here and the reader is referred to the book by Givoli[29] and the volume edited by Geers,[30] which gives access to recent developments. The papers in the Geers volume by Bielak, Givoli, Hagstrom, Hariharan, Higdon, Pinsky and Kallivokas should be consulted. An interesting development is the method of the perfectly matched layer (PML), or 'sponge layer', the idea being that the outgoing wave is not absorbed on a boundary, but in a domain which extends beyond the boundary. In this domain the wave is absorbed or damped in such a way that it does not return into the computational domain. See the papers by Monk and Collino, Hayder and Driscoll in reference 30.

8.9 Linking to exterior solutions

A general methodology for linking finite elements to exterior solutions was proposed by Zienkiewicz *et al.*,[33,34] following various *ad hoc* developments, and this is also discussed in Volume 1, particularly in Chapter 13. The exterior solution can take any form, and those chiefly used are (a) exterior series solutions and (b) exterior boundary integrals, although others are possible. The two main innovators in these cases were Berkhoff,[10,35] for coupling to boundary integrals, and Chen and Mei[36,37] for coupling to exterior series solutions. Although the methods proposed are quite different, it is useful to cast them in the same general form. More details of this procedure are given in reference 33. Basically the energy functional given in Eq. (8.23) is again used. If the functions used in the exterior automatically satisfy the wave equation, then the contribution on the boundary reduces to a line integral of the form

$$\Pi = \frac{1}{2}\int_{\Gamma} \eta \frac{\partial \eta}{\partial n} \, d\Gamma \tag{8.25}$$

It can be shown[15,33,34] that if the free parameters in the interior and exterior are **b** and **a** respectively, the coupled equations can be written

$$\begin{bmatrix} \mathbf{K} & \bar{\mathbf{K}}^{\mathrm{T}} \\ \bar{\mathbf{K}} & \breve{\mathbf{K}} \end{bmatrix} \begin{Bmatrix} \mathbf{a} \\ \mathbf{b} \end{Bmatrix} + \begin{Bmatrix} \mathbf{f} \\ \mathbf{0} \end{Bmatrix} = \begin{Bmatrix} \mathbf{0} \\ \mathbf{0} \end{Bmatrix} \tag{8.26}$$

where

$$\breve{K}_{ji} = \tfrac{1}{2}\int_{\Gamma} [(PN_j)N_i + N_j(PN_i)] \, d\Gamma \qquad \text{and} \qquad \bar{K}_{ji} = \int_{\Gamma} [(PN_j)(\bar{N}_i) \, d\Gamma \tag{8.27}$$

In the above P is an operator giving the normal derivative, i.e. $P \equiv \partial/\partial n$, $\bar{\mathbf{N}}$ is the finite element shape function, **N** is the exterior shape function, and **K** corresponds to the normal finite element matrix. The approach described above can be used with any suitable form of exterior solution, as we will see. All the nodes on the boundary become coupled.

8.9.1 Linking to boundary integrals

Berkhoff[10,35] adopted the simple expedient of identifying the nodal values of velocity potential obtained using the boundary integral, with the finite element nodal values. This leads to a rather clumsy set of equations, part symmetrical, real and banded, and part unsymmetrical, complex and dense. The direct boundary integral method for the Helmholtz equation in the exterior leads to a matrix set of equations

$$\mathbf{A}\tilde{\boldsymbol{\eta}} = \mathbf{B}\frac{\partial \tilde{\boldsymbol{\eta}}}{\partial n} \tag{8.28}$$

(The indirect boundary integral method can also be used.) The values of η and $\partial\eta/\partial n$ on the boundary are next expressed in terms of shape functions, so that

$$\eta \approx \hat{\eta} = \mathbf{N}\tilde{\boldsymbol{\eta}} \qquad \text{and} \qquad \frac{\partial \eta}{\partial n} \approx \frac{\partial \hat{\eta}}{\partial n} = \mathbf{M}\left\{\frac{\partial \tilde{\boldsymbol{\eta}}}{\partial n}\right\} \tag{8.29}$$

$\mathbf{N}$ and $\mathbf{M}$ are equivalent to $\mathbf{N}$ in the previous section. Using this relation, the integral for the outer domain can be written as

$$\Pi = \frac{1}{2}\int_\Gamma \frac{\partial \tilde{\boldsymbol{\eta}}}{\partial n}\mathbf{M}^{\mathrm{T}}\mathbf{N}\tilde{\boldsymbol{\eta}}\,\mathrm{d}\Gamma \tag{8.30}$$

where Γ is the boundary between the finite elements and the boundary integrals. The normal derivatives can now be eliminated, using the relation (8.28), and η can be identified with the finite element nodal values, η, to give

$$\Pi = \tfrac{1}{2}\mathbf{b}^{\mathrm{T}}(\mathbf{B}^{-1}\mathbf{A})^{\mathrm{T}}\int_\Gamma \mathbf{M}^{\mathrm{T}}\mathbf{N}\,\mathrm{d}\Gamma\mathbf{b} \tag{8.31}$$

Variations of this functional with respect to $\mathbf{b}$ can be set to zero, to give

$$\frac{\partial \Pi}{\partial \mathbf{b}} = \frac{1}{2}\left\{(\mathbf{B}^{-1}\mathbf{A})\int_\Gamma \mathbf{M}^{\mathrm{T}}\mathbf{N}\,\mathrm{d}\Gamma + \left[(\mathbf{B}^{-1}\mathbf{A})\int_\Gamma \mathbf{M}^{\mathrm{T}}\mathbf{N}\,\mathrm{d}\Gamma\right]^{\mathrm{T}}\right\}\mathbf{b} = \check{\mathbf{K}}\mathbf{b} \tag{8.32}$$

where $\check{\mathbf{K}}$ is a 'stiffness' matrix for the exterior region. It is symmetric and can be created and assembled like any other element matrix. The integrations involved must be carried out with care, as they involve singularities. Results obtained for the problem of waves refracted by a parabolic shoal are shown in Fig. 8.5.

8.9.2 Linking to series solutions

Chen and Mei[36,37] took the series solution for waves in the exterior, and worked out explicit expressions for the exterior and coupling matrices, $\bar{\mathbf{K}}$ and $\hat{\mathbf{K}}$, for piecewise linear shape functions, $\bar{\mathbf{N}}$, in the finite elements. The series used in the exterior consists of Hankel and trigonometric functions which automatically satisfy the Helmholtz equation and the radiation condition:

$$\eta = \sum_{j=0}^{m} H_j(kr)(\alpha_j \cos j\theta + \beta_j \sin j\theta) \tag{8.33}$$

The method described above leads to the following matrices.

$$\bar{\mathbf{K}}^{\mathrm{T}} = \frac{-knL_c}{2}\begin{bmatrix} 2H_0' & \cdots & H_n'(\cos n\theta_p + \cos n\theta_1) & H_n'(\sin n\theta_p + \sin n\theta_1) & \cdots \\ 2H_0' & \cdots & H_n'(\cos n\theta_1 + \cos n\theta_2) & H_n'(\sin n\theta_1 + \sin n\theta_2) & \cdots \\ 2H_0' & \cdots & H_n'(\cos n\theta_2 + \cos n\theta_3) & H_n'(\sin n\theta_2 + \sin n\theta_3) & \cdots \\ \vdots & \cdots & \vdots & \vdots & \ddots \\ 2H_0' & \cdots & H_n'(\cos n\theta_{p-1} + \cos n\theta_p) & H_n'(\sin n\theta_{p-1} + \sin n\theta_p) & \cdots \end{bmatrix} \tag{8.34}$$

$$\hat{\mathbf{K}} = \pi rkh\{\mathrm{diag}[2H_0H_0' \quad H_1'H_1 \quad H_1'H_1 \quad \cdots \quad H_s'H_s \quad H_s'H_s]\} \tag{8.35}$$

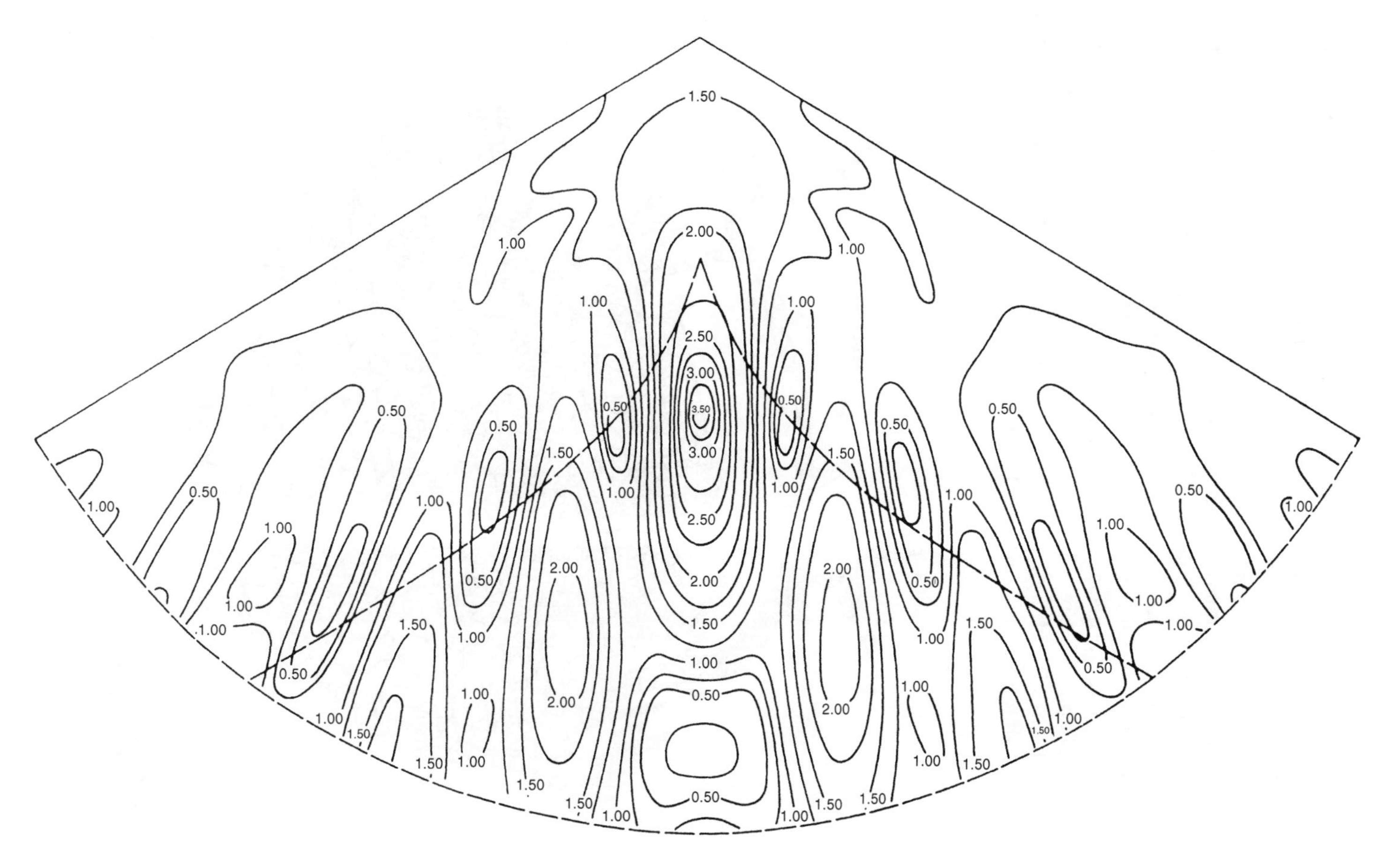

Fig. 8.5 Refraction–diffraction solution: lines of equal wave height, lines every 0.25 unit.

where m is the number of terms in the Hankel function series, r is the radius of the boundary, L_c is the distance between the equidistant nodes on Γ_C, p is the number of nodes, and H_n and H_n' are Hankel functions and derivatives.

Other authors have worked out the explicit forms of the above matrices for linear shape functions, and also it is possible to work them out for any type of shape function, using, if necessary, numerical integration. It will be noticed that the matrix $\hat{\mathbf{K}}$ is diagonal. This is because the boundary Γ_B is circular and the Hankel functions are orthogonal. If a non-circular domain is used, $\hat{\mathbf{K}}$ will become dense. Chen and Mei[36] applied the method very successfully to a range of problems, most notably that of resonance effects in an artificial offshore harbour, the results for which are shown in Volume 1, Chapter 17, Fig. 17.6.

The method was also utilized by Houston,[38] who applied it to a number of real problems, including resonance in Long Beach harbour, shown in Fig. 8.6.

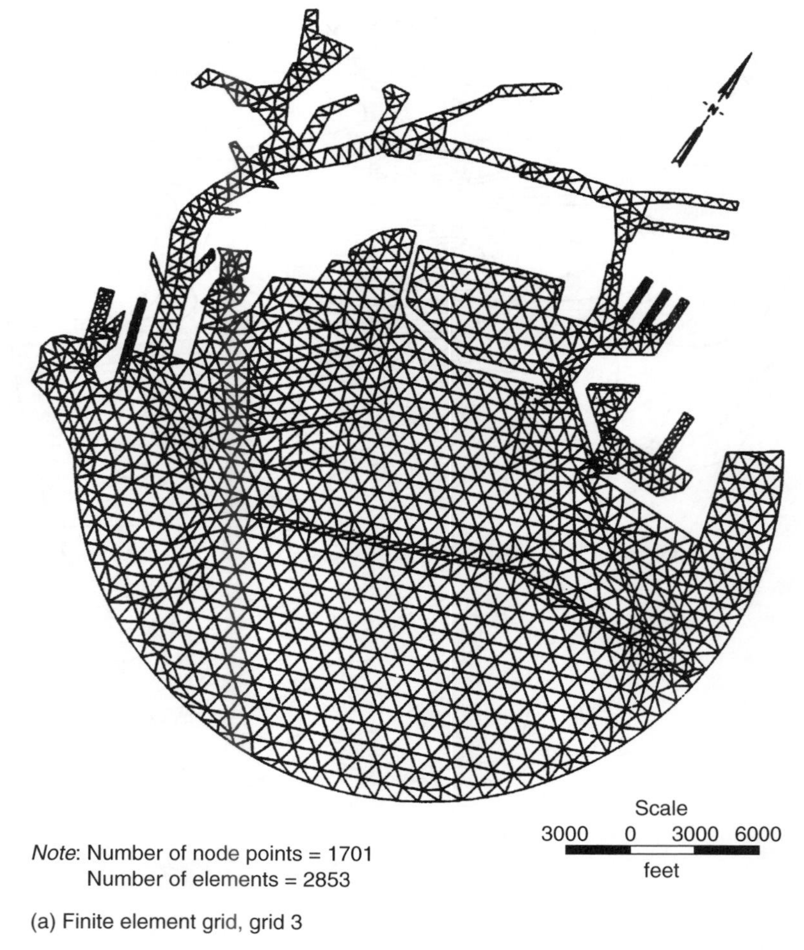

(a) Finite element grid, grid 3

Fig. 8.6 Finite element mesh and wave height magnification for Long Beach Harbour, Houston.[38]

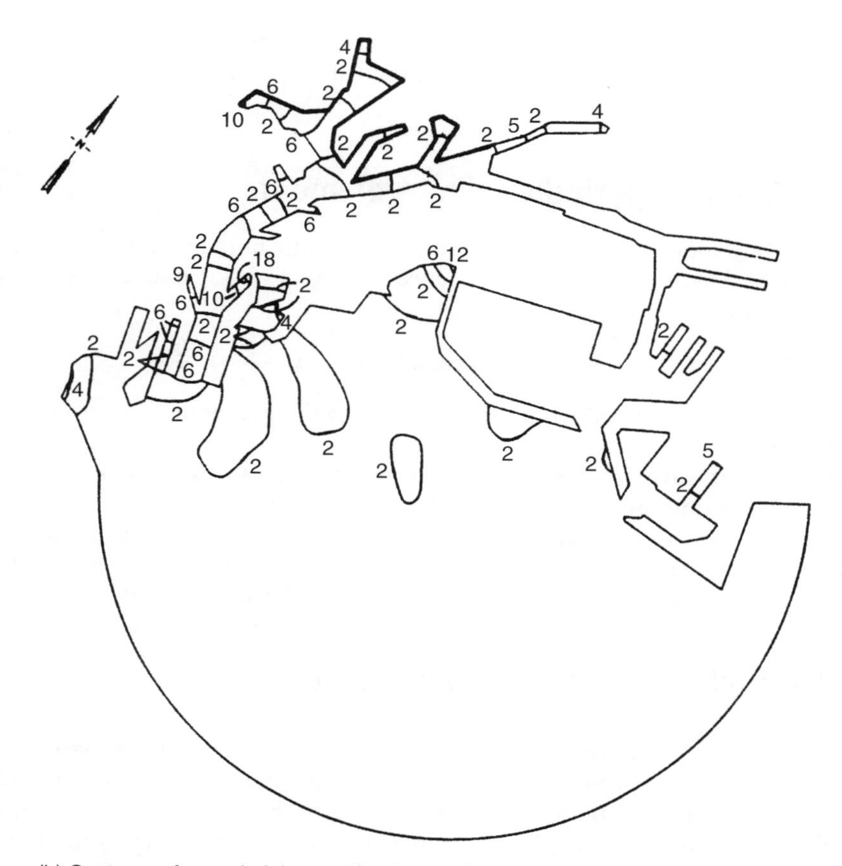

(b) Contours of wave height amplification, grid 3. 232-s wave period

Fig. 8.6 Continued.

8.9.3 DtN mapping

The approach described above has been re-invented by recent authors and given the title Dirichlet to Neumann (DtN) mapping. See a comparison by Astley.[39] A detailed survey of this approach, which goes beyond the wave equation, is given by Givoli.[29]

8.10 Infinite elements

Infinite elements are described in the book by Bettess,[40] which although somewhat out-of-date, still gives a useful introduction to the topic. More recent reviews are by Astley,[41] and Gerdes.[42] The methods described in Volume 1, Chapter 9, can be developed to include periodic effects. This was first done by Bettess and Zienkiewicz, using so-called 'decay function' procedures and they were very effective.[11,18] Comparison results with Chen and Mei[36,37] for the artificial island problem are shown in

Fig. 17.6 of Volume 1. Later 'mapped' infinite elements were developed for wave problems, and as these are more accurate than those using exponentials, they will be described here.

8.11 Mapped periodic infinite elements

The theory developed in Volume 1, Chapter 9 for static infinite elements, will not be repeated here. Details are given in references 28, 39, 43–47. Finite element polynomials of the form

$$P = \alpha_0 + \alpha_1\xi + \alpha_2\xi^2 + \ldots \quad \text{become} \quad P = \beta_0 + \frac{\beta_1}{r} + \frac{\beta_2}{r^2} + \ldots \tag{8.36}$$

where β_i can be determined from the α's and a. If the polynomial is zero at infinity then $\beta_0 = 0$.

Many exterior wave problems have solutions in which the wave amplitude decays radially like $1/r$ (and higher-order terms) and an advantage of this mapping is that such a decay can be represented exactly. In some cases, however, the amplitude decays approximately as $1/\sqrt{r}$, and this case needs a slightly different treatment. Accuracy can be increased by adding extra terms to the series (8.36).

8.11.1 Introducing the wave component

In two-dimensional exterior domains the solution to the Helmholtz equation can be described by a series of combined Hankel and trigonometric functions, the simplest solution to the Helmholtz equation being $H_0(kr)$. For large r the zeroth-order Hankel function oscillates roughly like $\cos(kr) + \mathrm{i}\sin(kr)$, while decaying in magnitude as $r^{-1/2}$. A series of terms $1/r$, $1/r^2$, etc., generated by the mapping, multiplied by $r^{1/2}$ and the periodic component $\exp(\mathrm{i}kr)$ will be used to model the $r^{-1/2}$ decay. The shape function is thus

$$N(\xi,\eta) = M(\xi,\eta)r^{1/2}\exp(\mathrm{i}kr) \tag{8.37}$$

where $r = A/(1-\xi)$. The shape function in Eq. (8.37) will now be, for compatibility with the finite elements,

$$N(\xi,\eta) = M(\xi,\eta)\left(\frac{2}{A}\right)^{1/2}\left(\frac{A}{1-\xi}\right)^{1/2}\exp\left(\frac{\mathrm{i}kA}{2}\right)\exp\left(\frac{\mathrm{i}kA}{1-\xi}\right) \tag{8.38}$$

In the improved version of this element,[28] the constant, A, varies within the element. A is now determined on each radial line from the positions of the nodes. It is interpolated between these values. The original mapped infinite element did not include the possibility of varying the mapping, so that the infinite elements had to be placed exterior to a sphere. There was also an uncertainty about the integrations in the infinite radial direction, which was resolved by Astley *et al.*[47] This arose because the boundary terms at infinity were incompletely stated, although the element, as presented in reference 46, is correct. Mapped wave envelope infinite elements were later developed, using the same methodology, but with a complex conjugate weighting. Later still Astley *et al.*,[48] Cremers and Fyfe and Coyette[49,50] generalized

Fig. 8.7 Real part of elevations of plane wave diffracted by an ellipse, of aspect ratio 2, Bettess.[28]

the mapping of these wave envelope infinite elements, so that it was no longer necessary to place them exterior to a sphere or cylinder. After this work Bettess and Bettess[28] generalized the original mapped wave infinite elements. Figure 8.7 shows some results from the diffraction of waves by an ellipse, for which there is an analytical solution.

8.12 Ellipsoidal type infinite elements of Burnett and Holford

Burnett with Holford[51–53] proposed a new type of infinite element for exterior acoustics problems. This uses prolate or oblate spheroidal coordinates, and separates the radial and angular coordinates. Burnett also further clarified the variational statement of the problem and explained in more detail the terms at the infinite boundary. It is known that a scattered wave exterior to a sphere can be written in spherical polar coordinates as

$$p = \frac{e^{-kr}}{r} \sum_{n=0}^{\infty} \frac{G_n(\theta, \phi, k)}{r^n} \tag{8.39}$$

This proof was generalized to the case when the coordinates θ, ϕ, r are not simply spherical, but prolate or oblate spheroidal or ellipsoidal. There are several benefits to using such coordinate systems:

- The volume integrals separate into radial and angular parts which can be carried out independently. This leads to economies in computation.
- The radial integration is identical for every such infinite element, so that the only integration which needs to be carried out for every infinite element is along the finite element interface.
- The radial integration is the only part containing the wave number.
- The ellipsoidal coordinates can be used to enclose a large variety of different geometries in the finite element interior, while still retaining a guarantee of convergence in 3D.

The angular shape functions are written in the conventional polynomial form. The radial shape functions take the form

$$N_i = \mathrm{e}^{-\mathrm{i}kr} \sum_{j=1}^{m} \frac{h_{ij}}{(kr)^j} \tag{8.40}$$

The coefficients h_{ij} are given from the condition of circumferential compatibility between adjacent infinite elements. There is effectively no difference between this radial behaviour and that originally proposed in the mapped infinite wave elements by Bettess *et al.*[45,46] The difference in the infinite element methodology lies in the fact that the radial variable, r, is now in ellipsoidal coordinates. Burnett and Holford[51–53] give the necessary detailed information for the element integrations and the programming of these infinite elements, together with some results. The analytical expressions are too long to include here. The elements have been used on submarine fluid–structure interaction problems, and substantial efficiencies over the use of boundary integral models for the scattered waves have been claimed. In one case Burnett states that the finite and infinite element model ran for 7 hours on a workstation. His projected time for the corresponding boundary element model was about 3000 hours, the infinite elements giving a dramatic improvement!

The Burnett elements have been tested up to very short-wave cases, up to $ka = 100$ for an elastic sphere diffraction problem, which is shown in Fig. 8.8.

8.13 Wave envelope infinite elements

Astley introduced a new type of finite element, in which the weighting function is the complex conjugate of the shape function.[19,20] The great simplification which this introduces is that the oscillatory function $\exp(\mathrm{i}kr)$ cancels after being multiplied by $\exp(-\mathrm{i}kr)$, and the remaining terms are all polynomials, which can be integrated using standard techniques, like Gauss–Legendre integration (see Volume 1, Chapter 9). This type of element was originally large (i.e. many wavelengths in extent), but not infinite. Figure 8.9 shows an example from acoustics, that of acoustical pressure in a hyperbolic duct. Good results were obtained despite using a relatively coarse mesh. Astley's shape function was of the form

$$N_i(r, \theta) \frac{r_i}{r} \mathrm{e}^{-\mathrm{i}k(r - r_i)} \tag{8.41}$$

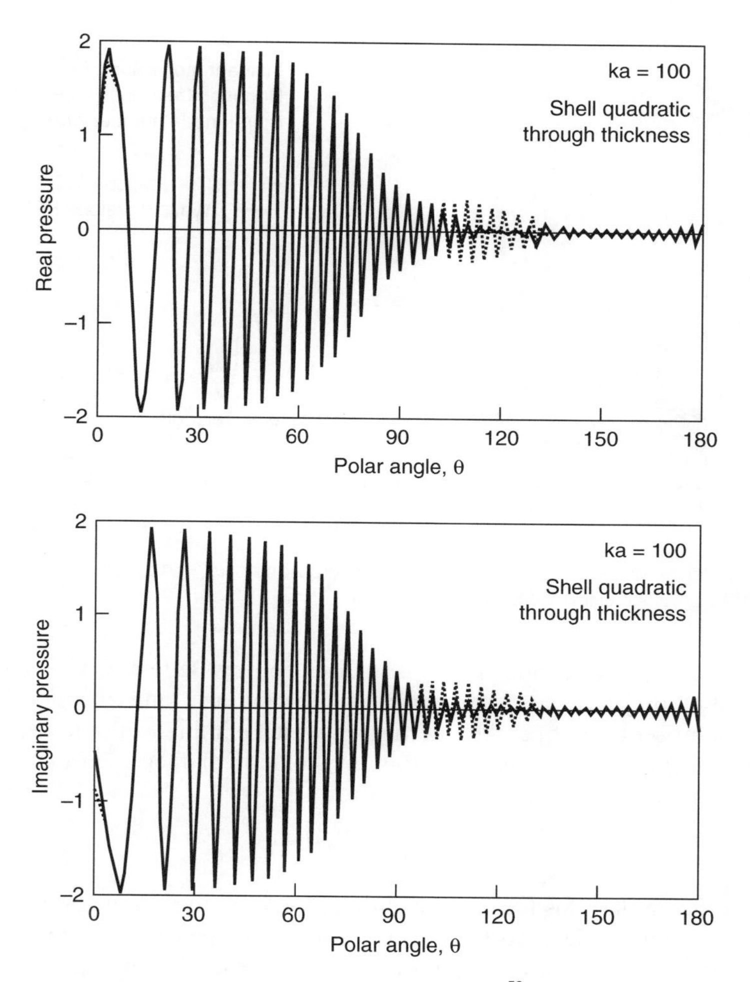

Fig. 8.8 Waves scattered by an elastic sphere for $ka = 100$, Burnett.[52]

where N_i is the standard shape function. The weighting function is thus

$$N_i(r,\theta)\frac{r_i}{r}\,\mathrm{e}^{\mathrm{i}k(r-r_i)} \tag{8.42}$$

Bettess[54] showed that for a one-dimensional synthetic wave-type equation the infinite wave envelope element recovers the exact solution. The element matrix is now hermitian rather than symmetric (though still complex), which necessitates a small alteration to the equation solver. (There are not usually any problems in changing standard profile or front solvers to deal with complex systems of equations.) Unfortunately the problem tackled by Bettess did not include the essential feature of physical waves, in two and three dimensions. Later workers applied the wave envelope concept to true wave problems. In this case it can be shown that if the weighting function is simply the complex conjugate of the shape function, terms

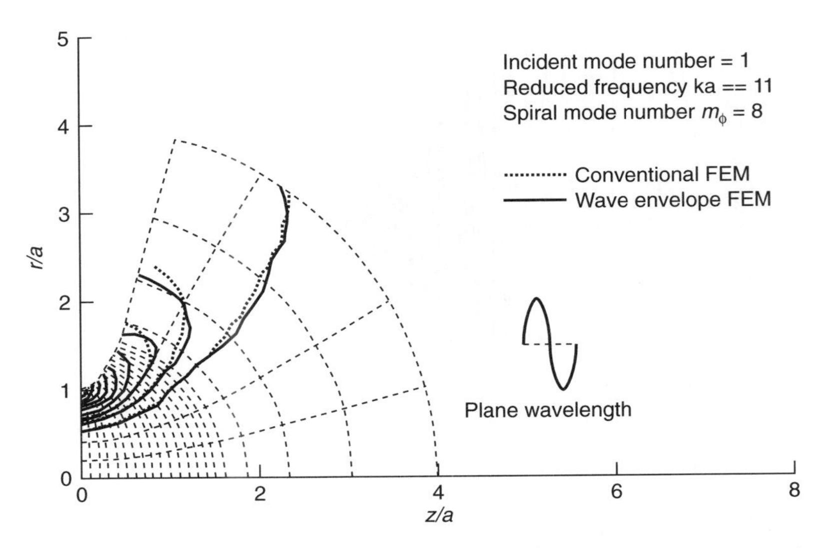

Fig. 8.9 Computed acoustical pressure contours for a hyperbolic duct ($\theta_0 = 70°$, $ka = 11$, $m_\phi = 8$). Conventional and wave envelope element solutions, Astley.[20]

arise on the boundary at infinity.† This is discussed by Bettess.[40] The terms can be evaluated, but they are not symmetrical (or hermitian), and therefore impose a change of solution technique. An alternative, which eliminates the terms at infinity, was proposed by Astley *et al.*[48] In this a 'geometrical factor' is included in the weighting function, which then takes the form

$$N_i(r, \theta)\left(\frac{r_i}{r}\right)^3 e^{ik(r - r_i)} \tag{8.43}$$

It has been shown that this form of weighting functions gives very good results. Such wave envelope infinite elements have been further developed by Coyette, Cremers and Fyfe.[49,50] These elements have incorporated a more general mapping than that in the original Zienkiewicz *et al.* mapped infinite wave element. Cremers and Fyfe allow the mapping to vary in the local ξ and η directions.

8.14 Accuracy of infinite elements

The use of a complex conjugate weighting in the wave envelope infinite elements means that the original variational statement, Eq. (8.22), must be changed to allow the use of the different weighting function. This gives rise to a number of issues relating to the nature of the weighted residual statement and the existence of various terms. These issues were touched on by Bettess,[40] but have been subsequently subjected to more detailed study. Gerdes and Demkowitz,[55,56] analysed the wave envelope elements, and subsequently the wave infinite elements.[57] Some of this work is restricted to

† Some writers, particularly mathematicians, prefer to call the usual wave infinite elements, unconjugated infinite elements, and the Astley type wave envelope infinite elements, conjugated infinite elements.

spherical scatterers. Other analysis is carried out by Shirron and Babuška,[58,59] who reveal a somewhat paradoxical result. The usual infinite elements give better results in the finite element mesh, but worse results in the infinite elements themselves. But the wave envelope elements give worse results in the finite elements, and better results in the far field. This result, which is ascribed to ill-conditioning, does seem to be counterintuitive. Astley[41] and Gerdes[42] have also surveyed current formulations and accuracies.

8.15 Transient problems

Recently Astley[60–63] has extended his wave envelope infinite elements, using the prolate and oblate spheroidal coordinates adopted by Burnett and Holford,[51–53] and has shown that they give accurate solutions to a range of periodic wave problems. With the geometric factor of Astley, which reduces the weighting function and eliminates the surface integrals at infinity, the stiffness, **K**, damping, **C**, and mass **M** matrices of the wave envelope infinite element become well defined and *frequency independent*, although unsymmetric. This makes it possible to apply such elements to unbounded transient wave problems. Figure 8.10 shows the transient response of a dipole.

More results from the application of infinite elements to transient problems are given by Cipolla and Butler,[64] who created a transient version of the Burnett infinite element. There appear to be more difficulties with such elements than with the wave envelope elements, and a consensus that the latter are better for transient problems seems to be emerging. Dampers and boundary integrals can also be used for transient problems. Space is not available to survey these fields, but the reader is directed, again, to Givoli[29] and Geers.[30] One set of interesting results was obtained using transient dampers by Thompson and Pinsky.[65]

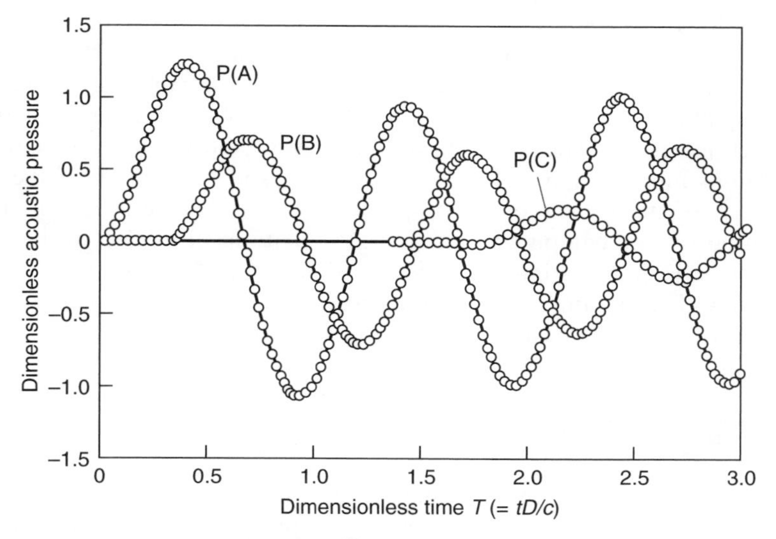

Fig. 8.10 Transient response of a dipole, Astley.[63]

8.16 Three-dimensional effects in surface waves

As has already been described, when the water is deep in comparison with the wavelength, the shallow-water theory is no longer adequate. For constant or slowly varying depth, Berkhoff's theory is applicable. Also the geometry of the problem may necessitate another approach. The flow in the body of water is completely determined by the conservation of mass, which in the case of incompressible flow reduces to Laplace's equation. The free surface condition is zero pressure. On using Bernoulli's equation and the kinematic condition, the free surface condition can be expressed, in terms of the velocity potential, ϕ, as

$$\frac{\partial^2 \phi}{\partial t^2} + g\frac{\partial \phi}{\partial t} + 2(\boldsymbol{\nabla}\phi)^{\mathrm{T}}\left[\boldsymbol{\nabla}\left(\frac{\partial \phi}{\partial t}\right)\right] + \tfrac{1}{2}(\boldsymbol{\nabla}\phi)^{\mathrm{T}}\boldsymbol{\nabla}[(\boldsymbol{\nabla}\phi)^{\mathrm{T}}\boldsymbol{\nabla}\phi] = 0 \tag{8.44}$$

where the velocities are $u_i = \partial\phi/\partial x_i$. This condition is applied on the free surface, whose position is unknown *a priori*. If only linear terms are retained, Eq. (8.44) becomes, for transient and periodic problems

$$\frac{\partial^2 \phi}{\partial t^2} + g\frac{\partial \phi}{\partial z} = 0 \qquad \text{or} \qquad \frac{\partial \phi}{\partial z} = \frac{\omega^2}{g}\phi \tag{8.45}$$

which is known as the *Cauchy–Poisson* free surface condition. It was derived in terms of pressure in Volume 1, Chapter 19 as Eq. (19.13). Three-dimensional finite elements can be used to solve such problems. The actual three-dimensional element is very simple, being a potential element of the type described in Volume 1 Chapter 7. The natural boundary condition is $\partial\phi/\partial n = 0$, where n is the outward normal, so to apply the free surface condition it is only necessary to add a surface integral to generate the ω^2/g term from the Cauchy–Poisson condition (see Eq. (19.13) of Volume 1). Two-dimensional elements in the far field can be linked to three-dimensional elements in the near field around the object of interest. Such models will predict velocity potentials, pressures throughout the fluid, and wave elevations. They can also be used to predict fluid–structure interaction. All the necessary equations are given in Volume 1 Chapter 19. More details of fluid–structure interactions of this type are given by Zienkiewicz and Bettess.[66] Essentially the fluid equations must be solved for incident waves, and for motion of the floating body in each of its degrees of freedom (usually six). The resulting fluid forces, masses, stiffnesses and damping are used in the equations of motion of the structure to determine its response. Figure 8.11 shows some results obtained by Hara *et al.*[67] using the WAVE program, for a floating breakwater. They obtained good agreement between the infinite elements and the methods of Sec. 8.9.

8.16.1 Large-amplitude water waves

There is no complete wave theory which deals with the case when η is not small in comparison with the other dimensions of the problem. Various special theories are invoked for different circumstances. We consider two of these, namely, large

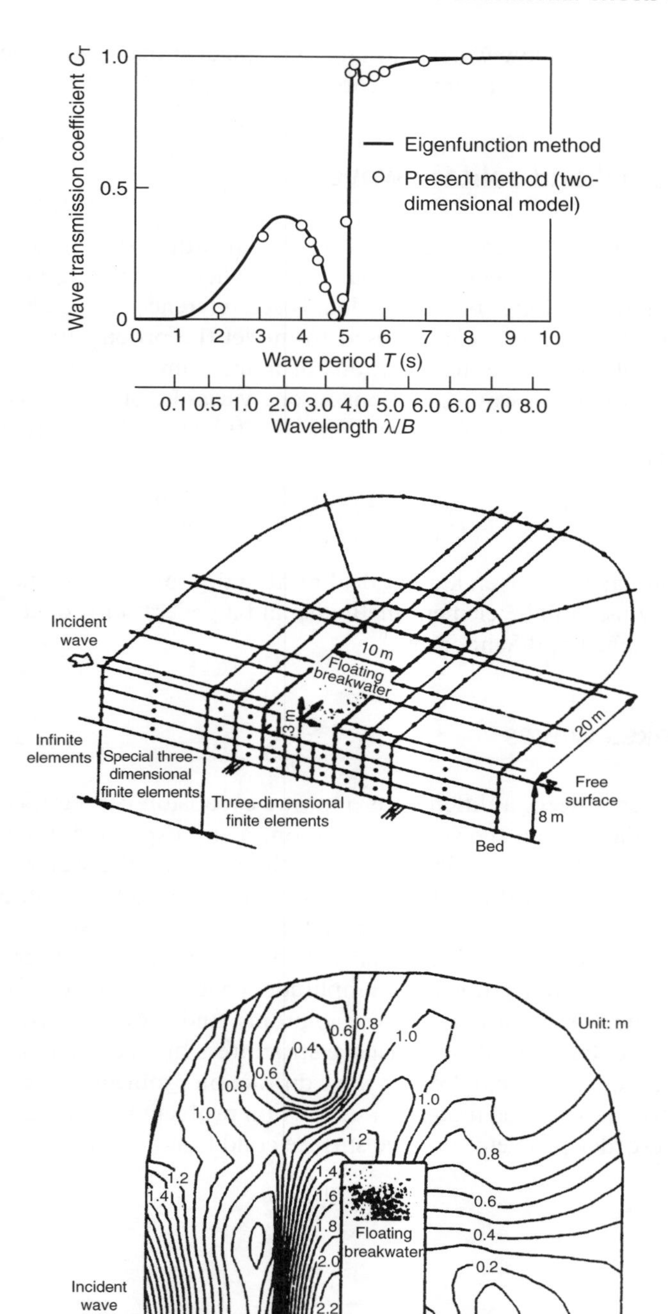

Fig. 8.11 Element mesh, contours of wave elevation and wave transmission coefficients for floating breakwater, Hara.[67]

wave elevations in shallow water and large wave elevations in intermediate to deep water. We have discussed a similar problem in Chapter 5.

8.16.2 Cnoidal and solitary waves

The equations modelled in Chapter 7 can deal with large-amplitude waves in shallow water. These are called cnoidal waves when periodic, and solitary waves when the period is infinite. For more details see references 1–4. The finite element methodology of Chapter 7, can be used to model the propagation of such waves. It is also possible to reduce the equations of momentum balance and mass conservation to corresponding wave equations in one variable, of which there are several different forms. One famous equation is the Korteweg–de Vries equation, which in physical variables is

$$\frac{\partial \eta}{\partial t} + \sqrt{gH}\left(1 + \frac{3\eta}{2h}\right)\frac{\partial \eta}{\partial x} + \frac{h^2}{6}\sqrt{gH}\frac{\partial^3 \eta}{\partial x^3} = 0 \tag{8.46}$$

This equation has been given a great deal of attention by mathematicians. It can be solved directly using finite element methods, and a general introduction to this field is given by Mitchell and Schoombie.[68]

8.16.3 Stokes waves

When the water is deep, a different asymptotic expansion can be used in which the velocity potential, ϕ, and the surface elevation, η, are expanded in terms of a small parameter, ε, which can be identified with the slope of the water surface. When these expressions are substituted into the free surface condition, and terms with the same order in ε are collected, a series of free surface conditions is obtained. The equations were solved by Stokes initially, and then by other workers, to very high orders, to give solutions for large-amplitude progressive waves in deep water. There is an extensive literature on these solutions, and they are used in the offshore industry for calculating loads on offshore structures. In recent years, attempts have been made to model the second-order wave diffraction problem, using finite elements, and similar techniques. The first-order diffraction problem is as described in Sec. 8.9. In the second-order problem, the free surface condition now involves the first-order potential.

First order:

$$\frac{\partial \phi^{(1)}}{\partial z} - \frac{\omega^2}{g}\phi^{(1)} = 0 \tag{8.47}$$

Second order:

$$\frac{\partial \phi^{(2)}}{\partial z} - \frac{\omega^2}{g}\phi^{(2)} = \alpha_D^{(2)} \tag{8.48}$$

$$\alpha_D^{(2)} = \alpha_{DI}^{(2)} + \alpha_{DD}^{(2)} \qquad \text{and} \qquad \nu = \frac{\omega^2}{g} \tag{8.49}$$

$$\alpha_{DI}^{(2)} = -\mathrm{i}\frac{\omega}{2g}\phi_D^{(1)}\left(\frac{\partial^2\phi_I^{(1)}}{\partial z^2} - \nu\frac{\partial\phi_I^{(1)}}{\partial z}\right) - \mathrm{i}\frac{\omega}{2g}\phi_I^{(1)}\left(\frac{\partial^2\phi_D^{(1)}}{\partial z^2} - \nu\frac{\partial\phi_D^{(1)}}{\partial z}\right) + \mathrm{i}\frac{2\omega}{g}\nabla\phi_I^{(1)}\nabla\phi_D^{(1)} \tag{8.50}$$

$$\alpha_{DD}^{(2)} = -\mathrm{i}\frac{\omega}{2g}\phi_D^{(1)}\left(\frac{\partial^2\phi_D^{(1)}}{\partial z^2} - \nu\frac{\partial\phi_D^{(1)}}{\partial z}\right) + \mathrm{i}\frac{2\omega}{g}(\nabla\phi_D^{(1)})^2 \tag{8.51}$$

The second-order boundary condition can be thought of as identical to the first-order problem, but with a specified pressure applied over the entire free surface, of value α. Now there is no *a priori* reason why such a pressure distribution should give rise to outgoing waves as in the first-order problem, and so the usual radiation condition is not applicable. The conventional procedure is to split the second-order wave into two parts, one the 'locked' wave, in phase with the first-order wave, and the other the 'free' wave, which is like the first-order wave but at twice the frequency, and with an appropriate wavenumber obtained from the dispersion relation. For further details of the theory, see Clark *et al.*[69] Figure 8.12 shows results for the second-order wave elevation around a circular cylinder, obtained by Clark *et al.* Although not shown, good agreement has been obtained with predictions made by boundary

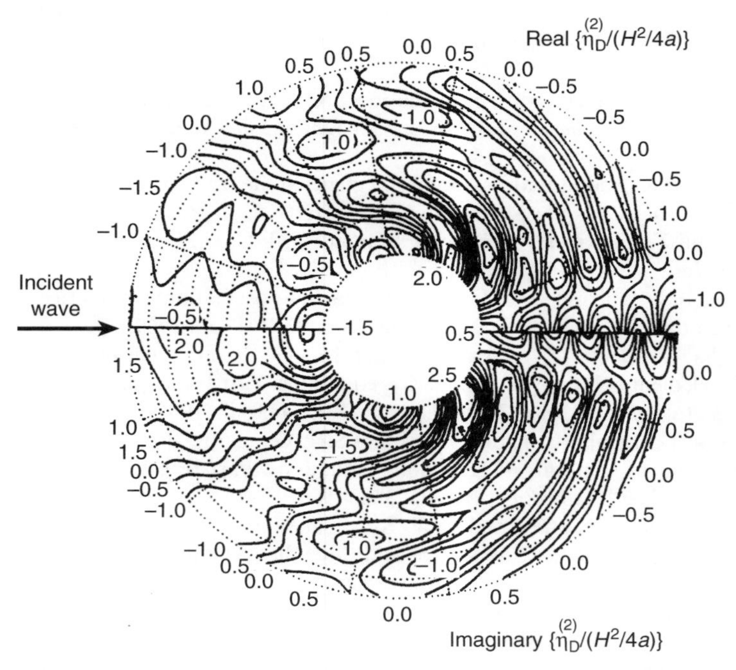

Fig. 8.12 Second order wave elevations around cylinder – real and imaginary parts Clark *et al.*[69]

integrals. Preliminary results, for wave forces only, have also been produced by Lau *et al.*[70] A much finer finite element mesh is needed to resolve the details of the waves at second order. The second-order wave forces can be very significant for realistic values of the wave parameters (those encountered in the North Sea for example). The first-order problem is solved first and the first-order potential is used to generate the forcing terms in Eqs (8.50) and (8.51). These values have to be very accurate. In principle the method could be extended to third and higher orders, but in practice the difficulties multiply, and in particular the dispersion relation changes and the waves become unstable.[4]

References

1. H. Lamb. *Hydrodynamics*, Cambridge University Press, Sixth edition, 1932.
2. G.B. Whitham. *Linear and Nonlinear Waves*, John Wiley, New York, 1974.
3. C.C. Mei. *The Applied Dynamics of Ocean Surface Waves*, Wiley, New York, 1983.
4. M.J. Lighthill. *Waves in Fluids*, Cambridge University Press, 1978.
5. G.M.L. Gladwell. A variational model of damped acousto-structural vibration. *Journal of Sound and Vibration*, **4**, 172–86, 1965.
6. O.C. Zienkiewicz and R.E. Newton. Coupled vibrations of a structure submerged in a compressible fluid. *Proc. International Symposium on Finite Element Techniques*, Stuttgart, pp. 360–78, 1–15 May, 1969.
7. A. Craggs. The transient response of a coupled plate acoustic system using plate and acoustic finite elements. *Journal of Sound and Vibration*, **15**, 509–28, 1971.
8. R.J. Astley. Finite elements in acoustics. *Sound and Silence: Setting the balance, Proc. Internoise 98*, Christchurch, New Zealand, Vol. 1, pp. 3–17, 16–18 November 1998.
9. P.M. Morse and H. Feshbach. *Methods of Theoretical Physics*, Volumes 1 and 2, McGraw-Hill, New York, 1953.
10. J.C.W. Berkhoff. Linear wave propagation problems and the finite element method, in *Finite Elements in Fluids*, 1, Eds. R.H. Gallagher *et al.*, Wiley, Chichester, pp. 251–80, 1975.
11. O.C. Zienkiewicz and P. Bettess. Infinite elements in the study of fluid structure interaction problems. *Proc. 2nd Int. Symp. on Comp. Methods Appl. Sci., Versailles*, 1975, also published in *Lecture Notes in Physics*, Volume 58, Eds. J. Ehlers *et al.*, Springer-Verlag, Berlin, 1976.
12. C. Taylor, B.S. Patil and O.C. Zienkiewicz. Harbour oscillation: a numerical treatment for undamped natural modes. *Proc. Inst. Civ. Eng.*, **43**, 141–56, 1969.
13. I. Babuška, F. Ihlenburg, T. Stroubolis and S.K. Gangaraj. *A posteriori* error estimation for finite element solutions of Helmholtz' equations. Part I: the quality of local indicators and estimators. *Int. J. Num. Meth. Eng.*, **40**(18), 3443–62, 1997.
14. I. Babuška, F. Ihlenburg, T. Stroubolis and S.K. Gangaraj. *A posteriori* error estimation for finite element solutions of Helmholtz' equations. Part II: estimation of the pollution error. *Int. J. Num. Meth. Eng.*, **40**(21), 3883–900, 1997.
15. O.C. Zienkiewicz, P. Bettess and D.W. Kelly. The finite element method for determining fluid loadings on rigid structures: two- and three-dimensional formulations, Chapter 4 of *Numerical Methods in Offshore Engineering*, Eds., O.C. Zienkiewicz, R.W. Lewis and K.G. Stagg, John Wiley, 1978.
16. K. Morgan, O. Hassan and J. Peraire. A time domain unstructured grid approach to the simulation of electromagnetic scattering in piecewise homogeneous media. *Comp. Methods Appl. Mech. Eng.*, **152**, 157–74, 1996.

17. K. Morgan, P.J. Brookes, O. Hassan and N.P. Weatherill. Parallel processing for the simulation of problems involving scattering of electromagnetic waves. *Comp. Methods Appl. Mech. Eng.*, **152**, 157–74, 1998.
18. P. Bettess and O.C. Zienkiewicz. Diffraction and refraction of surface waves using finite and infinite elements. *Int. J. Num. Meth. Eng.*, **11**, 1271–90, 1977.
19. R.J. Astley and W. Eversman. A note on the utility of a wave envelope approach in finite element duct transmission studies. *Journal of Sound and Vibration*, **76**, 595–601, 1981.
20. R.J. Astley. Wave envelope and infinite elements for acoustical radiation. *Int. J. Num. Meth. Fluids*, **3**, 507–26, 1983.
21. E. Chadwick, P. Bettess and O. Laghrouche. Diffraction of short waves modelled using new mapped wave envelope finite and infinite elements. *Int. J. Num. Meth. Eng.*, **45**, 335–54, 1999.
22. J.M. Melenk and I. Babuška. The partition of unity finite element method. Basic theory and applications. *Comp. Meths. Appl. Mech. Eng.*, **139**, 289–314, 1996.
23. J.M. Melenk and I. Babuška. The partition of unity finite element method. *Int. J. Num. Meth. Eng.*, **40**, 727–58, 1997.
24. O. Laghrouche and P. Bettess. Short wave modelling using special finite elements – towards an adaptive approach, presented at *10th Conference on the Mathematics of Finite Elements and Applications, Proceedings of Mafelap '99*, to appear, 2000.
25. O. Laghrouche and P. Bettess. Short wave modelling using special finite elements. *4th International Conference on Theoretical and Computational Acoustics*, Special Issue of *Journal of Computational Acoustics*, to appear, 2000.
26. A. Bayliss and E. Turkel. Radiation boundary conditions for wave-like equations, *ICASE Report No. 79-26*, 1979.
27. A. Bayliss, M. Gunzberger and E. Turkel. Boundary conditions for the numerical solution of elliptic equations in exterior regions. *ICASE Report No. 80–1*, 1980.
28. J.A. Bettess and P. Bettess. A new mapped infinite wave element for general wave diffraction problems and its validation on the ellipse diffraction problem. *Comp. Meth. Appl. Mech. Eng.*, **164**, 17–48, 1998.
29. D. Givoli. *Numerical Methods for Problems in Infinite Domains*, Elsevier, Amsterdam, 1992.
30. T.L. Geers. *Computational Methods for Unbounded Domains*, Kluwer Academic Publishers, Dordrecht, 1998.
31. K. Bando, P. Bettess and C. Emson. The effectiveness of dampers for the analysis of exterior wave diffraction by cylinders and ellipsoids. *Int. J. Num. Meth. Fluids*, **4**, 599–617, 1984.
32. B. Engquist and A. Majda. Radiation boundary conditions for the numerical simulation of waves. *Math. Comp.*, **31**, 629–52, 1977.
33. O.C. Zienkiewicz, D.W. Kelly and P. Bettess. The coupling of the finite element method and boundary solution procedures. *Int. J. Num. Meth. Eng.*, **11**(1), 355–75, 1977.
34. O.C. Zienkiewicz, D.W. Kelly and P. Bettess. Marriage à la mode – the best of both worlds (finite elements and boundary integrals), Chapter 5 of *Energy Methods in Finite Element Analysis*, Wiley, 1978.
35. J.C.W. Berkhoff. Computation of combined refraction–diffraction. *Proc. 13th Int. Conf. on Coastal Engineering*, Vancouver, July 10–14, 1972.
36. H.S. Chen and C.C. Mei. Oscillations and wave forces in an offshore harbor. *Parsons Laboratory, Massachusetts Institute of Technology, Report 190*, 1974.
37. H.S. Chen and C.C. Mei. Oscillations and wave forces in a man-made harbor in the open sea. *Proc. 10th. Symp. Naval Hydrodynamics, Office of Naval Research*, pp. 573–94, 1974.
38. J.R. Houston. Long Beach Harbor: numerical analysis of harbor oscillations. *US Army Engineering Waterways Experimental Station, Vicksburg, MS. Report 1, Misc. Paper H-76-20*, 1976.

39. R.J. Astley. FE mode matching schemes for the exterior Helmholtz problems and their relationship to the FE-DtN approach. *Comm. Num. Meth. Eng.*, **12**, 257–67, 1996.
40. P. Bettess. *Infinite Elements*, Penshaw Press, Sunderland, 1992.
41. R.J. Astley. Infinite elements for wave problems: a review of current formulations and an assessment of accuracy. *Int. J. Num. Meth. Eng.*, to appear, 1999.
42. K. Gerdes. Infinite elements for wave problems. *J. Comput. Acoustics*, **8**, 43–62, 2000.
43. O.C. Zienkiewicz, C. Emson and P. Bettess. A novel boundary infinite element. *Int. J. Num. Meth. Eng.*, **19**, 393–404, 1983.
44. O.C. Zienkiewicz, P. Bettess, T.C. Chiam and C. Emson. Numerical methods for unbounded field problems and a new infinite element formulation. *ASME, AMD*, **46**, 115–48, New York, 1981.
45. P. Bettess, C. Emson and T.C. Chiam. A new mapped infinite element for exterior wave problems, Chapter 17 of *Numerical Methods in Coupled Systems*, Eds. R.W. Lewis, P. Bettess and E. Hinton, John Wiley, Chichester, 1984.
46. O.C. Zienkiewicz, K. Bando, P. Bettess, C. Emson and T.C. Chiam. Mapped infinite elements for exterior wave problems. *Int. J. Num. Meth. Eng.*, **21**, 1229–51, 1985.
47. R.J. Astley, P. Bettess and P.J. Clark. Letter to the editor concerning Ref. 128. *Int. J. Num. Meth. Eng.*, **32**(1), 207–9, 1991.
48. R.J. Astley, G.J. Macaulay and J.P. Coyette. Mapped wave envelope elements for acoustical radiation and scattering. *J. Sound Vibration*, **170**(1), 97–118, 1994.
49. L. Cremers, K.R. Fyfe and J.P. Coyette. A variable order infinite acoustic wave envelope element. *J. Sound Vibration*, **171**(4), 483–508, 1994.
50. L. Cremers and K.R. Fyfe. On the use of variable order infinite wave envelope elements for acoustic radiation and scattering. *J. Acoust. Soc. Amer.*, **97**(4), 2028–40, 1995.
51. D.S. Burnett. A three-dimensional acoustic infinite element based on a prolate spheroidal multipole expansion. *J. Acoust. Soc. Amer.*, **95**(5), 2798–816, 1994.
52. D.S. Burnett and R.L. Holford. Prolate and oblate spheroidal acoustic infinite elements. *Comp. Meth. Appl. Mech. Eng.*, **158**, 117–41, 1998.
53. D.S. Burnett and R.L. Holford. An ellipsoidal acoustic infinite element. *Comp. Meth. Appl. Mech. Eng.*, **164**, 49–76, 1998.
54. P. Bettess. A simple wave envelope element example. *Comms. Applied Num. Meth.*, **3**, 77–80, 1987.
55. K. Gerdes and L. Demkowitz. Solution of 3D Laplace and Helmholtz equation in exterior domains of arbitrary shape, using HP-finite-infinite elements. *Comp. Meth. Appl. Mech. Eng.*, **137**, 239–73, 1996.
56. L. Demkowitz and K. Gerdes. Convergence of the infinite element methods for the Helmholtz equation in separable domains. *Numerische Mathematik*, **79**, 11–42, 1998.
57. L. Demkowitz and K. Gerdes. The conjugated versus the unconjugated infinite element method for the Helmholtz equation in exterior domains. *Comp. Meth. Appl. Mech. Eng.*, **152**, 125–45, 1998.
58. J.J. Shirron. Solution of exterior Helmholtz problems using finite and infinite elements, Ph.D. Thesis, University of Maryland, 1995.
59. J.J. Shirron and I. Babuška. A comparison of approximate boundary conditions and infinite element methods for exterior Helmholtz problems. *Comp. Meth. Appl. Mech. Eng.*, **164**, 121–39, 1998.
60. R.J. Astley. Transient wave envelope elements for wave problems. *J. Sound Vibration*, **192**(1), 245–61, 1996.
61. R.J. Astley, G.J. Macaulay, J.P. Coyette and L. Cremers. Three dimensional wave-envelope elements of variable order for acoustic radiation and scattering. Part 1. Formulation in the frequency domain. *J. Acoust. Soc. Amer.*, **103**(1), 49–63, 1998.

62. R.J. Astley, G.J. Macaulay, J.P. Coyette and L. Cremers. Three dimensional wave-envelope elements of variable order for acoustic radiation and scattering. Part 2. Formulation in the time domain. *J. Acoust. Soc. Amer.*, **103**(1), 64–72, 1998.
63. R.J. Astley. Mapped spheroidal wave-envelope elements for unbounded wave problems. *Int. J. Num. Meth. Eng.*, **41**, 1235–54, 1998.
64. J.L. Cipolla and M.J. Butler. Infinite elements in the time domain using a prolate spheroidal multipole expansion. *Int. J. Num. Meth. Eng.*, **43**(5), 889–908, 1998.
65. L.L. Thompson and P.M. Pinsky. A space-time finite element method for the exterior structural acoustics problem: Time dependent radiation boundary conditions in two space dimensions. *Int. J. Num. Meth. Eng.*, **39**, 1635–57, 1996.
66. O.C. Zienkiewicz and P. Bettess. Fluid–structure interaction and wave forces. An introduction to numerical treatment. *Int. J. Num. Meth. Eng.*, **13**, 1–16, 1978.
67. H. Hara, K. Kanehiro, H. Ashida, T. Sugawara and T. Yoshimura. Numerical simulation system for wave diffraction and response of offshore structures. *Mitsui Engineering and Shipbuilding Co., Technical Bulletin, TB 83–07*, October 1983.
68. A.R. Mitchell and S.W. Schoombie. Finite element studies of solitons, Chapter 16 of *Numerical Methods in Coupled Systems*, Eds. R.W. Lewis, P. Bettess and E. Hinton, John Wiley, Chichester, pp. 465–88, 1984.
69. P.J. Clark, P. Bettess, M.J. Downie and G.E. Hearn. Second order wave diffraction and wave forces on offshore structures, using finite elements. *Int. J. Num. Meth. Fluids*, **12**, 343–67, 1991.
70. L. Lau, K.K. Wong and Z. Tam. Nonlinear wave loads on large body by time-space finite element method. *Computers in Engineering, Proc. of the Int. Conf. on Computers in Engineering* **3**, Publ. by ASME, New York, pp. 331–37, 1987.

9

Computer implementation of the CBS algorithm

P. Nithiarasu*

9.1 Introduction

In this chapter we shall consider some essential steps in the computer implementation of the CBS algorithm on structured or unstructured finite element grids. Only linear triangular elements will be used and the notes given here are intended for a two-dimensional version of the program. The sample program listing and user manual along with several solved problems are available to down load from the publisher's web site http://www.bh.com/companions/fem free of charge.

The program discussed can be used to solve the following different categories of fluid mechanics problems:

1. Compressible viscous and inviscid flow problems
2. Incompressible viscous and inviscid flows
3. Incompressible flows with heat transfer
4. Porous media flows
5. Shallow-water problems.

With further simple modifications, many other problems such as turbulent flows, solidification, mass transfer, free surfaces, etc. can be solved. The procedures presented here are largely based on the computer implementation discussed in Chapter 20, Volume 1 of this book. Many programming aspects will not be discussed here in detail and the reader is referred back to Chapter 20, Volume 1. Here it is assumed that the reader is familiar with FORTRAN[1,2] and finite element procedures discussed in this volume as well as in Volume 1.[3]

We call the present program CBSflow since it is based on the CBS algorithm discussed in Chapter 3 of this volume. We prefer to keep the compressible and incompressible flow codes separate to avoid any confusion. However an experienced programmer can incorporate both parts into a single code without much memory loss. Each program listing is accompanied by some model problems which helps the reader to validate the codes. In addition to the model inputs to programs, a complete user manual is available to users explaining every part of the program in detail. Any error reported by readers will be corrected and the program will be continuously updated by the authors.

* Research Fellow, Department of Civil Engineering, University of Wales, Swansea, UK.

The modules are constructed essentially as in Chapter 20, Volume 1 starting with (1) the data input module with preprocessing and continuing with (2) the solution module and (3) the output module. However, unlike the generalized program of Chapter 20, Volume 1, the program CBSflow only contains the listing for solving transient Navier–Stokes (or Euler–Stokes) equations iteratively. Here there are many possibilities such as fully explicit forms, semi-implicit forms, quasi-implicit forms and fully implicit forms as discussed in Chapter 3 of this volume. We concentrate mainly on the first two forms which require small memory and simple solution procedures compared to other forms.

In both the compressible and incompressible flow codes, only non-dimensional equations are used. The reader is referred to the appropriate chapters of this volume (Chapters 3, 4 and 5) for different non-dimensional parameters.

In Sec. 9.2 we shall describe the essential features of data input to the program. Here either structured or unstructured meshes can be used to divide the problem domain into finite elements. Section 9.3 explains how the steps of the CBS algorithm are implemented. In that section, we briefly remark on the options available for shock capturing, various methods of time stepping and different procedures for equation solving. In Sec. 9.4, the output generated by the program and postprocessing procedures are considered. In the last section (Sec. 9.5) we shall consider the possibility of further extension of CBSflow to other problems such as mass transfer, turbulent flow, etc.

9.2 The data input module

This part of the program is the starting point of the calculation where the input data for the solution module are prepared. Here an appropriate input file is opened and the data are read from it. Unlike in Chapter 20, Volume 1, we have no mesh generator coupled with CBSflow. However an advancing front unstructured mesh generator and some structured mesh generators are provided separately. By suitable coupling, the reader can implement various adaptive procedures as discussed in Chapters 4 and 5. Either structured or unstructured mesh data can be given as input to the program. The general program structure and many more details can be found in Chapter 20, Volume 1.

9.2.1 Mesh data – nodal coordinates and connectivity

Once the nodal coordinates and connectivity of a finite element mesh are available from a mesh generator, they are allotted to appropriate arrays (for a detailed description on the mesh, numbering etc., see Chapter 20, Volume 1). Essentially the same arrays as described in Chapter 20, Volume 1 are used here. The coordinates are allotted to $X(i,j)$ with i defining the appropriate cartesian coordinates $x_1(i=1)$ and $x_2(i=2)$ and j defining the global node number. Similarly the connectivity is allotted to an array $IX(k,l)$. Here k is the local node number and l is the global element number. It should be noted that the material code normally used in heat conduction and stress analysis is not necessary.

Table 9.1 Non-dimensional parameters

Non-dimensional number	Symbol	Flow types
Conductivity ratio	k^*	Porous media flows
Darcy number	Da	Porous media flows
Mach number	M	Compressible flows
Prandtl number	Pr	Compressible, incompressible, thermal and porous media flows
Porosity	ε	Porous media flows
Rayleigh number	Ra	Natural convective flows
Reynolds number	Re	Compressible, incompressible, thermal and porous media flows
Viscosity ratio	ν	Porous media flows

If the structured meshes and banded solution are preferred by the user, a flag activated by the user calculates the half-bandwidth of the mesh and supplies it to the solution module. Alternatively, a diagonally preconditioned conjugate gradient solver can be used with an appropriate flag. These solvers are necessary only when the semi-implicit form of solution is used.

9.2.2 Boundary data

In general, the procedure discussed in Chapter 20, Volume 1 uses the boundary nodes to prescribe boundary conditions. However, in CBSflow we mostly use the edges to store the information on boundary conditions. Some situations require boundary nodes (e.g. pressure specified in a single node) and in such cases corresponding node numbers are supplied to the solution module.

9.2.3 Other necessary data and flags

In addition to the mesh data and boundary information, the user needs to input a few more parameters used in flow calculations. For example, compressible flow computations need the values of non-dimensional parameters such as the Mach number, Reynolds number, Prandtl number, etc. Here the reader may consult the non-dimensional equations and parameters discussed in Sec. 3.1, Chapter 3, and in Chapter 5, of this volume. The necessary parameters for different problems are listed in Table 9.1 for completeness.

Several flags for boundary conditions, shock capture, etc. need to be given as inputs. For a complete list of such flags, the reader is referred to the user manual and program listing at the publisher's web page.

9.2.4 Preliminary subroutines and checks

A few preliminary subroutines are called before the start of the time iteration loop. Establishing the surface normals, element area calculation (for direct integration),

```
            SUBROUTINE GETNRW(MXPOI,MBC,NPOIN,NBS,ISIDE,IFLAG,
     &                        COSX,COSY,ALEN,IWPOIN,WNOR,NWALL)

      IMPLICIT    NONE

      INTEGER     I,IB,IB2,IN,IW,J,JJ,MBC,MXPOI,NBS,NN,NPOIN,NWALL

      INTEGER     IFLAG(MXPOI), ISIDE(4,MBC), IWPOIN(3,MBC)

      REAL*8      ACH,ANOR,ANX1,ANY1
      REAL*8      ALEN(MBC),COSX(MBC),COSY(MBC) WNOR(2,MBC)

      DO I = 1,NPOIN
        IFLAG (I) = 0
      END DO ! I
      DO I = 1, NBS
        DO J = 1,3
          IWPOIN(J,I) = 0
        END DO ! J
      END DO ! I

      NWALL = 0

      DO IN = 1,2
        DO I = 1, NBS ! boundary sides.
c
c     flags on the wall points
c
          IF(ISIDE(4,I).EQ.2)THEN ! flag 2 for solid walls.
            NN = ISIDE(IN,I)
            JJ = IFLAG(NN)
            IF(JJ.EQ.0)THEN
              NWALL           = NWALL + 1
              IWPOIN(I,NWALL) = NN
              IWPOIN(2,NWALL) = I
              IFLAG(NN)       = NWALL
            ELSE
              IWPOIN(3,JJ)    = I
            ENDIF
          ENDIF
        END DO ! I
      END DO ! IN
c
      DO IW = 1, NWALL
        IB   = IWPOIN(2,IW)
        IB2  = IWPOIN(3,IW)
        ANX1 = ALEN(IB)*COSX(IB)
```

```
ANY1 = ALEN(IB)*COSY(IB)
ACH  = 0.0D00
IF(IB2.NE.0)THEN
  ANX1 = ANX1 + ALEN(IB2)*COSX(IB2)
  ANY1 = ANY1 + ALEN(IB2)*COSY(IB2)
  ACH  = COSX(IB)*COSX(IB2) + COSY(IB)*COSY(IB2)
ENDIF
ANOR        = DSQRT(ANX1*ANX1 + ANY1*ANY1)
ANX1        = ANX1/ANOR
ANY1        = ANY1/ANOR
WNOR(1,IW)  = ANX1
WNOR(2,IW)  = ANY1
IF(ACH.LT.-0.2) THEN
  WNOR(1,IW) = 0.0D00
  WNOR(2,IW) = 0.0D00
  WRITE(*,*)IWPOIN(1,IW),' is trailing edge' ! e.g. aerofoil.
ENDIF
END DO ! IW
END
```

Fig. 9.1 Subroutine calculating surface normals on the walls.

mass matrix calculation and lumping and some allocation subroutines are necessary before starting the time loop. The routine for establishing the surface normals is shown in Fig. 9.1. On sharp, narrow corners as at the trailing edge of an aerofoil, the boundary contributions are made zero by assigning a zero value for the surface normal as shown.

9.3 Solution module

Figure 9.2 shows the general flow diagram of CBSflow. As seen, the data from the input module are passed to the time loop and here several subprograms are used to solve the steps of the CBS algorithm. It should be noted that the semi-implicit form is used here only for incompressible flows and at the second step we only calculate pressure, as the density variation is here assumed negligible.

9.3.1 Time loop

The time iteration is carried out over the steps of the CBS algorithm and over many other subroutines such as the local time step and shock capture calculations. As mentioned in the flow chart, the energy can be calculated after the velocity correction. However, for a fully explicit form of solution, the energy equation can be solved in step 1 along with the intermediate momentum variable. Further details on different steps are given in Sec. 9.3.4 and the reader can refer to the theory discussed in Chapter 3 of this volume for a comprehensive review of the CBS algorithm.

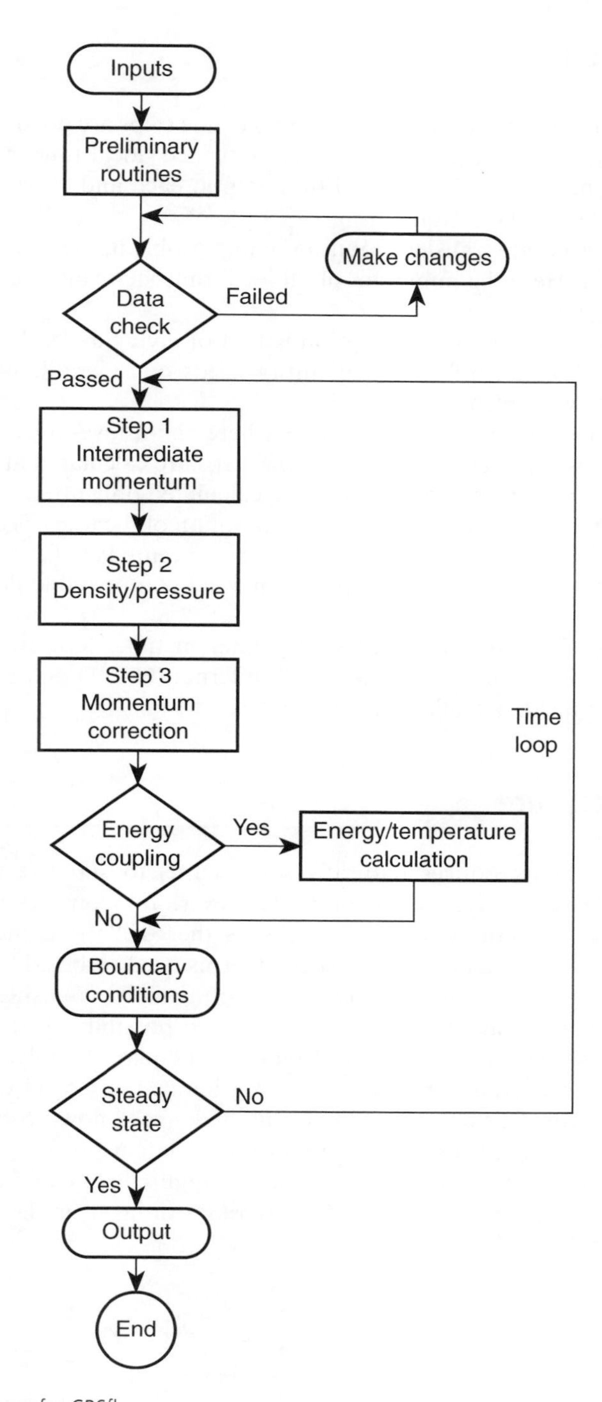

Fig. 9.2 Flow diagram for CBSflow.

9.3.2 Time step

In general, three different ways of establishing the time steps are possible. In problems where only the steady state is of importance, so-called 'local time stepping' is used (see Sec. 3.3.4, Chapter 3). Here a local time step at each and every nodal points is calculated and used in the computation.

When we seek accurate transient solution of any problem, the so-called 'minimum step' value is used. Here the minimum of all local time step values is calculated and used in the computation.

Another and less frequently used option is that of giving a 'fixed' user-prescribed time step value. Selection of such a quantity needs considerable experience from solving several flow problems.

The times loop starts with a subroutine where the above-mentioned time step options are available. In general the local time steps are calculated at every iteration for the initial few time steps and then they are calculated only after a certain number of iterations as prescribed by the user. If the last option of the user-specified fixed time step is used, the local time steps are not calculated. Figure 9.3 shows the subroutine used for calculating the local time steps for inviscid compressible flows with linear triangular elements.

As indicated in Sec. 4.3.3, Chapter 4, two different time steps are often useful in getting better stabilization procedures.[4] Such internal (`DELTI`) and external (`DELTP`) time stepping options are available in the routine of Fig. 9.3.

9.3.3 Shock capture

The CBS algorithm introduces naturally some terms to stabilize the oscillations generated by the convective acceleration. However, for compressible high-speed flows, these terms are not sufficient to suppress the oscillations in the vicinity of shocks and some additional artificial viscosity terms need to be added (see Sec. 6.5, Chapter 6). We have given two different forms of artificial viscosities based on the second derivative of pressure in the program. Another possibility is to use anisotropic shock capturing based on the residual of individual equations solved. However we have not used the second alternative in the program as the second derivative based procedures give quite satisfactory results for all high-speed flow problems.

In the first method implemented, we need to calculate a pressure switch (see Eq. (6.16), Chapter 6) from the nodal pressure values. Figure 9.4 gives a typical example of triangular elements inside and on the boundaries. For inside nodes (Fig. 9.4(a)) we calculate the nodal switch as

$$S_1 = \frac{|4p_1 - p_2 - p_3 - p_4 - p_5|}{|p_1 - p_2| + |p_1 - p_3| + |p_1 - p_4| + |p_1 - p_5|} \tag{9.1}$$

and for the boundary node (Fig. 9.4(b)) we calculate

$$S_1 = \frac{|5p_1 - 2p_2 - p_3 - 2p_4|}{2|p_1 - p_2| + |p_1 - p_3| + 2|p_1 - p_4|} \tag{9.2}$$

```
      SUBROUTINE TIMSTP(MXPOI,MXELE,NELEM,NPOIN,IALOT, IX, SFACT,
     &                  DTFIX,UNKNO,DELTP,DELTI,SONIC,PRES,GAMMA,
     &                  GEOME, X, NMAX,MAXCON,MODEL, NODEL)

c calculates the critical local time steps at nodes.
c calculates internal and external time steps.
c
      IMPLICIT  NONE

      IMPLICIT  MPOI
      PARAMETER(MPOI=9000)

      INTEGER   I,IALOT,IE,IP,IP1,IP2,IP3,MODEL,MXELE,MXPOI
      INTEGER   NELEM,NODEL,NPOIN

      INTEGER   IX(MODEL,MXELE),MAXCON(20,MIXPOI),NMAX(MXPOI)
c
      REAL*8    ALEN,ANX,ANY,CMAX, DTFIX, DTP, GAMMA, SFACT,TSTI
      REAL*8    TSTP,U,U1,U2,U3,V,V1,V2,V3,VN1,VN2,VN3,VELN,VSUM
c
      REAL*8    DELTI(MXPOI),DELTP(MXPOI), GEOME(7,MXELE)
      REAL*8    PRES(MXPOI), SONIC(MXPOI), UNKNO(4,MXPOI)
      REAL*8    X(2,MXPOI)
c
      REAL*8 PRS(MPOI),RHO(MPOI),VMAG(MPOI),VNORM(MPOI) ! local arrays

      IF(IALOT.EQ.-1)THEN
        CALL TIMFIL(MXPOI,DELTP,NPOIN,DTFIX)
        CALL TIMFIL(MXPOI,DELTI,NPOIN,DTFIX)
        RETURN
      ENDIF
c
c smoothing the variables
c
      DO I = 1, NPOIN
        VNORM(I)    = 0.00D+00
        RHO(I)      = 0.00D+00
        PRS(I)      = 0.00D+00
        U           = UNKNO(2,I)/UNKNO(1,I)
        V           = UNKNO(3,I)/UNKNO(1,I)
        VMAG(I)     = DSQRT(U**2+V**2)
        DO IP = 1,NMAX(I)
          IP1       = MAXCON(IP,I)
          VNORM(I) = VNORM(I) + VMAG(IP1)
          PRS(I)    = PRS(I) + PRES(IP1)
          RHO(I)    = RHO(I) + UNKNO(1,IP1)
        END DO ! IP
```

Fig. 9.3 Subroutine for time step calculation.

```
          VNORM(I) = VNORM(I)/FLOAT(NMAX(I))
          PRS(I)   = PRS(I)/FLOAT(NMAX(I))
          RHO(I)   = RHO(I)/FLOAT(NMAX(I))
          SONIC(I) = DSQRT(GAMMA*PRS(I)/RHO(I))
      END DO ! I
      DO IP = 1,NPOIN
        DELTP(IP) = 1.0d06
        SONIC(IP) = DSQRT(GAMMA*PRES(IP)/UNKNO(1,IP)) ! speed of sound
      END DO ! IP
c
c loop for calculation of local time steps
c
      DO IE = 1, NELEM
        IP1         = IX(1,IE)
        IP2         = IX(2,IE)                     ! connectivity
        IP3         = IX(3,IE)
        U1          = UNKNO(2,IP1)/UNKNO(1,IP1) ! u1 velocity
        V1          = UNKNO(3,IP1)/UNKNO(1,IP1) ! u2 velocity
        U2          = UNKNO(2,IP2)/UNKNO(1,IP2)
        V2          = UNKNO(3,IP2)/UNKNO(1,IP2)
        U3          = UNKNO(2,IP3)/UNKNO(1,IP3)
        V3          = UNKNO(3,IP3)/UNKNO(1,IP3)
        VN1         = DSQRT(U1**2 + U1**2)
        VN2         = DSQRT(U2**2 + U2**2)
        VN3         = DSQRT(U3**2 + U3**2)
        VELN        = MAX(VN1, VN2, VN3)
        CMAX        = MAX(SONIC(IP1), SONIC(IP2), SONIC(IP3))
        VSUM        = VELN + CMAX
c
        ANX         = GEOME(1,IE)                   ! shape function derivatives
        ANY         = GEOME(4,IE)
        ALEN        = 1.0/DSQRT(ANX**2 + ANY**2) ! element length at node 1
        TSTP        = ALEN/VSUM
        TSTI        = ALEN/VELN
        DELTP(IP1) = MIN(DELTP(IP1), TSTP)        ! external time step
        DELTI(IP1) = MIN(DELTI(IP1), TSTI)        ! internal time step
c
        ANX         = GEOME(2,IE)
        ANY         = GEOME(5,IE)
        ALEN        = 1.0/DSQRT(ANX**2 + ANY**2)
        TSTP        = ALEN/VSUM
        TSTI        = ALEN/VELN
        DELTP(IP2) = MIN(DELTP(IP2), TSTP)
        DELTI(IP1) = MIN(DELTI(IP1), TSTI)
```

Fig. 9.3 Continued.

```
c
        ANX          = GEOME(3,IE)
        ANY          = GEOME(6,IE)
        ALEN         = 1.0/DSQRT(ANX**2 + ANY**2)
        TSTP         = ALEN/VSUM
        TSTI         = ALEN/VELN
        DELTP(IP3) = MIN(DELTP(IP3), TSTP)
        DELTI(IP1) = MIN(DELTI(IP1), TSTI)
      END DO ! IE
      DO IP = 1, NPOIN
        DELTP(IP) = SFACT*DELTP(IP) ! SFACT - safety factor
      END DO ! IP
      IF(IALOT.EQ.0) THEN
        DTP    = 1.0d+06
        DO IP = 1,NPOIN
          DTP = MIN(DTP, DELTP(IP))
        END DO ! IP
          CALL TIMFIL(MXPOI,DELTP,NPOIN,DTP)
        ENDIF
      END
```

Fig. 9.3 Continued.

The nodal quantities calculated in a manner explained above are averaged over elements and used in the relations of Eq. (6.17), Chapter 6. Figure 9.5 shows the calculation of the nodal pressure switches for linear triangular elements.

In the next option available in the code, the second derivative of pressure is calculated from the smoothed nodal pressure gradients (see Sec. 4.5.1, Chapter 4) by averaging. Other approximations to the second derivative of pressure are described

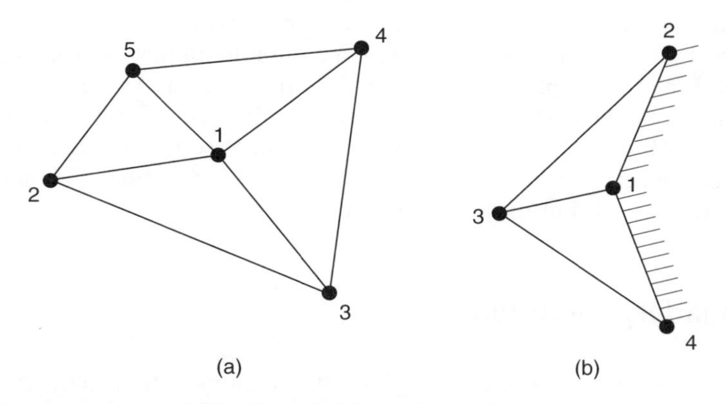

Fig. 9.4 Typical element patches (a) interior node (b) boundary node.

in Sec. 4.5.1, Chapter 4. The user can employ those methods to approximate the second derivative of pressure if desired.

9.3.4 CBS algorithm. Steps

Various steps involved in the CBS algorithm are described in detail in Chapter 3. There are three essential steps in the CBS algorithm (Fig. 9.2). First, an intermediate momentum variable is calculated and in the second step the density/pressure field is determined. The third step involves the introduction of density/pressure fields to obtain the correct momentum variables. In problems where the energy and other variables are coupled, calculation of energy is necessary in addition to the above three steps. In fully explicit form, however, the energy equation can be solved in the first step itself along with the intermediate momentum calculations.

In the subroutine `stepl` we calculate the temperature-dependent viscosity at the beginning according to Sutherland's relation (see Chapter 6). The averaged viscosity values over each element are used in the diffusion terms of the momentum equation and dissipation terms of the energy equation. The diffusion, convective and stabilization terms are integrated over elements and assembled appropriately to the RHS vector. The integration is carried out either directly or numerically. Finally the RHS vector is divided by the lumped mass matrices and the values of intermediate momentum variables are established.

In step two, in explicit form, the density/pressure values are calculated by the Eq. (3.53) (or Eq. (3.54)). The subroutine `step2` is used for this purpose. Here the option of using different values of θ_1 and θ_2 is available. In explicit form θ_2 is identically equal to zero and θ_1 varies between 0.5 and 1.0. For compressible flow computations, the semi-implicit form with θ_2 greater than zero has little advantage over the fully explicit form. For this reason we have not given the semi-implicit form for compressible flow problems in the program.

For incompressible flow problems, in general the semi-implicit form is used. In this θ_1, as before, varies between 0.5 and 1 and θ_2 is also in the same range. Now it is essential to solve the pressure equation in `step2` of the algorithm. Here in general we use a conjugate gradient solver as the coefficient matrix is not necessarily banded.

The third step is the one where the intermediate momentum variables are corrected to get the real values of the intermediate momentum. In all three steps, mass matrices are lumped if the fully explicit form of the algorithm is used. As mentioned in earlier chapters, this is the best way to accelerate the steady-state solution along with local time stepping. However, in problems where transient solutions are of importance, either a mass matrix correction as given in Sec. 2.6.3, Chapter 2 or simultaneous solution using a consistent mass matrix is necessary.

9.3.5 Boundary conditions

As explained before, the boundary edges are stored along with the elements to which they belong. Also in the same array `iside(i, j)` the flags necessary to inform the

```
      SUBROUTINE SWITCH(MXPOI, MXELE, MBC, NPOIN, NELEM, NBS, PRES,
     &                  CSHOCK,PSWTH,IX,DELUN,ISIDE,MODEL,ITYPE)
c
c this subroutine calculates the pressure switch at each node
c maximum value 1 and minimum value 0
c
      IMPLICIT      NONE

      INTEGER       IB,IELEM,IP,IP1,IP2,IP3,ITYPE,MBC,MODEL
      INTEGER       MXELE,MXPOI,NBS, NELEM, NPOIN

      INTEGER       ISIDE(4,MBC),IX(MODEL,MXELE)

      REAL*8        CSHOCK, PADD, P11, P22, P33,PS1,PS2,PS3
      REAL*8        XPS,XPD

      REAL*8        DELUN(MXPOI),PRES(MXPOI),PSWTH(MXPOI)
c
      DO IELEM = 1,NELEM
        IP1         = IX(1,IELEM)
        IP2         = IX(2,IELEM)
        IP3         = IX(3,IELEM)
        PS1         = PRES(IP1)
        PS2         = PRES(IP2)
        PS3         = PRES(IP3)
        PADD        = PS1+PS2+PS3
        P11         = (3.0d00*PS1 - PADD)
        P22         = (3.0d00*PS2 - PADD)
        P33         = (3.0d00*PS3 - PADD)
        PSWTH(IP1) = PSWTH(IP1) + P11
        PSWTH(IP2) = PSWTH(IP2) + P22
        PSWTH(IP3) = PSWTH(IP3) + P33
        DELUN(IP1) = DELUN(IP1) + DABS(PS1 - PS2) + DABS(PS1 - PS3)
        DELUN(IP2) = DELUN(IP2) + DABS(PS1 - PS2) + DABS(PS2 - PS3)
        DELUN(IP3) = DELUN(IP3) + DABS(PS3 - PS2) + DABS(PS1 - PS3)
      END DO ! IELEM
      DO IB = 1,NBS
        IP1         = ISIDE(1,IB)
        IP2         = ISIDE(2,IB)
        PS1         = PRES(IP1)
        PS2         = PRES(IP2)
        XPS         = PS1 + PS2
        XPD         = PS1 - PS2
        PSWTH(IP1) = PSWTH(IP1) + XPD
        PSWTH(IP2) = PSWTH(IP2) - XPD
        DELUN(IP1) = DELUN(IP1) + DABS(XPD)
        DELUN(IP2) = DELUN(IP2) + DABS(XPD)
```

```
END DO ! IB
DO IP = 1,NPOIN
  IF(DELUN(IP).LT.0.1*PRES(IP))DELUN(IP) = PRES(IP)
END DO ! IP
DO IP = 1,NPOIN
  PSWTH(IP) = CSHOCK*DABS(PSWTH(IP))/DELUN(IP)
END DO ! IP
END
```

Fig. 9.5 Calculation of nodal pressure switches for shock capturing.

solution module which type of boundary conditions are stored. In this array $i = 1, 2$ correspond to the node numbers of any boundary side of an element, $i = 3$ indicates the element to which the particular edge belongs and $i = 4$ is the flag which indicates the type of boundary condition (a complete list is given in the user manual available at the publisher's web page). Here j is the boundary edge number. A typical routine for prescribing the symmetry conditions is shown in Fig. 9.6.

9.3.6 Solution of simultaneous equations – semi-implicit form

The simultaneous equations need to be solved for the semi-implicit form of the CBS algorithm. Two types of solvers are provided. The first one is a banded solver which is effective when structured meshes are used. For this the half-bandwidth is necessary in order to proceed further. The second solver is a diagonal preconditioned conjugate gradient solver. The latter can be used to solve both structured and unstructured meshes. The details of procedures for solving simultaneous equations can be found in Chapter 20 of Volume 1.

9.3.7 Different forms of energy equation

In compressible flow computations only the fully conservative form of all equations ensures correct position of shocks. Thus in the compressible flow code, the energy equation is solved in its conservative form with the variable being the energy. However for incompressible flow computations, the energy equation can be written in terms of the temperature variable and the dissipation terms can be neglected. In general for compressible flows, Eq. (3.61) is used, and Eq. (4.6) is used for incompressible flow problems.

9.3.8 Thermal and porous media flows

As mentioned earlier the heat transfer and porous medium flows are also included in the incompressible flow code. Using the heat transfer part of the code, the user can solve forced, natural and mixed convection problems. Appropriate flags and

```
      SUBROUTINE SYMMET(MXPOI, MBC, NPOIN, NBS, UNKNO,ISIDE,RHOINF,
     &                  UINF,VINF, COSX,COSY)
c
c symmetric boundary conditions forced. one component of velocity
c forced to zero
c
      IMPLICIT   NONE

      INTEGER    I,IP,J,MBC,MXPOI,NBS,NPOIN

      INTEGER    ISIDE(4,MBC)

      REAL*8     ANX,ANY,RHOINF,UINF,US,VINF

      REAL*8     COSX(MBC), COSY(MBC), UNKNO(4,MXPOI)
c
      DO I = 1, NBS
        IF(ISIDE(4,I).EQ.4)THEN ! symmetry flag 4
        ANX    = COSX(I)
        ANY    = COSY(I)
        DO J = 1,2
          IP           = ISIDE(J,I)
          US           = -UNKNO(2,IP)*ANY + UNKNO(3,IP)*ANX
          UNKNO(2,IP) = - US*ANY
          UNKNO(3,IP) =   US*ANX
        END DO ! J
        ENDIF
      END DO ! I
      END
```

Fig. 9.6 Subroutine to impose symmetry conditions.

non-dimensional parameters need to be given as input. For the detailed discussion on these flows, the reader is referred to Chapter 5 of this volume.

9.3.9 Convergence

The residuals (difference between the current and previous time step values of parameters) of all equations are checked at every few user-prescribed number of iterations. If the required convergence (steady state) is achieved, the program stops automatically. The aimed residual value is prescribed by the user. The program calculates the maximum residual of each variable over the domain. The user can use them to fix the required accuracy. We give the routine used for this purpose in Fig. 9.7.

```
c
      SUBROUTINE RESID(MXPOI,NPOIN,ITIME,UNKNO,UNPRE,PRES,PRESN,IFLOW)
c
c     purpose : calculations of residuals.
c
      IMPLICIT   NONE

c
      INTEGER    I,ICON1,ICON2,ICON3,ICON4,IFLOW,ITIME,MXPOI,NPOIN

      REAL*8     EMAX1,EMAX2,EMAX3,EMAX4,ERR1,ERR2,ERR3,ERR4,ER1
      REAL*8     ER2,ER3,ER4

      REAL*8     PRES(MXPOI),PRESN(MXPOI),UNKNO(4,MXPOI)
      REAL*8     UNPRE(4,MXPOI)

          EMAX1 = 0.000d00
          EMAX2 = 0.000d00
          EMAX3 = 0.000d00
          EMAX4 = 0.000d00

      DO I = 1,NPOIN
        ERR1 = UNKNO(1,I) - UNPRE(1,I) ! density or pressure
        ERR2 = UNKNO(2,I) - UNPRE(2,I) ! u1 velocity or mass flux
        ERR3 = UNKNO(3,I) - UNPRE(3,I) ! u2 velocity or mass flux
        ERR4 = UNKNO(4,I) - UNPRE(4,I) ! energy or temperature
        ER1  = DABS(ERR1)
        ER2  = DABS(ERR2)
        ER3  = DABS(ERR3)
        ER4  = DABS(ERR4)
        IF (ER1.GT.EMAX1) THEN
          EMAX1 = ER1
          ICON1 = I
        ENDIF
        IF (ER2.GT.EMAX2) THEN
          EMAX2 = ER2
          ICON2 = I
        ENDIF
        IF (ER3.GT.EMAX3) THEN
          EMAX3 = ER3
          ICON3 = I
        ENDIF
        IF (ER4.GT.EMAX4) THEN
          EMAX4 = ER4
          ICON4 = I
        ENDIF
      END DO ! I
      END
```

Fig. 9.7 Subroutine to check convergence rate.

9.4 Output module

If the imposed convergence criteria are satisfied then the output is written into a separate file. The user can modify the output according to the requirements of post-processor employed. Here we recommend the education software developed by CIMNE (GiD) for post and preprocessing of data.[5] The facilities in GiD include two- and three-dimensional mesh generation and visualization.

9.4.1 Stream function calculation

The stream function value is calculated from the following equation:

$$\frac{\partial^2 \psi}{\partial x_1^2} + \frac{\partial^2 \psi}{\partial x_2^2} = \frac{\partial u}{\partial x_2} - \frac{\partial v}{\partial x_1} \qquad (9.3)$$

This equation is derived from the definition of stream function in terms of the velocity components. We again use the finite element method to solve the above equation.

9.5 Possible extensions to CBSflow

As mentioned earlier, there are several possibilities for extending this code. A simple subroutine similar to the temperature equation can be incorporated to solve mass transport. Here another variable 'concentration' needs to be solved.[6]

Another subject which can be incorporated and studied is that of a 'free surface' given in Chapter 5 of this volume. Here another equation needs to be solved for the surface waves.[7]

The phase change problems need appropriate changes in the energy equation.[8–12] The liquid, solid and mushy regions can be accounted for in the equations by simple modifications. The latent heat also needs to be included in phase change problems.

The turbulent flow requires solution of another set or sets of equations similar to the momentum or energy equations as explained in Chapter 5. For the κ–ε model the reader is referred to reference 13.

The program CBSflow is an educational code which can be modified to suit the needs of the user. For instance, the modification of this program to incorporate a 'command language' could make the code very efficient and compact.

References

1. I. Swith and D.V. Griffiths. *Programming the Finite Element Method*, Third Edition, Wiley, Chichester, 1998.
2. D.R. Willé. *Advanced Scientific Fortran*, Wiley, Chichester, 1995.
3. O.C. Zienkiewicz and R.L. Taylor. *The Finite Element Method, Vol. 1, The Basics*, 5th Edition, Arnold, London, 2000.
4. P. Nithiarasu and O.C. Zienkiewicz. On stabilization of the CBS algorithm. Internal and external time steps. *Int. J. Num. Meth. Eng.*, **48**, 875–80, 2000.

5. GiD. International Center for Numerical Methods in Engineering, Universidad Politécnica de Cataluña, 08034, Barcelona, Spain.
6. P.Nithiarasu, K.N. Seetharamu and T. Sundararajan. Double-diffusive natural convection in an enclosure filled with fluid saturated porous medium – a generalised non-Darcy approach. *Numerical Heat Transfer, Part A, Applications*, **30**, 413–26, 1996.
7. I.R. Idelsohn, E. Oñate and C. Sacco. Finite element solution of free surface ship wave problems. *Int. J. Num. Meth. Eng.*, **45**, 503–28, 1999.
8. K. Morgan. A numerical analysis of freezing and melting with convection. *Comp. Meth. Appl. Mech. Eng.*, **28**, 275–84, 1981.
9. A.S. Usmani, R.W. Lewis and K.N. Seetharamu. Finite element modelling of natural convection controlled change of phase. *Int. J. Num. Meth. Fluids*, **14**, 1019–36, 1992.
10. S.K. Sinha, T. Sundararajan and V.K. Garg. A variable property analysis of alloy solidification using the anisotropic porous medium approach. *Int. J. Heat Mass Transfer*, **35**, 2865–77, 1992.
11. R.W. Lewis, K. Morgan, H.R. Thomas and K.N. Seetharamu. *The Finite Element Method for Heat Transfer Analysis*, Wiley, Chichester, 1996.
12. P. Nithiarasu. An adaptive finite element procedure for solidification problems. *Heat and Mass Transfer* (to appear, 2000).
13. O.C. Zienkiewicz, B.V.K.S. Sai, K. Morgan and R. Codina. Split characteristic based semi-implicit algorithm for laminar/turbulent incompressible flows. *Int. J. Num. Meth. Fluids*, **23**, 1–23, 1996.

Appendix A

Non-conservative form of Navier–Stokes equations

To derive the Navier–Stokes equations in their non-conservative form, we start with the conservative form.

Conservation of mass:

$$\frac{\partial \rho}{\partial t}+\frac{\partial(\rho u_i)}{\partial x_i}=\frac{\partial \rho}{\partial t}+\rho\frac{\partial u_i}{\partial x_i}+u_i\frac{\partial \rho}{\partial x_i}=0 \tag{A.1}$$

Conservation of momentum:

$$\frac{\partial(\rho u_i)}{\partial t}+\frac{\partial(u_j\rho u_i)}{\partial x_j}-\frac{\partial \tau_{ij}}{\partial x_j}+\frac{\partial p}{\partial x_i}=0 \tag{A.2}$$

Conservation of energy:

$$\frac{\partial(\rho E)}{\partial t}+\frac{\partial(u_j\rho E)}{\partial x_j}-\frac{\partial}{\partial x_i}\left(k\frac{\partial T}{\partial x_i}\right)+\frac{\partial(u_j p)}{\partial x_j}-\frac{\partial(\tau_{ij}u_j)}{\partial x_j}=0 \tag{A.3}$$

Rewriting the momentum equation with terms differentiated as

$$\rho\frac{\partial u_i}{\partial t}+u_i\left(\frac{\partial \rho}{\partial t}+\rho\frac{\partial u_j}{\partial x_j}+u_j\frac{\partial \rho}{\partial x_j}\right)+\rho u_j\frac{\partial u_i}{\partial x_j}-\frac{\partial \tau_{ij}}{\partial x_j}+\frac{\partial p}{\partial x_i}=0 \tag{A.4}$$

and substituting the equation of mass conservation (Eq. A.1) into the above equation gives the reduced momentum equation

$$\frac{\partial u_i}{\partial t}+u_j\frac{\partial u_i}{\partial x_j}-\frac{1}{\rho}\frac{\partial \tau_{ij}}{\partial x_j}+\frac{1}{\rho}\frac{\partial p}{\partial x_i}=0 \tag{A.5}$$

Similarly as above, the energy equation (Eq. A.3) can be written with differentiated terms as

$$E\left(\frac{\partial \rho}{\partial t}+\rho\frac{\partial u_j}{\partial x_j}+u_j\frac{\partial \rho}{\partial x_j}\right)+\rho\frac{\partial E}{\partial t}+\rho u_j\frac{\partial E}{\partial x_j}-\frac{\partial}{\partial x_i}\left(k\frac{\partial T}{\partial x_i}\right)$$
$$+\frac{\partial(u_i p)}{\partial x_i}-\frac{\partial(\tau_{ij}u_j)}{\partial x_i}=0 \tag{A.6}$$

Again substituting the continuity equation into the above equation, we have the reduced form of the energy equation

$$\frac{\partial E}{\partial t} + u_j \frac{\partial E}{\partial x_j} - \frac{1}{\rho} \frac{\partial}{\partial x_i} \left(k \frac{\partial T}{\partial x_i} \right) + \frac{1}{\rho} \frac{\partial (u_i p)}{\partial x_i} - \frac{1}{\rho} \frac{\partial (\tau_{ij} u_j)}{\partial x_i} \qquad \text{(A.7)}$$

Some authors use Eqs. (A.1), (A.5) and (A.7) to study compressible flow problems. However these non-conservative equations can result in multiple or incorrect solutions in certain cases. This is true especially for high-speed compressible flow problems with shocks. The reader should note that such non-conservative equations are not suitable for simulation of compressible flow problems.

Appendix B

Discontinuous Galerkin methods in the solution of the convection–diffusion equation*

In Volume 1 of this book we have already mentioned the words 'discontinuous Galerkin' in the context of transient calculations. In such problems the discontinuity was introduced in the interpolation of the function in the time domain and some computational gain was achieved.

In a similar way in Chapter 13 of Volume 1, we have discussed methods which have a similar discontinuity by considering appropriate approximations in separate element domains linked by the introduction of Lagrangian multipliers or other procedures on the interface to ensure continuity. Such *hybrid* methods are indeed the precursors of the discontinuous Galerkin method as applied recently to fluid mechanics.

In the context of fluid mechanics the advantages of applying the discontinuous Galerkin method are:

- the achievement of complete flux conservation for each element or cell in which the approximation is made;
- the possibility of using higher-order interpolations and thus achieving high accuracy for suitable problems;
- the method appears to suppress oscillations which occur with convective terms simply by avoiding a prescription of Dirichlet boundary conditions at the flow exit; this is a feature which we observed to be important in Chapter 2.

To introduce the procedure we consider a model of the steady-state *convection–diffusion* problem in one dimension of the form

$$u\frac{\mathrm{d}\phi}{\mathrm{d}x} - \frac{\mathrm{d}}{\mathrm{d}x}\left(k(x)\frac{\mathrm{d}\phi}{\mathrm{d}x}\right) = f \qquad 0 \leqslant x \leqslant L \tag{B.1}$$

where u is the convection velocity, $k = k(x)$ the diffusion (conduction) coefficient (always bounded and positive), and $f = f(x)$ the source term. We add boundary conditions to Eq. (B.1); for example,

$$\phi(L) = \bar{\phi} \qquad \text{and} \qquad k(0)\frac{\mathrm{d}\phi(0)}{\mathrm{d}x} = g \tag{B.2}$$

* J.T. Oden, personal communication, 1999.

As usual the domain $\Omega = (0, L)$ is partitioned into a collection of N elements (intervals) $\Omega_e = (x_{e-1}, x_e), e = 1, 2, \ldots, m$. In the present case, we consider the special weak form of Eqs (B.1) and (B.2) defined on this mesh by

$$\sum_{e=1}^{m} \int_{x_{e-1}}^{x_e} \left(k\frac{\mathrm{d}\phi}{\mathrm{d}x}\frac{\mathrm{d}v}{\mathrm{d}x} - \frac{\mathrm{d}(uv)}{\mathrm{d}x}\phi \right) \mathrm{d}x + \sum_{e=1}^{m} \left\{ \left\langle k\frac{\mathrm{d}v}{\mathrm{d}x} \right\rangle [\phi] - \left\langle k\frac{\mathrm{d}\phi}{\mathrm{d}x} \right\rangle [v] \right\}(x_e)$$

$$+ \left(k\frac{\mathrm{d}v}{\mathrm{d}x} \right)\phi(L) - v\left(k\frac{\mathrm{d}\phi}{\mathrm{d}x} \right)(L) + v\phi^{-}u(0)$$

$$= \sum_{e=1}^{m} \int_{x_{e-1}}^{x_e} fv \, \mathrm{d}x + k\frac{\mathrm{d}v}{\mathrm{d}x}(L)\bar{\phi} + v(0)g - uv(0)\bar{\phi} \tag{B.3}$$

for arbitrary weight functions v. Here $\langle . \rangle$ denotes (flux) averages

$$\langle kv' \rangle (x_e) = \frac{kv'(x_{e+}) + kv'(x_{e-})}{2} \tag{B.4}$$

and $[.]$ denote jumps

$$[\phi](x_e) = \phi(x_{e+}) - \phi(x_{e-}) \tag{B.5}$$

it being understood that $x_{e\pm} = \lim_{\varepsilon \to 0}(x_e \pm \varepsilon)$, $v' = \mathrm{d}v/\mathrm{d}x$ etc.

The particular structure of the weak statement in Eq. (B.3) is significant. We make the following observations concerning it:

1. If $\phi = \phi(x)$ is the exact solution of Eqs (B.1) and (B.2), then it is also the (one and only) solution of Eq. (B.3); i.e. Eqs (B.1) and (B.2) imply the problem given by Eq. (B.3).
2. The solution of Eqs (B.1) and (B.2) satisfies Eq. (B.3) because ϕ is continuous and the fluxes $k\,\mathrm{d}\phi/\mathrm{d}x$ are continuous:

$$[\phi](x_e) = 0 \qquad \text{and} \qquad \left\langle k\frac{\mathrm{d}u}{\mathrm{d}x} \right\rangle (x_e) = 0 \tag{B.6}$$

3. The Dirichlet boundary conditions (an inflow condition) enter the weak form on the left-hand side, an uncommon property, but one that permits discontinuous weight functions at relevant boundaries.
4. The signs of the second term on the left side $(\sum_e\{\langle kv'[\phi]\rangle - \langle k\phi'\rangle[v]\})$ can be changed without affecting the equivalence of Eq. (B.3) and Eqs (B.1) and (B.2), but the particular choice of signs indicated turns out to be crucial to the stability of the discontinuous Galerkin method (DGM).
5. We can consider the conditions of continuity of the solution and of the fluxes at interelement boundaries, conditions (B.6), as *constraints* on the true solution. Had we used Lagrange multipliers to enforce these constraints then, instead of the second sum on the left-hand side of Eq. (B.3), we would have terms like

$$\sum_{e=1}^{m} \{\lambda[\phi] + \mu\langle k \, \mathrm{d}\phi \, \mathrm{d}x \rangle\}(x_e) \tag{B.7}$$

where λ and μ are the multipliers. A simple calculation shows that the multipliers can be identified as average fluxes and interface jumps:

$$\lambda = \left\langle k\frac{\mathrm{d}v}{\mathrm{d}x} \right\rangle (x_e); \qquad \mu = -[v](x_e) \tag{B.8}$$

Introducing Eq. (B.8) into Eq. (B.7) gives the second term on the left hand side of Eq. (B.3). Incidently, had we constructed independent approximations of λ and μ, a setting for the construction of a *hybrid* finite element approximation of Eq. (B.1) and Eq. (B.2) would be obtained (see Chapter 13, Volume 1).

We are now ready to construct the approximation of Eqs (B.1) and (B.2) by the DGM. Returning to Eq. (B.3), we introduce over each element Ω_e a polynomial approximation of ϕ;

$$\phi \simeq \hat{\phi} = \sum_{k=0}^{p_e} a_k^e N_k^e(x) \tag{B.9}$$

where the a_k^e are undetermined constants and $N_k^e = x^k$ are monomials (shape functions) of degree k each associated only with Ω_e. Introducing Eq. (B.9) into (B.3) and using, for example, complete polynomials N_e of degree p_e for weight functions in each element, we arrive at the discrete system

$$\begin{aligned}
&\sum_{e=1}^{m}\sum_{k=0}^{p_e}\left\{\int_{x_{e-1}}^{x_e}\left[k\frac{\mathrm{d}N_k^e}{\mathrm{d}x}\frac{\mathrm{d}N_j^e}{\mathrm{d}x} - N_k^e\frac{\mathrm{d}(uN_j^e)}{\mathrm{d}x}\right]\mathrm{d}x + \left\langle k\frac{\mathrm{d}N_j^e}{\mathrm{d}x}\right\rangle[N_k^e](x_e)\right\}a_k^e \\
&-\left\{\left\langle k\frac{\mathrm{d}N_k^e}{\mathrm{d}x}\right\rangle[N_j^e]\right\}a_k^e + \left\{\left\langle k\frac{\mathrm{d}N_k^e}{\mathrm{d}x}\right\rangle N_k^e(L) - N_j^e\left\langle k\frac{\mathrm{d}N_k^e}{\mathrm{d}x}\right\rangle(L) + N_j^e(0)u(0)N_k^{e-}(0)\right\}a_k^e \\
&\quad = \sum_{e=1}^{N}\int_{x_{e-1}}^{x_e} fN_j^e\,\mathrm{d}x + k\frac{\mathrm{d}N_j^e}{\mathrm{d}x}(L)\bar{\phi} + gN_j^e(0) - u(0)N_j^e(0)\bar{\phi}; \\
&\qquad j = 1,2,\ldots,p_e, \quad e = 1,2,\ldots,m
\end{aligned} \tag{B.10}$$

This is the DGM approximation of Eq. (B.3). Some properties of Eq. (B.10) are noteworthy:

1. The shape functions N_k^e need not be the usual nodal based functions; *there are no nodes in this formulation*. We can take N_k^e to be any monomial we please (representing, for example, complete polynomials up to degree p_e for each element Ω_e and even orthogonal polynomials). The unknowns are the coefficients a_k^e which are not necessarily the values of $\hat{\phi}$ at any point.
2. We can use different polynomial degrees in each element Ω_e; thus Eq. (B.10) provides a natural setting for *hp*-version finite element approximations.
3. Suppose $u = 0$. Then the operator in Eq. (B.1) is symmetric. Even so, the formulation in Eq. (B.10) leads to an unsymmetric stiffness matrix owing to the presence of the jump terms and averages on the element interfaces. However, it can be shown that the resulting matrix is always positive definite, the choice of signs in the boundary and interface terms being critical for preserving this property.
4. In general, the formulation in Eq. (B.10) involves more degrees of freedom than the conventional continuous (conforming) Galerkin approximation of Eqs (B.1) and (B.2) owing to the fact that the usual dependencies produced in enforcing continuity across element interfaces are now not present. However, the very localized nature of the discontinuous approximations contributes to the surprising robustness of the DGM.

5. While the piecewise polynomial basis $\{N_1^1, \ldots, N_{p_1}^n, \ldots, N_1^n, \ldots, N_{p_n}^n\}$ contains complete polynomials from degree zero up to $p = p\min_e p_e$, numerical experiments indicate that stability demands $p \geqslant 2$, in general.
6. The DGM is elementwise conservative while the standard finite element approximation is conservative only in element patches. In particular, for any element Ω_e, we always have

$$\int_{\Omega_e} f\,\mathrm{d}x + k\frac{\mathrm{d}\hat{\phi}}{\mathrm{d}x}\bigg|_{x_{e-1}}^{x_e} = 0 \tag{B.11}$$

This property holds for arbitrarily high-order approximations p_e.

The DGM is robust and essentially free of the global spurious oscillations of continuous Galerkin approximations when applied to convection–diffusion problems.

We now consider the solution to a convection–diffusion problem with a turning point in the middle of the domain. The Hemker problem is given as follows:

$$k\frac{\mathrm{d}^2\phi}{\mathrm{d}x_2} + x\frac{\mathrm{d}\phi}{\mathrm{d}x} = -k\pi^2\cos(\pi x) - \pi x\sin(\pi x) \qquad \text{on} \quad [0, 1]$$

with $\phi(-1) = -2$, $\phi(1) = 0$. Exact solution for above shows a discontinuity of

$$\phi(x) = \cos(\pi x) + \mathrm{erf}(x/\sqrt{2k})/\mathrm{erf}(1/\sqrt{2k})$$

Figures B.1 and B.2 show the solutions to the above problem ($k = 10^{-10}$ and $h = 1/10$) obtained with the continuous and discontinuous Galerkin method, respectively. Extension to two and three dimensions is discussed in references given in Chapter 2.

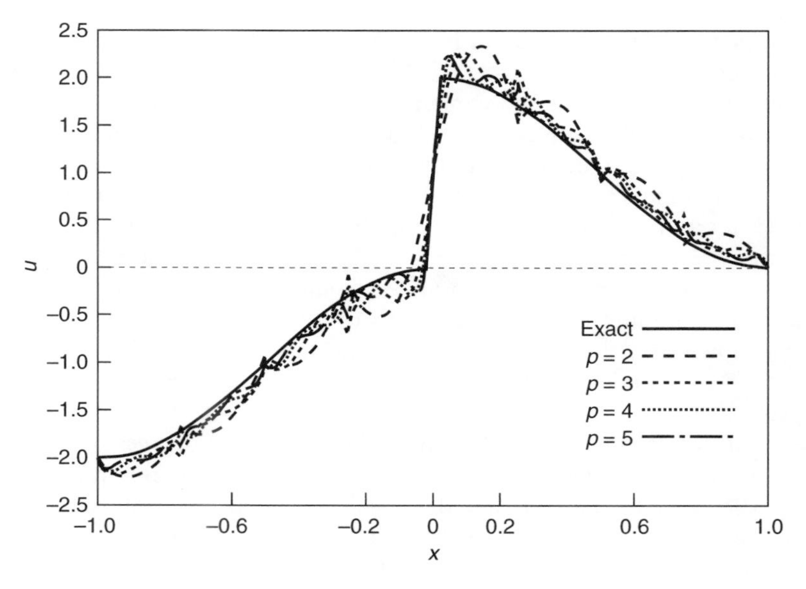

Fig. B1. Continuous Galerkin approximation.

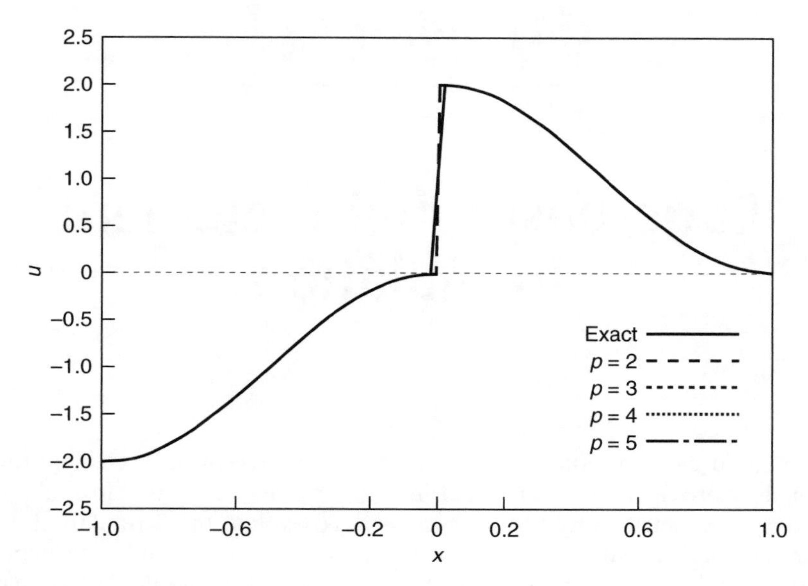

Fig. B2. Discontinuous Galerkin approximation.

Appendix C

Edge-based finite element formulation

The edge-based data structure has been used in many recent finite element formulations for flow problems. As mentioned in Sec. 6.8, Chapter 6, this formulation has many advantages such as smaller storage, etc. To explain the formulation we shall consider the Euler equations and a few assembled linear triangular elements on a two-dimensional finite element mesh as shown in Fig. C.1. From Eq. (1.24) we rewrite the following Euler equations

$$\frac{\partial \boldsymbol{\Phi}}{\partial t} + \frac{\partial \mathbf{F}_i}{\partial x_i} = 0 \tag{C.1}$$

where $\boldsymbol{\Phi}$ are the conservative variables. If the element-based formulation for the above equation omits the stabilization terms, the weak form can be written as

$$\int_\Omega N^k \frac{\Delta \boldsymbol{\Phi}}{\Delta t}\,\mathrm{d}\Omega = -\int_\Omega (N^k)^{\mathrm{T}} \frac{\partial \mathbf{F}_i}{\partial x_i}\,\mathrm{d}\Omega \tag{C.2}$$

In a fully explicit form of solution procedure, the left-hand side becomes $\mathbf{M}(\Delta\boldsymbol{\Phi}/\Delta t)$ and here $\mathbf{M}$ is the consistent mass matrix (see Chapter 3). We can write the RHS of the above equation for an interior node I (Fig. C.1(a)) by interpolating $\mathbf{F}_i$ in each element and after applying Green's theorem as

$$\sum_{E\in I}\int_{A_E} \frac{\partial N^I}{\partial x_i}(N^k \mathbf{F}_i^k)\,\mathrm{d}\Omega = \sum_{E\in I}\left[\frac{A_E}{3}\frac{\partial N_I}{\partial x_i}\right]_E (\mathbf{F}_i^I + \mathbf{F}_i^J + \mathbf{F}_i^K) \tag{C.3}$$

where A_E is the area and I, J and K are the three nodes of the element (triangle) E. This is an acceptable added approximation which is frequently used in the Taylor–Galerkin method (see Chapter 2). In another form, the above RHS can be written as (Fig. C.1(a))

$$\frac{A_1}{3}\frac{\partial N_I}{\partial x_i}(\mathbf{F}_I + \mathbf{F}_1 + \mathbf{F}_2) + \frac{A_2}{3}\frac{\partial N_I}{\partial x_i}(\mathbf{F}_I + \mathbf{F}_2 + \mathbf{F}_3) + \frac{A_3}{3}\frac{\partial N_I}{\partial x_i}(\mathbf{F}_I + \mathbf{F}_3 + \mathbf{F}_1) \tag{C.4}$$

where A_1, A_2 and A_3 are the areas of elements 1, 2 and 3 respectively. For integration over the boundary on the RHS, we can write the following in the element formulation

$$\sum_{B\in I}\int_{\Gamma_B} N^I (N^k \mathbf{F}_i^k)\,\mathrm{d}\Gamma \mathbf{n}_B = \sum_{B\in I}\left[\frac{\Gamma_B}{6}(2\mathbf{F}_i^I + \mathbf{F}_i^J)\mathbf{n}\right]_B \tag{C.5}$$

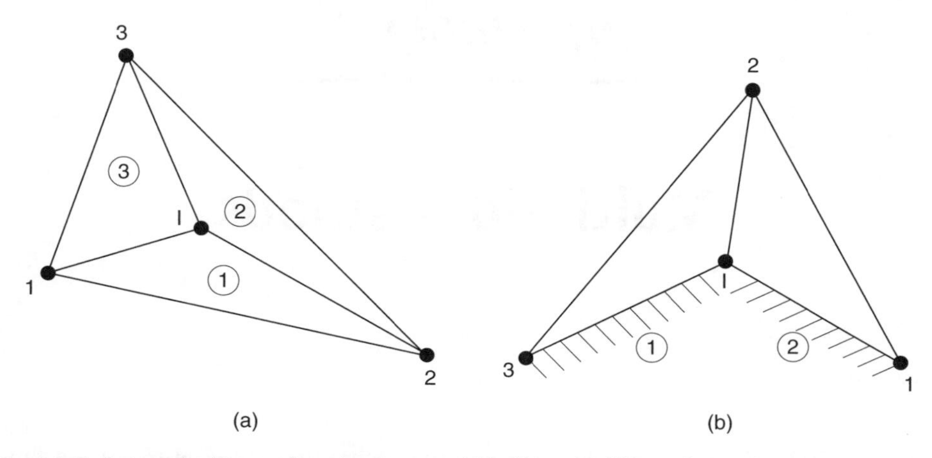

Fig. C.1 Typical patch of linear triangular elements: (a) inside node; (b) boundary node.

where $\mathbf{n}$ is the boundary normal. The above equation can be rewritten for the node I in Fig. C.1(b) as

$$\frac{\Gamma_{B1}}{6}(2\mathbf{F}_i^I + \mathbf{F}_i^3)\mathbf{n}_1 + \frac{\Gamma_{B2}}{6}(2\mathbf{F}_i^I + \mathbf{F}_i^1)\mathbf{n}_2 \tag{C.6}$$

where Γ_{B1} and Γ_{B2} are appropriate edge lengths.

The above equations (C.3) and (C.5) can be reformulated for an edge-based data structure. In such a procedure, Eq. (C.3) can be rewritten as (for an interior node I)

$$\sum_{E \in I} \int_{\Omega_E} \frac{\partial N^I}{\partial x_i}(N^k \mathbf{F}_i^k)\, \mathrm{d}\Omega = \sum_{S=1}^{m_s} \left\{ \sum_{E \in II_s} \left[\frac{A_E}{3} \frac{\partial N_I}{\partial x_i} \right]_E (\mathbf{F}_i^I + \mathbf{F}_i^{I_s}) \right\} \tag{C.7}$$

where m_s is the number of edges in the mesh which are directly connected to the node I and the summation $\sum_{E \in II_s}$ extends over those elements that contain the edges II_s. The user can readily verify that the above equation is identically equal to the standard element formulation of Eq. (C.4) if we consider the node I in Fig. C.1(a). The inclusion of boundary sides is direct from Eqs (C.5) and (C.6).

Appendix D

Multigrid methods

It is intuitively obvious that whenever iterative techniques are used to solve a finite element or finite difference problem it is useful to start from a coarse mesh solution and then to use this coarse mesh solution as a starting point for iteration in a finer mesh. This process repeated on many meshes has been used frequently and obviously accelerates the total convergence rate. This acceleration is particularly important when a hierarchical formulation of the problem is used. We have indeed discussed such hierarchical formulations in Chapter 8 of the first volume and the advantages are pointed out there.

The simple process which we have just described involves going from coarser meshes to finer ones. However it is not useful if no return to the coarser mesh is done. In hierarchical solutions such returning is possible as the coarser mesh matrix is embedded in the finer one with the same variables and indeed the iteration process can be described entirely in terms of the fine mesh solution. The same idea is applied to the multigrid form of iteration in which the coarse and fine mesh solution are suitably linked and use is made of the fact that the fine mesh iteration converges very rapidly in eliminating the higher frequencies of error while the coarse mesh solution is important in eliminating the low frequencies.

To describe the process let us consider the problem of

$$L\phi = f \quad \text{in} \quad \Omega \tag{D.1}$$

which we discretize incorporating the boundary conditions suitably. On a coarse mesh the discretization results in

$$\mathbf{K}^c\tilde{\boldsymbol{\phi}}^c = \mathbf{f}^c \tag{D.2}$$

which can be solved directly or iteratively and generally will converge quite rapidly if $\tilde{\boldsymbol{\phi}}^c$ is not a big vector. The fine mesh discretization is written in the form

$$\mathbf{K}^f\tilde{\boldsymbol{\phi}}^f = \mathbf{f}^f \tag{D.3}$$

and we shall start the iteration after the solution has been obtained on the coarse mesh. Here we generally use a *prolongation* operator which is generally an interpolation from which the fine mesh values at all nodal points are described in terms of the coarse mesh values. Thus

$$\boldsymbol{\phi}_i^f = \mathbf{P}\boldsymbol{\phi}_{i-1}^c + \Delta\boldsymbol{\phi}_i^f \tag{D.4}$$

where $\Delta\boldsymbol{\phi}_i^f$ is the increment obtained in direct iteration. If the meshes are nesting then of course the matter of obtaining $\mathbf{P}$ is fairly simple but this can be done quite generally by interpolating from a coarser to a finer mesh even if the points are not coincident. Obviously the values of the matrices $\mathbf{P}$ will be close to unity whenever the fine mesh points lie close to the coarse mesh ones. This leads to an almost hierarchical form. Once the prolongation to $\boldsymbol{\phi}^f$ has been established at a particular iteration i the fine mesh solutions can be attempted by solving

$$\mathbf{K}^f \Delta\tilde{\phi}^f = \mathbf{f}^f - \mathbf{R}_i^f \tag{D.5}$$

where the residual $\mathbf{R}$ is easily evaluated from the actual equations. We note that the solution need not be complete and can well proceed for a limited number of cycles after which a return to the coarse mesh is again made to cancel out major low-frequency errors. At this stage it is necessary to introduce a matrix $\mathbf{Q}$ which transforms values from the fine mesh to the coarse mesh. We now write for instance

$$\tilde{\boldsymbol{\phi}}_i^c = \mathbf{Q}\tilde{\boldsymbol{\phi}}_i^f \tag{D.6}$$

where one choice for $\mathbf{Q}$ is, of course, $\mathbf{P}^{\mathrm{T}}$. In a similar way we can also write

$$\mathbf{R}_i^c = \mathbf{Q}\mathbf{R}_i^f \tag{D.7}$$

where $\mathbf{R}_i$ are residuals. The above interpolation of residuals is by no means obvious but is intuitively at least correct and the process is self-checking as now we shall start a coarse mesh solution written as

$$\mathbf{K}^c(\tilde{\boldsymbol{\phi}}_{i+1}^c - \tilde{\boldsymbol{\phi}}_i^c) = \mathbf{R}_i^c \tag{D.8}$$

At this stage we solve for $\tilde{\boldsymbol{\phi}}_{i+1}^c$ using the values of previous iterations of $\tilde{\boldsymbol{\phi}}_i^c$ and putting the collected residuals on the right-hand side. This way of transferring residuals is by no means unique but has established itself well and the process is rapidly convergent.

In general more than two mesh subdivisions will be used and suitable operators $\mathbf{P}$ and $\mathbf{Q}$ have to be established for transition between each of the stages. The total process of solution is vastly accelerated and proceeds well as shown by the many papers cited in Chapter 6.

Appendix E

Boundary layer–inviscid flow coupling

A few references on the topic of boundary layer–inviscid flow coupling are given in Chapter 6. In this appendix we shall briefly explain a simple procedure of this flow coupling procedure. To understand the process of coupling the Euler and integral boundary solutions we shall consider a typical flow pattern around a wing as shown in Fig. E.1. Both turbulent and laminar regimes are shown in this figure.

We summarize the procedure as follows.

Step 1 Solve the Euler equations in the domain considered around the aerofoil. Here any mesh can be used independently of the mesh used for the boundary layer solution. The solution thus obtained will give a pressure distribution on the surface of the wing.

Step 2 Solve the boundary layer using an integral approach over an independently generated surface mesh. If the surface nodes do not coincide with the Euler mesh, the pressure needs to be interpolated to couple the two solutions. The laminar portion near the boundary (Fig. E.1) is calculated by the 'Thwaites compressible' method and the turbulent region is predicted by the 'lag-entrainment' integral boundary layer model.

Step 3 The Euler and integral solutions are coupled by transferring the outputs from one solution to the other. As indicated in Fig. E.1, direct and semi-inverse couplings

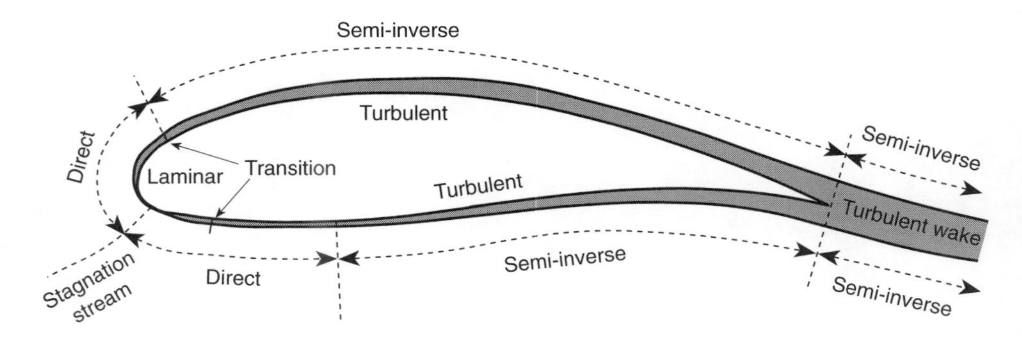

Fig. E1. Flow past an aerofoil. Typical problem for boundary layer–inviscid flow coupling.

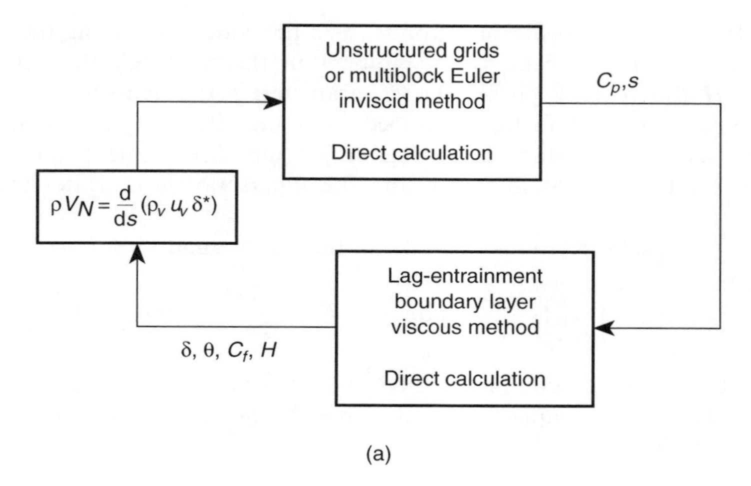

(a)

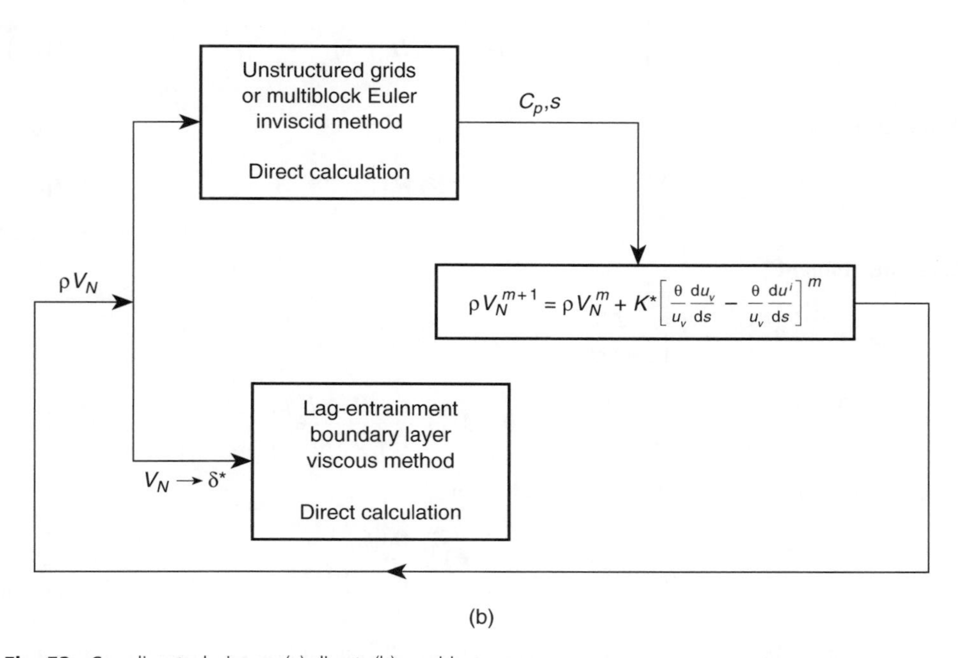

(b)

Fig. E2. Coupling techniques: (a) direct; (b) semi-inverse.

can be used for different regions. The semi-inverse coupling is introduced here mainly to stabilize the solution in the turbulent region close to separation. Figure E.2 shows the flow diagrams for the present boundary layer–inviscid coupling.

Further details on the Thwaites compressible method and semi-inverse coupling can be found in the references discussed in Sec. 6.12, Chapter 6 (Le Balleur and coworkers).

In Fig. E.2, C_p is the coefficient of pressure; s the coordinate along the surface; δ the boundary layer thickness; θ the momentum thickness; C_f the skin friction coefficient; H the velocity profile shape parameter; ρ the density; V_N the transpiration velocity; K^* is a factor developed from stability analysis; the subscript v marks the viscous boundary layer region; δ^* the displacement thickness; the superscript i indicates inviscid region and the superscript m indicates the current iteration.

Following are useful relations for some of the above quantities:

$$H = \frac{\delta^*}{\theta}, \qquad \delta^* = \int_0^\infty \left(1 - \frac{\rho u}{\rho_v u_v}\right) \mathrm{d}n, \qquad K^* = \frac{\beta\lambda}{2\pi\theta}, \qquad \beta = \sqrt{1 - M^2} \tag{E.1}$$

where n is the normal direction from the wing surface.

We have the following equations to be solved in the integral boundary layer lag-entrainment model.

Continuity

$$\theta\frac{\mathrm{d}\bar{H}}{\mathrm{d}s} = \frac{\mathrm{d}\bar{H}}{\mathrm{d}H}\left[C_e - H_1\left(\frac{C_f}{2} - (H+1)\frac{\theta}{u_v}\frac{\mathrm{d}u_v}{\mathrm{d}s}\right)\right] \tag{E.2}$$

Momentum

$$\frac{\mathrm{d}\theta}{\mathrm{d}s} = \frac{C_f}{2} - (H + 2 - M^2)\frac{\theta}{u_v}\frac{\mathrm{d}u_v}{\mathrm{d}s} \tag{E.3}$$

Lag-entrainment

$$\theta\frac{\mathrm{d}C_e}{\mathrm{d}s} = F\left[\frac{2.8}{H + H_1}\left((C_\tau)^{0.5}_{\mathrm{EQ_o}} - \lambda C_f^{0.5}\right) + \left(\frac{\theta}{u_v}\frac{\mathrm{d}u_v}{\mathrm{d}s}\right)_{\mathrm{EQ}} - \frac{\theta}{u_v}\frac{\mathrm{d}u_v}{\mathrm{d}s}(1 + 0.075M^2)\frac{(1 + 0.2M^2)}{(1 + 0.1M^2)}\right] \tag{E.4}$$

where F is a function of C_e and C_f and given as

$$F = \frac{\left(0.02C_e + C_e^2 + \dfrac{0.8C_f}{3}\right)}{(0.01 + C_e)} \tag{E.5}$$

In the above equations, $\bar{H}$ and H_1 are the velocity profile shape parameters defined as

$$\bar{H} = \frac{1}{\theta}\int_0^\infty \left(1 - \frac{u}{u_v}\right) \mathrm{d}n, \qquad H_1 = \frac{\delta - \delta^*}{\theta} \tag{E.6}$$

C_e is the entrainment coefficient; u_v the mean component of the streamwise velocity at the edge of the boundary layer; M the Mach number; C_τ the shear stress coefficient; λ the scaling factor on the dissipation length; the subscripts EQ and $\mathrm{EQ_o}$ denote respectively the equilibrium conditions and equilibrium conditions in the absence of secondary influences on the turbulence structure.

Once the above equations are solved, the transpiration velocity V_N is calculated as shown in Fig. E.2 and is added to the standard Euler boundary conditions on the wall and plays the role of a surface source. The coupling continues until convergence. In practice, in one coupling cycle, several Euler iterations are carried out for each boundary layer solution.

Author index

Page numbers in **bold** refer to the list of references at the end of each chapter.

Subject index